Edited by
Kim Ekroos

Lipidomics

Further Titles of Interest

Jelinek, R. (Ed.)

Lipids and Cellular Membranes in Amyloid Diseases

2011

ISBN: 978-3-527-32860-4

Lämmerhofer, M., Weckwerth, W. (Eds.)

Metabolomics in Practice

Successful Strategies to Generate and Analyze Metabolic Data

2012

ISBN: 978-3-527-33089-8

Devaux, P. (Ed.)

Transmembrane Dynamics of Lipids

Series: Wiley Series in Protein and Peptide Science

2011

ISBN: 978-0-470-38845-7

Gompper, G., Schick, M. (Eds.)

Soft Matter

Volume 4: Lipid Bilayers and Red Blood Cells
Series: Soft Matter (Volume 4)

2008

ISBN: 978-3-527-31502-4

Van Eyk, J. E., Dunn, M. J. (Eds.)

Clinical Proteomics

From Diagnosis to Therapy

2008

ISBN: 978-3-527-31637-3

Edited by Kim Ekroos

Lipidomics

Technologies and Applications

WILEY-VCH Verlag GmbH & Co. KGaA

The Editor

Dr. Kim Ekroos
Zora Biosciences Oy
Biologinkuja 1
02150 Espoo
Finland

Library of Congress Card No.: applied for

British Library Cataloguing-in-Publication Data
A catalogue record for this book is available from the British Library.

Bibliographic information published by the Deutsche Nationalbibliothek
The Deutsche Nationalbibliothek lists this publication in the Deutsche Nationalbibliografie; detailed bibliographic data are available on the Internet at http://dnb.d-nb.de.

Print ISBN: 978-3-527-33098-0
ePDF ISBN: 978-3-527-65597-7
ePub ISBN: 978-3-527-65596-0
mobi ISBN: 978-3-527-65595-3
oBook ISBN: 978-3-527-65594-6

Cover Design Adam-Design, Weinheim

Typesetting Thomson Digital, Noida, India

Printing and Binding Markono Print Media Pte Ltd, Singapore

Printed on acid-free paper

Contents

Preface

This is the first advanced textbook on *lipidomics*, which has been written with two major objectives in mind. One is to provide an advanced textbook covering the present state-of-the-art technologies in lipidomics, whereas the second objective is to provide a clear summary of applications in lipid cell biology, lipoprotein metabolism, lipid-related diseases, and pharmaceutical drug discovery.

The contributors are key professional leaders in their field of research. This textbook is unique in that it not only covers the analytical aspect of this field, but also covers the assimilation of the technological advancements and the biological and medical research topics to collectively advance research and human health. Therefore, the chapters within this book are written for a broad audience of readers. It is not a series of exhaustive reviews of the various topics, but rather is a current, readable, and critical summary of these areas of research. This textbook should satisfy the need for a general reference and review book for students and scientists studying lipids, lipoproteins, membranes, and lipid-related diseases in combination with lipidomics. This book should simultaneously allow scientists to become familiar with recent developments related to their own research interests, and should also aid clinical researchers and medical scientists keep abreast of developments in basic and applied sciences that are important for subsequent clinical advances.

All the chapters have been extensively revised. This volume comprises of 16 chapters altogether, covering analytical technologies, such as shotgun and targeted lipidomics, and numerous applications in basic lipid research, in lipid-related disease research, and in drug discovery. The editor and contributors assume full responsibility for the content of the various chapters.

I am indebted to all contributors and many other people who have made this book possible. In particular, I extend my thanks to my supporting wife Marika.

Espoo, Finland
May 2012

Kim Ekroos

List of Contributors

Makoto Arita
University of Tokyo
Graduate School of Pharmaceutical Sciences
Department of Health Chemistry
7-3-1 Hongo
Bunkyo-ku
Tokyo 113-0033
Japan

and

PRESTO
Japan Science and Technology Agency
Saitama
Japan

Robert M. Barkley
University of Colorado Denver
Department of Pharmacology
Mail Stop 8303
12801 East 17th Avenue
Aurora
CO 80045
USA

Stephen J. Blanksby
University of Wollongong
School of Chemistry
NSW
Australia

Jan Borén
Sahlgrenska Academy
Sahlgrenska Center for Cardiovascular and Metabolic Research
Gotenburg
Sweden

Simon H.J. Brown
University of Wollongong
School of Chemistry
NSW
Australia

Laurent Camont
National Institute for Health and Medical Research (INSERM)
Dyslipidemia, Inflammation and Atherosclerosis Research Unit (UMR939)
Paris 75013
France

and

Universite Pierre et Marie Curie-Paris 6
Paris 75013
France

and

AP-HP
Groupe Hospitalier Pitie-Salpetriere
Paris 75013
France

M. John Chapman
National Institute for Health and
Medical Research (INSERM)
Dyslipidemia, Inflammation and
Atherosclerosis Research Unit
(UMR939)
Paris 75013
France

and

Universite Pierre et Marie
Curie-Paris 6
Paris 75013
France

and

AP-HP
Groupe Hospitalier Pitie-Salpetriere
Paris 75013
France

Yanfeng Chen
Georgia Institute of Technology
School of Chemistry and
Biochemistry
Atlanta
GA 30332-0400
USA

Jeffrey H. Chuang
Boston College, Department of
Biology
140 Commonwealth Avenue
Chestnut Hill
MA 02467
USA

Christer S. Ejsing
University of Southern Denmark
Department of Biochemistry and
Molecular Biology
5230 Odense
Denmark

Kim Ekroos
Zora Biosciences Oy
Biologinkuja 1
02150 Espoo
Finland

Xianlin Han
Washington University School of
Medicine
Division of Bioorganic Chemistry
and Molecular Pharmacology
Department of Medicine
660 S. Euclid Ave
St. Louis
MO 63110-1093
USA

and

Sanford-Burnham Medical Research
Institute
Diabetes and Obesity Research
Center
Lake Nona
Orlando
FL 32827
USA

Joseph A. Hankin
University of Colorado Denver
Department of Pharmacology
Mail Stop 8303
12801 East 17th Avenue
Aurora
CO 80045
USA

Peter Husen
University of Southern Denmark
Department of Biochemistry and
Molecular Biology
5230 Odense
Denmark

Kazutaka Ikeda
The University of Tokyo
Graduate School of Medicine
Department of Metabolome
Tokyo Daigaku
7-3-1 Hongo
Bunkyo-ku
Tokyo 113-0033
Japan

and

Keio University
Institute for Advanced Biosciences
246-2 Mizukami
Kakuganji
Tsuruoka
Yamagata 997-0052
Japan

Yosuke Isobe
University of Tokyo
Graduate School of Pharmaceutical
Sciences
Department of Health Chemistry
7-3-1 Hongo
Bunkyo-ku
Tokyo 113-0033
Japan

Ryo Iwamoto
University of Tokyo
Graduate School of Pharmaceutical
Sciences
Department of Health Chemistry
7-3-1 Hongo
Bunkyo-ku
Tokyo 113-0033
Japan

Minna T. Jänis
Zora Biosciences Oy
Biologinkuja 1
02150 Espoo
Finland

Hui Jiang
Washington University School of
Medicine
Division of Cardiology
St. Louis
MO 63110
USA

Michael A. Kiebish
Washington University School of
Medicine
Division of Bioorganic Chemistry
and Molecular Pharmacology
Department of Medicine
660 S. Euclid Ave
St. Louis
MO 63110-1093
USA

Daniel A. Kirschner
Boston College
Biology Department
Chestnut Hill
MA 02467
USA

Christian Klose
Max Planck Institute of Molecular
Cell Biology and Genetics
Pfotenhauerstr. 108
01307 Dresden
Germany

Anatol Kontush
National Institute for Health and Medical Research (INSERM)
Dyslipidemia, Inflammation and Atherosclerosis Research Unit (UMR939)
Paris 75013
France

and

Universite Pierre et Marie Curie-Paris 6
Paris 75013
France

and

AP-HP
Groupe Hospitalier Pitie-Salpetriere
Paris 75013
France

Reijo Laaksonen
Zora Biosciences Oy
Biologinkuja 1
02150 Espoo
Finland

Marie C. Lhomme
National Institute for Health and Medical Research (INSERM)
Dyslipidemia, Inflammation and Atherosclerosis Research Unit (UMR939)
Paris 75013
France

and

Universite Pierre et Marie Curie-Paris 6
Paris 75013
France

and

AP-HP
Groupe Hospitalier Pitie-Salpetriere
Paris 75013
France

Lynette Lim
National University of Singapore
Centre for Life Sciences
Yong Loo Lin School of Medicine
Department of Biochemistry
28 Medical Drive
Singapore 117456
Singapore

Ying Liu
Emory University School of Medicine
Department of Human Genetics
Atlanta
GA 30322
USA

Todd W. Mitchell
University of Wollongong
School of Health Sciences
NSW
Australia

Robert C. Murphy
University of Colorado Denver
Department of Pharmacology
Mail Stop 8303
12801 East 17th Avenue
Aurora
CO 80045
USA

Hiroki Nakanishi
The University of Tokyo
Graduate School of Medicine
Department of Metabolome
Tokyo Daigaku
7-3-1 Hongo
Bunkyo-ku
Tokyo 113-0033
Japan

and

Research Center for Biosignal
Akita University
1-1-1 Hondo
Akita-city
Akita 010-8543
Japan

Yoshinori Satomi
Takeda Pharmaceutical Co. Ltd.
Pharmaceutical Research Division
Biomolecular Research Laboratories
2-26-1 Muraokahigashi
Fujisawa
Kanagawa 251-0012
Japan

Thomas N. Seyfried
Boston College, Department of
Biology
140 Commonwealth Avenue
Chestnut Hill
MA 02467
USA

Guanghou Shui
National University of Singapore
Centre for Life Sciences
Life Sciences Institute
28 Medical Drive
Singapore 117456
Singapore

M. Mobin Siddique
Duke National University of
Singapore
Graduate Medical School
8 College Road #8-15
Singapore 169857
Singapore

Kai Simons
Max Planck Institute of Molecular
Cell Biology and Genetics
Pfotenhauerstr. 108
01307 Dresden
Germany

Marcus Ståhlman
Sahlgrenska Academy
Sahlgrenska Center for
Cardiovascular and Metabolic
Research
Gotenburg
Sweden

M. Cameron Sullards
Georgia Institute of Technology
School of Chemistry and
Biochemistry
Atlanta
GA 30332-0400
USA

and

Georgia Institute of Technology
School of Biology and the Petit Institute for Bioengineering and
Bioscience
Atlanta
GA 30332-0400
USA

Scott A. Summers
Duke University Medical Center
Sarah W. Stedman Nutrition and
Metabolism Center
Durham
NC
USA

and

Duke National University of
Singapore
Graduate Medical School
8 College Road #8-15
Singapore 169857
Singapore

Michal Surma
Max Planck Institute of Molecular
Cell Biology and Genetics
Pfotenhauerstr. 108
01307 Dresden
Germany

Ryo Taguchi
Chubu University
College of Life and Health Sciences
Department of Biomedical Sciences
1200 Matsumoto-cho
Kasugai
Aichi 487-8501
Japan

and

The University of Tokyo
Graduate School of Medicine
Department of Metabolome
Tokyo Daigaku
7-3-1 Hongo
Bunkyo-ku
Tokyo 113-0033
Japan

Kirill Tarasov
Zora Biosciences Oy
Biologinkuja 1
02150 Espoo
Finland

Markus R. Wenk
National University of Singapore
Centre for Life Sciences
Yong Loo Lin School of Medicine
Department of Biochemistry
28 Medical Drive
Singapore 117456
Singapore

and

National University of Singapore
Centre for Life Sciences
Department of Biological Sciences
28 Medical Drive
Singapore 117456
Singapore

and

National University of Singapore
Centre for Life Sciences
Life Sciences Institute
28 Medical Drive
Singapore 117456
Singapore

Karin A. Zemski Berry
University of Colorado Denver
Department of Pharmacology
Mail Stop 8303
12801 East 17th Avenue
Aurora
CO 80045
USA

Lu Zhang
Boston College
Department of Biology
140 Commonwealth Avenue
Chestnut Hill
MA 02467
USA

1
Lipidomics Perspective: From Molecular Lipidomics to Validated Clinical Diagnostics

Kim Ekroos

1.1
Introduction

Lipids are recognized as extremely diversified molecules, with nearly 10^4 different structures of lipids currently being stored in the most comprehensive lipid structure database (LIPID MAPS, http://www.lipidmaps.org). The complexity is confounded by the fact that the absolute quantity of individual molecular lipids can differ among lipid species up to several million-fold depending on the matrix of origin. These features unprecedentedly complicate a precise assessment of the actual number of lipid entities and their identities and quantities making up a lipidome of a biological system. The accomplishment of this task now relies on lipidomics. Cutting edge lipidomics has already demonstrated its supremacy by revealing, for example, over 500 lipid species in human plasma [1, 2] and 250 lipid species in yeast [3]. With the currently ongoing meticulous developments in this field, it is highly anticipated that it will in the coming years facilitate delivery of a close to complete lipidomic content and precisely determined. If this increases the total number of species, for instance, in the human plasma lipidome to comprise thousand lipid species or more still remains to be seen.

The lipidome of eukaryotic cells is believed to contain thousands of lipid entities that structurally and chemically regulate cell membranes, store energy, or become precursors to bioactive metabolites [4, 5]. Lipids primarily reside in cellular membranes. The individual membranes of a cell have unique lipid compositions, required for serving their vital biological functions. For example, the free cholesterol (FC) to total phospholipid (PL) ratio in the endoplasmic reticulum (ER) membrane in mammals has been shown to be 0.15, whereas in the plasma membrane the ratio is 1 [6]. Although ER is the main site of lipid synthesis, it is the local lipid metabolism that is the prime determinant of the unique compositions of organelles. Moreover, further membrane specialization is orchestrated by lateral organization to form dynamic platforms, that is, lipid rafts, within the cellular membranes serving as functional assemblies for diverse processes such as signal transduction, membrane trafficking, and cell adhesion [7]. The physiological response and bioactive output of such lipid raft domain or cellular membrane will

Lipidomics, First Edition. Edited by Kim Ekroos.

collectively be defined by the present molecular lipid structures, their local concentrations, and spatial distributions [8]. In addition, several studies have demonstrated and highlighted the importance and specificity of single-molecule lipid structures in determining the biofunctionality. Thus, based on these facts, it is highly anticipated that a defect in the underlying lipid regulation can lead to deleterious effects on the cell or organism and assist in the pathophysiology of diseases.

Precise determination of molecular lipid species becomes a prerequisite not only to gain their biological functions that might vary depending on the localization but also to gain their roles in a lipid collective. It is highly envisioned that this opens up new avenues in cell biology, biochemistry, and biophysics as it will untie the organization and function of the complex lipid metabolism machinery and its association with the construction and formation of unique cellular membranes. An essential aspect is that this information will accelerate our understanding of human diseases. This applies not only to pharmaceutical drug discovery programs but also to nutrition programs. It is known that major diseases such as atherosclerosis, infectious diseases, Alzheimer's disease, and cancer all have a lipid component in their epidemiology. Through precisely defined maps of the lipid metabolism and its regulation, more targeted delineations of the underlying dysfunctional metabolic pathways and cellular events can be obtained that are likely to result in the discovery of the culprit(s) associated with a particular disease. As it will unravel the mechanisms of action, one can envision that this will advance the discovery of new drug targets and efficacy and diagnostic biomarkers. In the field of nutrition research, the gained know-how will facilitate innovation of healthier food formulas. Finally, as the valid drug efficacy and disease diagnostic lipid biomarkers are discovered, they need to be transferred into a regulatory environment. Not only will this demand stringent analytical quality fulfilling the regulatory guidelines, but will also require the assays to be cost-effective and high-throughput oriented.

This chapter describes the types of lipid information the currently applied analytical platforms produce and how these differently assist in understanding biology. Future viewpoints of lipidomics in respect to its expected deliveries and upcoming challenges will be given. Special focus has been put on molecular lipids as these hold the answers in lipid biology.

1.2
Hierarchical Categorization of the Analytical Lipid Outputs

The applied analytical approach determines the level of details of delivered lipid information. According to the currently available techniques, the following hierarchical categorization can be made: (i) lipid class, (ii) sum compositions, (iii) molecular lipids and its related category, and (iv) structurally defined molecular lipids (Figure 1.1). Following this hierarchy, the number of entries belonging to phosphatidylcholine (PC) in human red blood cells would, for instance, be 1 lipid class, 18 sum compositions, more than 40 molecular lipids, and finally more than 100 structurally defined lipids [9].

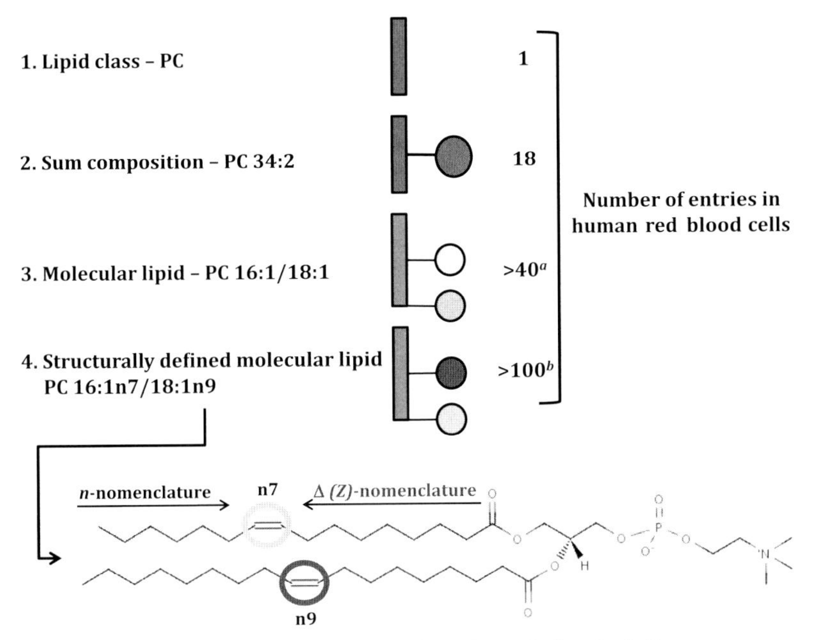

Figure 1.1 Hierarchical categorization of lipid outputs based on the analytical approach. The number of entries per category is based on PC of human red blood cells [9]. The structure of PC 16:1n7/18:1n9 is shown. [a]Expected number of entries including all positional isomers. [b]Likelihood of number of entries, although it still remains unknown.

1.2.1 Lipid Class

The first lipid information level is lipid class. This output principally originates from the use of traditional techniques such as thin-layer chromatography (TLC) and normal-phase liquid chromatography (NPLC). These techniques together with gas chromatography (GC) have been the principal tools for assessing lipid measurements over decades. Profound descriptions of their principles and applications exist in literature. The benefit of these techniques is their capability to separate lipids into respective classes. This ability has been enormously utilized for exploring the lipid content of biological tissues and biofluids, results that strongly impacted the evolution of the lipid biology framework. However, their major drawbacks are recognized in the incapability of elucidating individual lipid entities, low detection sensitivity, and time ineffectiveness.

Lipid class measurements can also be obtained using the current lipidomic techniques. This can, for example, be achieved using shotgun lipidomics by monitoring lipid class selected fragments. For instance, precursor ion scanning (PIS) analyses of m/z 184.1 in positive ion mode selectively detects phosphorylcholine containing lipids such as phosphatidylcholine (PC) and sphingomyelin (SM) [10]. The corresponding PC or SM lipid class levels would be obtained by adding up all signals of the identified species, for example, 50. We have previously proven this approach to be valid. Here, we monitored the molecular composition of lipid species analyzed by shotgun lipidomic analysis of total HepG2 lipid extracts before and after NPLC fractionation [11]. The molecular species composition was not affected by the NPLC separation. Moreover and most importantly, direct analysis of total lipid extracts by PIS estimated the total amount of cholesteryl ester (CE) in HepG2 cells to be 57 nmol/mg protein. In comparison, quantification using evaporative light scattering (ELS) detection together with the NPLC fractionation determined the total amount of CE to be 53 nmol/mg protein. Thus, two independent approaches and detector systems were shown to produce identical outputs. This strongly indicates that the applied lipidomic method is both qualitatively and quantitatively valid. Notably, factors such as methodological approach, selection of internal standards, and isotopic correction will influence the quantification accuracy.

Global or untargeted lipidomics, such as liquid chromatography-based full scan mass spectrometry (MS) analyses (i.e., LC-MS), could also in theory be utilized to determine total lipid class content. However, it has been recognized that ion suppression, which strongly influences quantification, might be more complex during LC-MS analysis. Since ion suppression is likely to vary during the chromatographic run due to the difference in the eluting mobile-phase composition (for gradient LC-MS methods) and sample matrix, it can lead to unequal signal responses of the different lipid species of the same lipid class even though they are present at equimolar concentrations. Data supporting this idea were recently published [1]. Optionally, the collision energy could be optimized for each analyte to correct the suppression effects. However, this becomes difficult or even unpractical as the settings are likely to be different, depending on, for example, sample matrix (different background) and LC conditions. This issue is best solved by using stable isotope-labeled lipid standards that are structurally similar to the endogenous species. Under such circumstances, targeted LC-MS approaches have shown to be superior for absolute quantification. However, since synthetic standards for each endogenous lipid species are still unavailable, lipid class quantification in absolute amounts by summing multiple various lipid species of the same class is not feasible by this method. This has been described in greater detail in Chapter 5. Thus, the quantification accuracy related to absolute lipid class content from reverse-phase LC-MS-based lipidomic data still awaits to be proven. Until then, it is recommended that the available published results should only be considered as estimates.

1.2.2 Sum Compositions

The level following lipid class is sum composition or brutto lipids. Common lipidomic approaches are capable of elucidating lipids with different sum compositions, for example, phosphatidylethanolamine (PE) 36:4, where 36 represent the total number of carbon and 4 the total number of double bonds in the attached fatty acids [10]. In a full mass spectrometry analysis, such type of information can already be obtained. Since no selective analysis modes are usually required, a profile of the sum lipid composition can be very rapidly acquired, either in conjunction with LC or with direct infusion approaches. For instance, in the latter, by taking advantage of the high mass-resolving power of instruments, such as an orbitrap or a Fourier transform ion cyclotron resonance mass spectrometer, a broad profile of brutto lipids can be readily identified and quantified in only minutes from unresolved samples [12]. Here, the high mass accuracy is used to separate the actual lipid peaks from the chemical noise. The simplicity, reliability, and the speed of such methods assisted by the lipid software advancements [13] have become not only attractive for standard lipidomic analyses but also very appealing for high-throughput lipidomic screenings. However, the grave weakness of this approach is that the results are still difficult to biologically interpret due to the missing details of the molecular lipids. This is more thoroughly discussed below.

1.2.3 Molecular Lipids

After sum compositions follow molecular lipids in the hierarchy (Figure 1.1). Targeted or focused lipidomic approaches such as LC-MRM and shotgun-based PIS and neutral loss scanning (NLS) are well-established techniques for the identification and quantification of molecular lipids. Their common lipid output could, for example, be PC 16:0/18:1, where the information on the type of fatty acids and their positions attached to the glycerol backbone making up the particular lipid molecule are revealed. Alternatively, this could be output as PC 16:0–18:1, where " – " describes that the positions of the fatty acids are not determined. The basis of these approaches is to monitor the lipid characteristic fragmentation ions, for example, head groups and acyl anions, to delineate the molecular species. MRM, PIS, and NLS techniques are described in greater detail in other chapters of this book and therefore only a brief overview is given here.

In MRM, m/z of both precursor and fragment ions are specified. Precursors of interests are isolated in quadrupole Q1 and subjected to fragmentation in quadrupole Q2. Subsequently, selected fragment ions are set to pass in Q3, and the abundance of the specified fragment ions is monitored by the detector allowing quantification of targeted molecular lipids. In conjunction with LC, this approach becomes highly selective as the latter system facilitates a vast and reproducible sample cleanup prior to MRM. Reduction of sample complexity prior to MS analysis improves not only the success rate of monitoring molecular lipids but also the sensitivity of the method. However, a drawback of this approach is the limited transitions, that is, number of molecular lipids, which can be covered during an analysis run due to insufficient chromatographic peak collection caused by the limited acquisition speed of the MS.

In contrast, a shotgun lipidomics-based PIS and NLS analysis is typically not limited to acquisition time, as it has been shown that minute sample extracts can be robustly infused for an hour or even more [11]. Therefore, the lipid coverage can be significantly greater with this approach. On a quadrupole time-of-flight (QTOF) instrumentation, we previously demonstrated the possibility to simultaneously acquire 40–50 PIS using multiple precursor ion scanning (MPIS) [14]. Using the recent QTRAP technology, we can rapidly and sensitively acquire a total of 70–80 PIS and 20–30 NLS that cover both fatty acids and lipid head group fragment ions within the quadrupole Q3, while the quadrupole Q1 is scanning lipid precursors [1]. Typically such an analysis identifies and quantifies several hundred different molecular species in approximately 30 min. These shotgun lipidomic methods have proven suitable for high-throughput lipidomic screenings.

Molecular lipid information could also be retrieved by, for instance, fragmenting (i.e., MS/MS) all eluting peaks during a chromatographic run or all precursors detected in a direct infusion full scan (i.e., MS) analysis. An example of the latter is the recently described technique sequential precursor ion fragmentation [15]. Here, precursors in a selected mass range are stepwise isolated (1 amu) in Q1 at unit-based resolution and subjected to collision-induced dissociation (CID) in Q2, while collecting more than a thousand MS/MS spectra covering every precursor in the mass range of each cycle. The power of this methodology is that it collects full MS and MS/MS of every precursor, and therefore nothing is left behind. Utilizing the high acquisition speed and mass accuracy of the recent QTOF technology, a complete lipidomic analysis covering over 400 molecular lipids in human plasma could be accomplished in less than 12 min including positive and negative polarities [15]. This is the best performance of a molecular lipidomic methodology at present. Evidently, this type of emerging instrumental technologies in combination with matching software tools will create new opportunities in molecular lipidomics as it amends the extensive acquisition times and maintains outstanding data quality and comprehensiveness. Thus, the outlooks are most promising and positively will open up new solutions for high-throughput screenings.

1.2.4
Structurally Defined Molecular Lipids

The molecular lipid information will immensely facilitate the untying of the unknown knowledge in lipid biology. Further advancements will be achieved once more structural information of the particular molecular species can be determined, such as the double bond position determination in the attached fatty acids. Analytical approaches for elucidating this type of information have recently emerged. Although the technologies are still rather immature, they deliver essential biological information. Therefore, the final level in the hierarchy is defined as structurally defined molecular lipids (Figure 1.1).

A most promising technique is OzID, which is described in greater detail in Chapter 6. The basis of this technology is that ozone vapor is introduced to the collision cell of the mass spectrometer, which will react with double bonds, for example, of fatty acids, and selectively dissociate them. This process therefore generates characteristic fragment ions that facilitate determination of the double bond position [16, 17]. This technique applies in principle to all types of double bonds.

For example, Mitchell and colleagues have shown that OzID facilitates proper identification of ether lipids, that is, containing alkyl and alkylen bonds, which are typically difficult to assess by conventional MS approaches [18]. Preliminary results also suggest that this technique could distinguish *cis*- and *trans*-bonds, however this still needs to be proven. Thus, the emergence of completely structurally defined molecular lipids is awaited in the near future. Clearly, this evolution will depend on OzID and other similar nascent techniques.

1.3 The Type of Lipid Information Delivers Different Biological Knowledge

It is critical to underscore that the various detail levels represent different implications in biology. For example, lipid class information does not reveal the detailed composition of a plasma membrane, whereas molecular lipid species information is required to fulfill this task. In contrast, triacylglycerol (TAG) level in human plasma enables us to better understand the health condition. For instance, high TAG levels in human plasma (hypertriglyceridemia), has been identified as a risk factor for coronary artery disease (CAD). Thus, the different types of lipid outputs guide us to understand a biological system from diverse angles.

As mentioned above, our current know-how in lipid biology has been strongly impacted by the extensive measurements of lipid classes performed over decades. Lipid class information produces an essential overview of a biological system. For example, it is well known that high level of cholesterol in low-density lipoprotein (LDL) is a hallmark for increased risk of atherosclerosis. Another example would be monitoring of membrane fluidity by measuring the PC to PE ratio. Here, it has been implicated that a decrease in the ratio might induce a loss of membrane integrity [19, 20]. The repertoire of biological examples is extensive and well documented. Thus, substantial biological understanding has already been gained through studying deviations in lipid classes from their normal levels. Undoubtedly, this will remain as an essential asset for upcoming lipid research.

Information on brutto lipids has been rapidly emerging during the recent years. Sum composition information has been, for example, utilized to estimate the lipid content of cellular membranes [21], isolated viruses [22], and cells [23, 24]. The available information on total double bonds has further been used to determine the degree of saturation, that is, saturation index, which is useful for studying membrane behavior. However, sum compositions to a great extent have been assessed in studies related to diseases or dysfunctions. Here, the main objective has been to identify diagnostic or prognostic biomarkers based on observed deviations in brutto species between healthy controls and cases. Many studies of this kind have been described over the recent years. However, less focus has been put on elucidating the underlying biological mechanisms causing the observed changes. A prime reason for this is that the obtained results are normally difficult to interpret as such sum compositions do not exist in biological systems, rather it represents a collection of lipids. Moreover, there is a high risk that such a collection

Δ5	Δ6	Δ9	Δ11	Δ13	Δ15
n15	**n12; n10**	**n5; n7; n9; n11**	**n7; n9; n11**	**n9**	**n7; n9**
16:0/20:1n15/16:0	16:0/18:0/18:1n12	14:1n5/20:0/18:0	16:0/18:0/18:1n7	14:0/22:1n9/16:0	14:0/22:1n7/16:0
16:0/16:0/20:1n15	16:0/18:1n12/18:0	14:1n5/18:0/20:0	16:0/18:1n7/18:0	14:0/16:0/22:1n9	14:0/16:0/22:1n7
14:0/20:1n15/18:0	18:1n12/16:0/18:0	20:0/14:1n5/18:0	18:1n7/16:0/18:0	22:1n9/14:0/16:0	22:1n7/14:0/16:0
14:0/18:0/20:1n15	14:0/20:0/18:1n12	14:1n5/22:0/16:0	14:0/20:0/18:1n7		14:0/24:1n9/14:0
18:0/14:0/20:1n15	14:0/18:1n12/20:0	16:0/14:1n5/22:0	14:0/18:1n7/20:0		14:0/14:0/24:1n9
	20:0/14:0/18:1n12	22:0/16:0/14:1n5	20:0/14:0/18:1n7		
	16:1n10/18:0/18:0	14:0/24:0/14:1n5	14:0/20:1n9/18:0		
	18:0/16:1n10/18:0	24:0/14:0/14:1n5	14:0/18:0/20:1n9		
	16:0/20:0/16:1n10	24:0/14:1n5/14:0	18:0/14:0/20:1n9		
	16:0/16:1n10/20:0	16:1n7/18:0/18:0	16:0/20:1n9/16:0		
	16:1n10/16:0/20:0	18:0/16:1n7/18:0	16:0/16:0/20:1n9		
	14:0/22:0/16:1n10	16:0/20:0/16:1n7	16:0/20:1n11/16:0		
	14:0/16:1n10/22:0	16:0/16:1n7/20:0	16:0/16:0/20:1n11		
	22:0/14:0/16:1n10	14:0/22:0/16:1n7	14:0/22:1n11/16:0		
		14:0/16:1n7/22:0	14:0/16:0/22:1n11		
		22:0/14:0/16:1n7	22:1n11/14:0/16:0		
		16:0/18:0/18:1n9			
		16:0/18:1n9/18:0			
		18:1n9/16:0/18:0			
		14:0/20:0/18:1n9			
		14:0/18:1n9/20:0			
		20:0/14:0/18:1n9			
		14:0/20:1n11/18:0			
		14:0/18:0/20:1n11			
		18:0/14:0/20:1n11			

Figure 1.2 Possible molecular TAG species corresponding to brutto TAG 52:1 in mammalians. Delta (Δ) and n nomenclatures describes the double bond positions. Delta nomenclature is used to assign the position of an individual double bond or specificity of enzyme inserting it, whereas n nomenclature is used to assign individual fatty acids within a family of structurally related lipids [25].

could produce misleading results due to influence of contaminating species that are not associated with the actual study topic. For instance, as shown in Figure 1.2, in mammalians the TAG 52:1 could comprise nearly 90 different structurally defined molecular species. Thus, this aspect hinders both the biological interpretation and the success rate of the biomarker discovery. Despite this drawback, sum composition information can offer a valuable biological insight, but notably at a more general level. How to most optimally use sum composition information still remains blurred. Deconvolution strategies to outline the underlying species could be a way forward to gain insights into the biological mechanisms as this would facilitate metabolic mapping explorations.

Most promise in lipid research rests on the molecular lipids and structurally defined lipids as these are highly expected to open up new paths in biomedical research. As described above, biology is not regulated at the lipid class or sum composition level, but rather at the level of actual molecular lipid species. As already mentioned, the likelihood of misidentification significantly increases in a sum composition analysis, due to which molecular lipid species information is masked in these data. This becomes evident from Figure 1.2, recognizing the potential of vast arsenal of underlying species of a single brutto entity. Undoubtedly, reliable analysis of the molecular lipid species is of utmost importance as they may have very well-defined functional roles. Therefore, the molecular lipid species information (together with structurally defined molecular species) should give us the highest success rate in identifying the culprits in the causal lipid metabolic networks leading to metabolic dysfunction states. This is discussed in more detail in the following section.

1.4 Untying New Biological Evidences through Molecular Lipidomic Applications

Chapter 2 intriguingly reviews the multifaceted lipid architecture in cells with emphasis on the capability of lipids to form morphologically different membrane structures for maintaining the cell function. Fascinatingly, a functional human red blood cell requires over 100 distinct PC molecules at individual concentrations and distribution in its plasma membrane (Figure 1.1). Why such a high number of individual molecules is required still remains unknown. However, it demonstrates the biological complexity of lipids. Moreover, the authors describe that platforms, that is, lipid rafts, existing within membranes have highly specific functions. However, the knowledge of which lipids make up the lipid rafts is still limited. In addition, the platforms are likely to undergo reformations in their compositions, both in lipids and proteins, to alter their function. Obviously, the molecular lipid composition of this platform plays a vital role and therefore needs to be delineated in great detail to understand its function. Only molecular lipidomics accompanied with structurally defined molecular species can address this type of questions. This fulfills precise identification and quantification of each present lipid species, although issues such as insufficient sensitivity will remain as the sample amounts will be extremely small. Undoubtedly, this is an enormous challenge that lipidomics will be facing in the coming years; hence, lipidomics should be orchestrated with biophysical and biochemistry experiments to answer the cell biological questions.

Several studies have demonstrated the importance and specificity of single-molecule lipid structures rather than a lipid (and protein) collective in determining the biofunctionality. An impressive work has been shown by Shinzawa-Itoh *et al.*. They showed by sophisticated experiments that the oxygen transfer mechanism in cytochrome c oxidase requires a specific phosphatidylglycerol molecular lipid with palmitate (C16:0) and vaccenate (C18:1n7) at the *sn*-1 and *sn*-2 positions, respectively, on the glycerol backbone [26]. Altering the double position from n7 to n9 (from vaccenate to oleate) inactivated the function. Thus, a small conformational change in the attached fatty acid is sufficient to inactivate the oxygen transfer mechanism. Moreover, Menuz *et al.* showed that C24–C26 carbon ceramides mediated the death of a *Caenorhabditis elegans* mutant that failed to resist asphyxia, whereas ceramides with shorter chains had the opposite effect [27]. We have recently been able to obtain similar findings in mammalians. Here, we showed that the C24 carbon ceramide induced ER stress, whereas the shorter chain (C20–C22 and C16) ceramides had no effect on HL-1 cardiomyocyte cells [28]. Finally, Ewers *et al.* recently showed that the structure of the ganglioside GM1 determines the simian virus 40 (SV40)-induced membrane invagination and infection [29]. They demonstrated that GM1 molecules with long acyl chains facilitated entering of SV40 through the host cell plasma membrane, while GM1 molecular species with short hydrocarbon chains failed to support the invagination and endocytosis and infection. Evidently, these examples already underscore the essence of molecular lipids and technologies for their discoveries.

Exhaustive research lies behind our current knowledge of lipid metabolic pathways. Although the lipid metabolism is well characterized, information about the molecular lipid metabolism still remains unknown. The main reason for this is that the applied technologies were capable of determining only those lipids that are at the lipid class level. Even though gas chromatography-based analyses have directly allowed tracking of the fatty acid and oxidized fatty acid metabolisms, there still remain a significant number of unknowns in their metabolisms, especially of the latter. This is mainly due to lack of sensitivity and specificity of the applied approaches.

Much promise now relies on lipidomics to untie the lipid metabolism at molecular level (and structural defined). Lipidomics exhibits the analytical preferences for this quest. We have recently demonstrated the power of lipidomics in elucidating the lipid metabolism in yeast [3]. Although the pathways are outlined in the form of lipid classes, the experiments identified that the individual classes comprise of unique lipid species, thus indicating that the metabolism is regulated at the molecular level. However, there is still no solution in respect to how to connect the molecular lipids in the metabolism. An utmost challenge lies in the measurements of single metabolic events. It can be expected that a whole-cell measurement represents the total sum of events that concurrently occur in the cell. Thus, how to separate, for instance, a local metabolism of PC in the plasma membrane from a coexisting PC synthesis in ER or metabolism in other organelles remains unclear. Fortunately, enlightening approaches tackling this are emerging. One approach is to perform metabolic flux experiments using stable isotopes that incorporate selectively and efficiently into lipids of interest. The approach will then be to specifically measure the labeled lipids during a time course to produce a kinetic readout of the lipids of interest. In this way, the synthesis or catabolism rates of lipids could be established. Haynes *et al.* recently described the use of a stable isotope-labeled precursor ([U-^{13}C]palmitate) to analyze *de novo* sphingolipid biosynthesis by tandem mass spectrometry [30]. Moreover, Pynn *et al.* incorporated a deuterated methyl-D9-labeled choline chloride to quantify biosynthesis fluxes through both the PC synthetic pathways *in vivo* in human volunteers and compared these fluxes with those in mice [31]. In conjunction with sophisticated lipidomics, they were able for the first time to show that phosphatidylethanolamine-*N*-methyltransferase (PEMT) pathway in human liver is selective for polyunsaturated PC species, especially those containing docosahexaenoic acid. Kuerschner *et al.* utilized a highly unique isotopic label facilitating detailed lipidomics-assisted tracking of labeled molecular lipids and in concert with their cellular localization by fluorescent microscopy [32]. The label does not necessary need to be detectable by advanced microcopy techniques. For instance, a general (nonfluorescent) labeled lipid could be precisely determined by imaging lipidomic techniques (described in Chapter 7). Very recently, this labeling approach in conjunction with lipidomics lent to the discovery of exclusively one sphingomyelin species, namely, d18:1/18:0, interacting directly and being highly specific with the transmembrane domain (TMD) of the COPI machinery protein p24 [33]. The results demonstrate that the exclusive molecular sphingomyelin acts as cofactor to regulate the function of a transmembrane protein and thus again point out why biological membranes are assembled from such a large variety of different lipids and the essence of single entities. Taken together, labeling experiments in concert with molecular and imaging lipidomics compose the right ingredients for tackling the delineation of the molecular lipid metabolism. Together with subcellular dissection of the cell, enzymatic silencing or inhibitory experiments, and supported bioinformatics tools, these show the most promise for the discovery of the biological roles of molecular lipids and mapping of the molecular lipid pathways.

1.5 Molecular Lipidomics Approaches Clinical Diagnostics

The foreseeable biological specificity residing in molecular lipids make them prime candidates for drug and biomarker discovery. However, the success rate will substantially depend on the selected experimental design and its accomplishment. The use of isotopic tracers, lipid imaging, and subcellular dissections are likely to be key assets. These types of experiments are rather trivial to perform *in vitro*, however, for instance, isotopic labeling *in vivo* is still a very challenging task. Another asset is the integration of genomic and proteomic data with the molecular lipidomic data set. However, it still remains blurry in how to mine such large data sets. How biological representative the currently existing results are remain very unclear, bearing in mind that the applied lipidomics results are mainly based on lipid sum compositions. No clear evidences exist demonstrating that the "omic" mining results correspond to the results from basic biochemistry and biology. Thus, the risk for misinterpretation can still be rather high in such hypothesis-driven experiments. A final strength is the combination of biochemistry, analytical chemistry, biophysics, biology, bioinformatics, and medicine know-how and expertise. Taken together, the receipt for discovery of new drugs and lipid biomarkers rests on how to retrieve most out of the above-described assets (and others) alone and/or collectively, and in most accurate way. Convincingly, an optimal component setup will take us beyond our current understanding in lipid biochemistry and biology. It is anticipated that the localization and function of the lipid metabolic machinery, including its active components and interactive companions, to be inclusively illuminated. The resulting novel lipid maps will foster the discovery of novel mechanisms of action (MoA), drug targets and drug efficacy and disease diagnostic biomarkers.

As clearly illuminated in other chapters, lipids are highly awaited serving as drug efficacy and disease diagnostic and prognostic biomarkers. A main reason for this is that lipids are physiological readouts of the complex gene-driven system that is affected by environmental factors. Molecular lipidomics in combination with the appropriate clinical samples and biobank material can therefore be highly considered for escalating the improvement of disease diagnostics and prediction. This applies not only to a certain disease but also to many therapeutic areas, including cardiovascular diseases, neurological states, cancers, metabolic diseases such as diabetes, and inflammatory processes. It is anticipated that a single or up to a handful of molecular lipids rather than 20–50 different biological molecules fulfill the diagnostic purposes. Moreover, as lipids are considered as intermediate phenotypes that are actually much closer to the disease state in question than for instance genetic information, they could also serve as candidates for companion diagnostics in the pharmaceutical arena, which is moving increasingly toward the specialized therapeutics model. Similarly, as lipids have been highly preserved throughout the evolution of life, that is, highly similar lipid contents throughout the mammalian species, they can be highly considered for the assessment of translational medicine and thereby help in identifying the optimal experimental animal model most closely mimicking the human disease. A promising example has recently been described by Chan *et al.* [34]. They could identify a correlating behavior of GM3 and CE in certain Alzheimer's transgenic mouse models and in Alzheimer's disease patients. Although these results are most encouraging, they need to be further proved considering that the transgenic mice displayed highly dramatic lipid changes that were not seen in humans.

Once novel drug efficacy and prognostic and diagnostic molecular lipid biomarkers are discovered, the next step will be to move their monitoring into clinical laboratories. However, before this can take place it is required that the biomarkers are thoroughly validated. For instance, the identity and quantity of the discovered lipid biomarkers should be verified simultaneously as the analysis of other independent cohorts should validate the findings. Optimally, the biomarker validation should be performed in different clinical or diagnostic laboratories. In this cascade of the biomarker development, it will be required to adapt the lipidomic assay according to the regulatory requirements following the US Food and Drug Administration (FDA) guidelines. Consequently, the lipidomic assay has to undergo a thorough validation procedure, which includes, for instance, the determination of the method accuracy, precision, lower limit of quantification (LLOQ), long- and short-term stability, freeze–thaw cycles, and robustness. An example of a calibration curve of ceramide d18:1/17:0 including corresponding quality controls (QC) is shown in Figure 1.3. Here, the performance of the used LC-MRM method can be verified by the observed linear instrument response and performance of minimum three different QC samples in accordance with the FDA guidelines. Thus, this type of results confidently indicates that validated methods for measuring molecular lipids such as ceramides can be established. Indeed, Scherer *et al.* recently demonstrated a rapid and validated LC-MS assay for the measurement of plasma sphingosine 1-phosphate, sphinganine-1-phosphate, and lysophosphatidic acid [35]. Bearing in mind the labiality and the analysis difficulty of these lipids, these results prosperously demonstrate the feasibility of using lipids as validated diagnostic endpoints. The upcoming challenges, however, include the availabilities of qualified

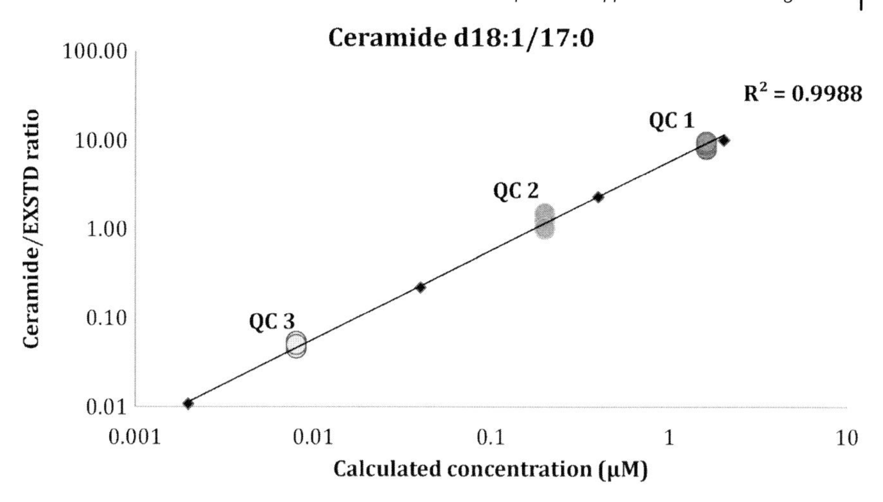

Figure 1.3 Calibration curve with corresponding QCs. Samples prespiked with various amounts of the analyte ceramide 18:1/17:0 and extracted followed by a postspiked external standard served for constructing the calibration curve ($n = 7$). Extracted individual samples of equivalent biological matrix containing known concentrations of the analyte (low QC, QC 3; middle QC, QC 2; and high QC, QC 1) served as quality controls ($n = 6$ per QC). All samples were measured by LC-MRM. A linear regression coefficient R^2 of 0.9988 demonstrates a linear instrument response inside the selected concentration range. From this, the lower limit of quantification (LLOQ) is determined. The mean values of the measured QCs are within ±15%, thus fulfilling the accuracy and precision requirements set in the FDA guidelines. The obtained results justify the FDA acceptance criteria and are essential parts in the full method validation process. *x*-axis represents the calculated concentration of the analyte (μM) and *y*-axis the measured analyte to external standard ratio.

internal standards, qualified sample handling, high-throughput adaptation, and FDA-approved MS-based setups.

Finally, but not the least, molecular lipidomics is recognized as an essential toolkit in nutrition research. Here, the scope will be to assist in optimizing food recipes aiming for improving human health. For instance, an interest is to reduce the omega-6/omega-3 ratio as a high level, such as 15: 1, in the diet of today's Western world, which promotes the pathogenesis of many diseases, including cardiovascular disease, cancer, and inflammatory and autoimmune diseases syndrome. It has been shown that a ratio of 2.5: 1 reduces rectal cell proliferation in patients with colorectal cancer and a ratio of 2–3: 1 suppresses inflammation in patients with rheumatoid arthritis [36]. As the current studies are typically based on total fatty acid analyses, scarce or even no insights are produced on the underlying lipid metabolisms. Such delineation by molecular lipidomics and especially the emerging OzID lipidomics would produce a detailed overview of the fatty acid and lipid metabolic machinery and how this is regulated. An example of such delineation is given by Ståhlman *et al.* where they pinpointed selected metabolic shifts in the TAG metabolism of dyslipidemic subjects [37]. They showed that dyslipidemia was associated with elevation in TAG molecular species containing vaccenic acid, especially

TAG 16:0/18:1n7/16:0, indicating an involvement of delta-9 desaturase and elovl-5. It needs to be noted that this is the first time complex endogenous lipid species have been identified and quantified at this detailed level. In the case of omega-6/omega-3 ratio, a target of a similar outlining could be to identify and utilize the metabolic switches that favor the omega-3 production or other beneficial metabolites.

1.6 Current Roadblocks in Lipidomics

The success of a MoA, drug, or biomarker study decidedly depends on the bioanalytical quality. Since the stability information of molecular lipids in various matrices or milieus remains scarce or unknown, the sample handling should be performed with precaution. Here, not only storage of the samples but also the sample preparation practices and processes throughout the workflow should be considered. Careful sample collection is required where samples are preferably quickly frozen and stored at the appropriate storage conditions if they cannot directly be subjected to lipidomic analysis. This has recently been reviewed by Jung *et al.* [1]. It has been shown that certain biological matrices can be safely stored for years at −80 °C [38]. However, the stability can dramatically vary depending on the time, type of lipid, type of sample, and type of storage material and solvent. For instance, Hammad *et al.* recently showed that ethylenediaminetetraacetic acid (EDTA) is the preferred anticoagulant for sphingolipids [39]. This indicates that the material should carefully be selected to avoid unspecific reactions and interferences with the lipids of interest. Furthermore, the materials should also be resistant to the harsh solvent treatment, as typically a lipid extraction is performed in hazardous organic solvents such as chloroform, which itself can unfavorably react with lipids [40]. An example of the stability of lactosyl ceramide d18:1/24:0 from human serum is shown in Figure 1.4. This sphingolipid is shown to be rather stable when stored (10 days) in chloroform:methanol (1: 2, v/v) at −20 °C in standard eppendorf tubes (Eppendorf AG). Unexpectedly, its concentration is more than threefold elevated when stored in 96-well plates from the same vendor. The reason for this remains unknown. However, this shows that a small change in the storage condition can have fatal consequences and thus points out the essence in sample handling. Moreover, it is recognized that the number of freeze and thaw cycles can influence the end result [41]; however, two freeze–thaw cycles did not affect the levels of this sphingolipid.

Another challenge is that the end results are strongly influenced by the chosen extraction methodology and lipidomic instrumentations, leading to deviations in the final lipidomic outputs. Although much effort is currently put in making of synthetic standards, a substantial lack of proper nonendogenous standards is still a facto. Therefore, users have been stranded using the available synthetic standards, which regrettably in most cases have been insufficient for absolute quantification of monitored lipids. Thus, most of the currently available lipidomic reports are based on relative or semiabsolute lipid outputs, which can be difficult to compare due to

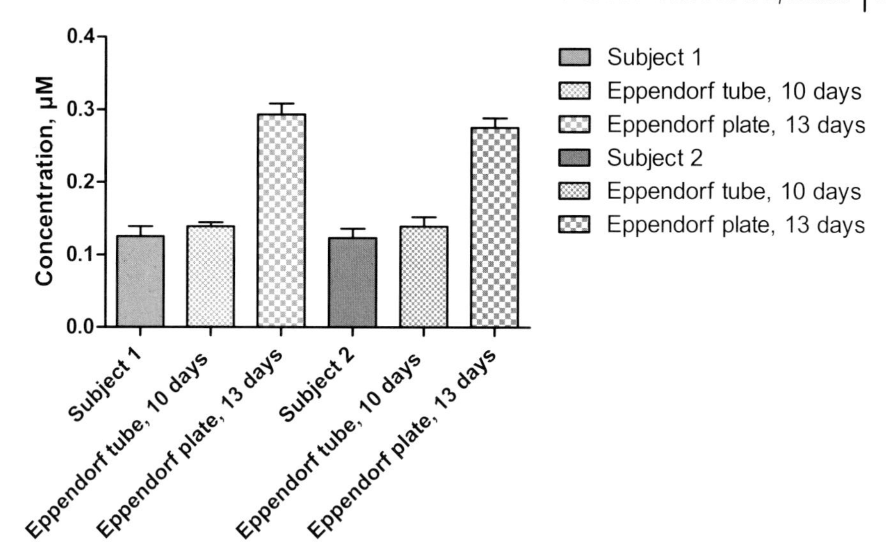

Figure 1.4 Effect of storage material on the stability of lactosyl ceramide d18:1/24:0. The sphingolipid was extracted as described in Ref. [1] from serum of two healthy donors. The total chloroform:methanol (1: 2, v/v) extract was analyzed by LC-MRM [28] fresh (solid bars), after 10 days of storage in standard 2 ml eppendorf tubes (small squared bars), and after 13 days of storage in eppendorf 96-well microtiter plates (large squared bars) at −20 °C. Independent samples were analyzed and error bars indicate standard deviation ($n = 6$).

deviations in both applied methodology and applied internal standards. A relative or semiabsolute lipidomic measurement can be considered valid for retrieving the first glims of lipidomic differences. However, once the focus is turned to determine, for example, the precise size of cellular membranes or lipoproteins and the lipid biomarker in clinical diagnostics, absolute measurements becomes a prerequisite.

Indisputably, lipidomics suffers from the lack of standardization. Naturally users apply their own preferred protocols and methodologies to assess their lipidomic quests. Regrettably, deviations in, for example, sample handling, extractions, synthetic standards, instrumentations, and data processing tools between laboratories lead to inconsistencies in the lipidomic results, which in turn complicate further data comparison and combination efforts. Transparent data collections will be a fundamental foundation triggering the direction of future lipidoimcs. Rigorous standardization and validation of the lipidomic processes are therefore urgently needed. Once established, new generation of lipidomic assays can set off and widely be adapted to clinical practices. This on the other hand leads to new challenges since such an adaptation typically demands high-throughput performance assays. We have recently demonstrated the feasibility of such attempts in discovery work [1]. The virtue of this high-throughput molecular lipidomic workflow relies on its high reproducibility and controllability, gained through robot-assisted sample preparation and lipid extraction and multiple lipidomic platforms integrated with a sophisticated bioinformatics system. Currently, it offers the highest throughput in delivering simultaneously the most comprehensive and quantitative lipidomic outputs at the molecular lipid level. For instance, approximately 5 days are required to determine the concentration of over 500 molecular lipids in 20 different lipid classes of 96 human plasma samples. Thus, this setup illuminates and demonstrates the first attempts toward high-throughput quantitative molecular lipidomics and, although not confirmed, prosperously supports its suitability in a regulatory setting.

1.7 Conclusions

The lipidomics era is currently occurring. Applications in basic and applied research have clearly pointed out and demonstrated its indispensable value. It will open up new avenues in the biomedical research community, with high expectations that this discovery toolkit will enhance biomarker discovery and provide novel information to target discovery programs as it will prospectively shed new light onto the affected metabolic and signaling pathways. Undoubtedly, it is the delineation of molecular lipids and their precise determination that will lead the way forward and accelerate our understanding of molecular lipids, the integrated lipidomic networks, and decoding the coordinately regulated pathways.

New attempts will be taken to overcome the challenges lipidomics currently faces. The standardization of sample preparation and analytical and bioinformatic procedures has to be properly addressed. Once solved, this launches the transition of lipidomics into clinical laboratories. In parallel, the demand for high-throughput technologies that do not compromise on the data quality is required. Emerging new MS technologies and methodologies already show promises by offering quantitative information on over 400 molecular lipid species obtained in less than 12 min [15]. If these platforms fulfill the regulatory requirements still needs to be seen.

It is now the time for researchers in biochemistry, analytical chemistry, biophysics, biology, bioinformatics, and medicine to gather and utilize lipidomics in the most productive way. It is the collective wisdom that will take us beyond our current know-how in lipid biology.

References

1 Jung, H.R., Sylvanne, T., Koistinen, K.M., Tarasov, K., Kauhanen, D., and Ekroos, K. (2011) High throughput quantitative molecular lipidomics. *Biochim. Biophys. Acta*, **1811**, 925–934.

2 Quehenberger, O., Armando, A.M., Brown, A.H., Milne, S.B., Myers, D.S., Merrill, A.H., Bandyopadhyay, S., Jones, K.N., Kelly, S., Shaner, R.L., Sullards, C.M., Wang, E., Murphy, R.C., Barkley, R.M., Leiker, T.J., Raetz, C.R., Guan, Z., Laird, G.M., Six, D.A., Russell, D.W., McDonald, J.G., Subramaniam, S., Fahy, E., and Dennis, E.A. (2010) Lipidomics reveals a remarkable diversity of lipids in human plasma. *J. Lipid. Res.*, **51**, 3299–3305.

3 Ejsing, C.S., Sampaio, J.L., Surendranath, V., Duchoslav, E., Ekroos, K., Klemm, R. W., Simons, K., and Shevchenko, A. (2009) Global analysis of the yeast lipidome by quantitative shotgun mass spectrometry. *Proc. Natl. Acad. Sci. USA*, **106**, 2136–2141.

4 Shevchenko, A. and Simons, K. (2010) Lipidomics: coming to grips with lipid diversity. *Nat. Rev. Mol. Cell Biol.*, **11**, 593–598.

5 van Meer, G. (2005) Cellular lipidomics. *EMBO J.*, **24**, 3159–3165.

6 van Meer, G., Voelker, D.R., and Feigenson, G.W. (2008) Membrane lipids: where they are and how they behave. *Nat. Rev. Mol. Cell Biol.*, **9**, 112–124.

7 Balasubramanian, N., Scott, D.W., Castle, J.D., and Casanova, J.E., and Schwartz, M.A. (2007) Arf6 and microtubules in adhesion-dependent trafficking of lipid rafts. *Nat. Cell Biol.*, **9**, 1381–1391.

8 Lingwood, D., Binnington, B., Rog, T., Vattulainen, I., Grzybek, M., Coskun, U., Lingwood, C.A., and Simons, K. (2011) Cholesterol modulates glycolipid conformation and receptor activity. *Nat. Chem. Biol.*, **7**, 260–262.

9 Ekroos, K., Ejsing, C.S., Bahr, U., Karas, M., Simons, K., and Shevchenko, A. (2003) Charting molecular composition of phosphatidylcholines by fatty acid scanning and ion trap MS3 fragmentation. *J. Lipid Res.*, **44**, 2181–2192.

10 Brugger, B., Erben, G., Sandhoff, R., Wieland, F.T., and Lehmann, W.D. (1997) Quantitative analysis of biological membrane lipids at the low picomole level by nano-electrospray ionization tandem mass spectrometry. *Proc. Natl. Acad. Sci. USA*, **94**, 2339–2344.

11 Stahlman, M., Ejsing, C.S., Tarasov, K., Perman, J., Boren, J., and Ekroos, K. (2009) High-throughput shotgun lipidomics by quadrupole time-of-flight mass spectrometry. *J. Chromatogr. B. Analyt. Technol. Biomed. Life Sci.*, **877**, 2664–2672.

12 Schuhmann, K., Almeida, R., Baumert, M., Herzog, R., Bornstein, S.R., and Shevchenko, A. (2012) Shotgun lipidomics on a LTQ orbitrap mass spectrometer by successive switching between acquisition polarity modes. *J. Mass Spectrom.*, **47**,, 96–104.

13 Herzog, R., Schuhmann, K., Schwudke, D., Sampaio, J.L., Bornstein, S.R., Schroeder, M., and Shevchenko, A. (2012) LipidXplorer: a software for consensual cross-platform lipidomics. *PLoS One*, **7**, e29851.

14 Ekroos, K., Chernushevich, I.V., Simons, K., and Shevchenko, A. (2002) Quantitative profiling of phospholipids by multiple precursor ion scanning on a hybrid quadrupole time-of-flight mass spectrometer. *Anal. Chem.*, **74**, 941–949.

15 Simons, B., Kauhanen, D., Sylvänne, T., Tarasov, K., Duchoslav, E., and Ekroos, K. (2012) Shotgun lipidomics by sequential precursor ion fragmentation on a hybrid quadrupole time-of-flight mass spectrometer. *Metabolites*, **2**, 195–213.

16 Thomas, M.C., Mitchell, T.W., Harman, D.G., Deeley, J.M., Nealon, J.R., and Blanksby, S.J. (2008) Ozone-induced dissociation: elucidation of double bond position within mass-selected lipid ions. *Anal. Chem.*, **80**, 303–311.

17 Poad, B.L., Pham, H.T., Thomas, M.C., Nealon, J.R., Campbell, J.L., Mitchell, T. W., and Blanksby, S.J. (2010) Ozone-induced dissociation on a modified tandem linear ion-trap: observations of different reactivity for isomeric lipids. *J. Am. Soc. Mass Spectrom.*, **21**, 1989–1999.

18 Deeley, J.M., Thomas, M.C., Truscott, R.J., Mitchell, T.W., and Blanksby, S.J. (2009) Identification of abundant alkyl ether glycerophospholipids in the human lens by tandem mass spectrometry techniques. *Anal. Chem.*, **81**, 1920–1930.

19 Li, Z., Agellon, L.B., Allen, T.M., Umeda, M., Jewell, L., Mason, A., and Vance, D.E. (2006) The ratio of phosphatidylcholine to phosphatidylethanolamine influences membrane integrity and steatohepatitis. *Cell Metab.*, **3**, 321–331.

20 Sergent, O., Ekroos, K., Lefeuvre-Orfila, L., Rissel, M., Forsberg, G.B., Oscarsson, J., Andersson, T.B., and Lagadic-Gossmann, D. (2009) Ximelagatran increases membrane fluidity and changes membrane lipid composition in primary

human hepatocytes. *Toxicol. In Vitro*, **23**, 1305–1310.

21 Zech, T., Ejsing, C.S., Gaus, K., de Wet, B., Shevchenko, A., Simons, K., and Harder, T. (2009) Accumulation of raft lipids in T-cell plasma membrane domains engaged in TCR signalling. *EMBO J.*, **28**, 466–476.

22 Gerl, M.J., Sampaio, J.L., Urban, S., Kalvodova, L., Verbavatz, J.M., Binnington, B., Lindemann, D., Lingwood, C.A., Shevchenko, A., Schroeder, C., and Simons, K. (2012) Quantitative analysis of the lipidomes of the influenza virus envelope and MDCK cell apical membrane. *J. Cell Biol.*, **196**, 213–221.

23 Dennis, E.A., Deems, R.A., Harkewicz, R., Quehenberger, O., Brown, H.A., Milne, S. B., Myers, D.S., Glass, C.K., Hardiman, G., Reichart, D., Merrill, A.H., Jr., Sullards, M.C., Wang, E., Murphy, R.C., Raetz, C.R., Garrett, T.A., Guan, Z., Ryan, A.C., Russell, D.W., McDonald, J.G., Thompson, B.M., Shaw, W.A., Sud, M., Zhao, Y., Gupta, S., Maurya, M.R., Fahy, E., and Subramaniam, S. (2010) A mouse macrophage lipidome. *J. Biol. Chem.*, **285**39976–39985.

24 Sampaio, J.L., Gerl, M.J., Klose, C., Ejsing, C.S., Beug, H., Simons, K., and Shevchenko, A. (2011) Membrane lipidome of an epithelial cell line. *Proc. Natl. Acad. Sci. USA*, **108**, 1903–1907.

25 Vance, D.E. and Vance, J.E. (1996) *Biochemistry of Lipids, Lipoproteins, and Membranes*, Elsevier, Amsterdam.

26 Shinzawa-Itoh, K., Aoyama, H., Muramoto, K., Terada, H., Kurauchi, T., Tadehara, Y., Yamasaki, A., Sugimura, T., Kurono, S., Tsujimoto, K., Mizushima, T., Yamashita, E., Tsukihara, T., and Yoshikawa, S. (2007) Structures and physiological roles of 13 integral lipids of bovine heart cytochrome c oxidase. *EMBO J.*, **26**, 1713–1725.

27 Menuz, V., Howell, K.S., Gentina, S., Epstein, S., Riezman, I., Fornallaz-Mulhauser, M., Hengartner, M.O., Gomez, M., Riezman, H., and Martinou, J.C. (2009) Protection of *C. elegans* from anoxia by HYL-2 ceramide synthase. *Science*, **324**, 381–384.

28 Perman, J.C., Bostrom, P., Lindbom, M., Lidberg, U., StAhlman, M., Hagg, D., Lindskog, H., Scharin Tang, M., Omerovic, E., Mattsson Hulten, L., Jeppsson, A., Petursson, P., Herlitz, J., Olivecrona, G., Strickland, D.K., Ekroos, K., Olofsson, S.O., and Boren, J. (2011) The VLDL receptor promotes lipotoxicity and increases mortality in mice following an acute myocardial infarction. *J. Clin. Invest.*, **121**, 2625–2640.

29 Ewers, H., Romer, W., Smith, A.E., Bacia, K., Dmitrieff, S., Chai, W., Mancini, R., Kartenbeck, J., Chambon, V., Berland, L., Oppenheim, A., Schwarzmann, G., Feizi, T., Schwille, P., Sens, P., Helenius, A., and Johannes, L. (2010) GM1 structure determines SV40-induced membrane invagination and infection. *Nat. Cell Biol.*, **12**, 11–18.

30 Haynes, C.A., Allegood, J.C., Wang, E.W., Kelly, S.L., Sullards, M.C., and Merrill, A. H., Jr. (2011) Factors to consider in using [U-C]palmitate for analysis of sphingolipid biosynthesis by tandem mass spectrometry. *J. Lipid Res.*, **52**, 1583–1594.

31 Pynn, C.J., Henderson, N.G., Clark, H., Koster, G., Bernhard, W., and Postle, A.D. (2011) Specificity and rate of human and mouse liver and plasma phosphatidylcholine synthesis analyzed *in vivo*. *J. Lipid Res.*, **52**, 399–407.

32 Kuerschner, L., Ejsing, C.S., Ekroos, K., Shevchenko, A., Anderson, K.I., and Thiele, C. (2005) Polyene-lipids: a new tool to image lipids. *Nat. Methods*, 239–45.

33 Contreras, F.X., Ernst, A.M., Haberkant, P., Bjorkholm, P., Lindahl, E., Gonen, B., Tischer, C., Elofsson, A., von Heijne, G., Thiele, C., Pepperkok, R., Wieland, F., and Brugger, B. (2012) Molecular recognition of a single sphingolipid species by a protein's transmembrane domain. *Nature*, **481**, 525–529.

34 Chan, R.B., Oliveira, T.G., Cortes, E.P., Honig, L.S., Duff, K.E., Small, S.A., Wenk, M.R., Shui, G., and Di Paolo, G. (2012) Comparative lipidomic analysis of mouse and human brain with Alzheimer disease. *J. Biol. Chem.*, **287**, 2678–2688.

35 Scherer, M., Schmitz, G., and Liebisch, G. (2009) High-throughput analysis of sphingosine 1-phosphate, sphinganine

1-phosphate, and lysophosphatidic acid in plasma samples by liquid chromatography-tandem mass spectrometry. *Clin. Chem.*, **55**, 1218–1222.

36 Simopoulos, A.P. (2002) The importance of the ratio of omega-6/omega-3 essential fatty acids. *Biomed. Pharmacother.*, **56**, 365–379.

37 Stahlman, M., Pham, H.T., Adiels, M., Mitchell, T.W., Blanksby, S.J., Fagerberg, B., Ekroos, K., and Boren, J. (2012) Clinical dyslipidaemia is associated with changes in the lipid composition and inflammatory properties of apolipoprotein-B-containing lipoproteins from women with type 2 diabetes. *Diabetologia*, **55** (4), 1156–1166.

38 Matthan, N.R., Ip, B., Resteghini, N., Ausman, L.M., and Lichtenstein, A.H. (2010) Long-term fatty acid stability in human serum cholesteryl ester, triglyceride, and phospholipid fractions. *J. Lipid Res.*, **51**, 2826–2832.

39 Hammad, S.M., Pierce, J.S., Soodavar, F., Smith, K.J., Al Gadban, M.M., Rembiesa, B., Klein, R.L., Hannun, Y. A., Bielawski, J., and Bielawska, A. (2010) Blood sphingolipidomics in healthy humans: impact of sample collection methodology. *J. Lipid Res.*, **51**, 3074–3087.

40 Owen, J.S., Wykle, R.L., Samuel, M.P., and Thomas, M.J. (2005) An improved assay for platelet-activating factor using HPLC-tandem mass spectrometry. *J. Lipid Res.*, **46**, 373–382.

41 Zivkovic, A.M., Wiest, M.M., Nguyen, U.T., Davis, R., Watkins, S.M., and German, J.B. (2009) Effects of sample handling and storage on quantitative lipid analysis in human serum. *Metabolomics*, **5**, 507–516.

2
Lipids in Cells

Kai Simons, Christian Klose, and Michal Surma

2.1
Introduction

The oldest valid molecular model in biology is the lipid bilayer, proposed in 1925 for the organization of cell membranes [1]. Membrane research was for a long time dominated by the lipids; proteins were barely considered. There were even postulates that ion transport across cell membranes would occur through lipid pores. In the Danielli–Davson unit model of cell membrane structure, the proteins were proposed to be plastered on both sides of the bilayer with no proteins spanning the membrane [2]. Mark Bretscher was the first to demonstrate in 1971 that transmembrane proteins with a fixed orientation existed in the erythrocyte plasma membrane [3]. With recombinant DNA technology at hand to clone the cDNAs encoding membrane proteins, the pendulum of membrane research swung toward the proteins, which have ever since received most of the attention in the field. Cell membranes are crowded with proteins and about 20% of the surface (depending on the membrane) is occupied by proteins. An increasing number of membrane protein structures are being solved [4]. Many of the different processes in cells take place membrane bound, reflected by the fact that about 30% of the eukaryotic genome encodes membrane proteins and many other proteins spend part of their lives bound to either side of a cell membrane, taking part in numerous membrane activities. However, the lipids cannot be simply ignored. Cells use ~5% of their genes to synthesize their lipids, generating a diversity of thousands of different lipid species [5]. This complex lipid mixture not only forms the bilayer matrix, but is involved in shaping cellular architecture and tissue formation, storing energy, mediating membrane trafficking, regulating membrane protein activity, facilitating signal transduction, and forming the basis for creating dynamic subcompartments within membranes.

In this chapter we will give an overview of how membrane lipids are distributed within the cell and how this distribution contributes to cellular function.

Lipidomics, First Edition. Edited by Kim Ekroos.

2.2
Basis of Cellular Lipid Distribution

The main lipid biosynthetic organelle is the endoplasmic reticulum (ER). Here most glycerophospholipids and sterols (e.g., cholesterol, which is the major animal sterol) are produced [5]. Cholesterol although synthesized in the ER is rapidly moved out from this organelle, heading toward the plasma membrane [6]. Therefore, the lipid composition of the ER is dominated by glycerophospholipids. Sphingolipids, another major category of lipids, are mainly produced in the Golgi apparatus and are therefore low in abundance in the ER [6].

The ER is the starting station for biosynthetic membrane traffic of proteins and lipids, from where membrane constituents are trafficked to the Golgi apparatus and from there further to the cell surface and other destinations. In the Golgi apparatus, sphingolipids, like sphingomyelin in animals, and glycosphingolipids are produced from the ceramide backbone, which itself is originally synthesized in the ER [7, 8]. The concentration gradient of sterols and sphingolipids increases along the biosynthetic pathway to reach the highest in abundance in the plasma membrane (PM), where sterols constitute a stunning 40–50 mol% [9–11]. From the PM, endocytosis routes move membrane to early endosomes that are similar in composition to that of the PM [12]; however, when they mature into late endosomes, a decrease of sterols is observed. Also a characteristic for late endosomes, a new lipid, bis(monoacylglycerol)phosphate, is generated during endocytosis [13]. The end station of endocytosis, the lysosomes, are kept low in both sterols and sphingolipids [14, 15]

The different organelles in biosynthetic and endocytic trafficking are marked by their "own" phosphoinositides (PIPs) derived from phosphatidylinositol (PI) made in the ER [16, 17]. In the Golgi apparatus, on the way toward the *trans*-side, PI is phosphorylated to PtdIns4P, while PtdIns(4,5)P_2 and PtdIns(3,4,5)P_3 dominate the PM. PtdIns3P is in the early endosomal membranes and PtdIns(3,5)P_2 in late endosomes. This distribution is maintained by a network of kinases and phosphatases that contribute to "lipid coding" of these organelles for specific protein interactions.

Organelles outside the ER–PM circuit, for instance, the mitochondria, have their own specific lipid composition [6, 18]. For example, cardiolipin is specific for the inner membrane of mitochondria, which is also depleted of cholesterol [19]. Cholesterol is present in the outer mitochondrial membrane, but in a lower concentration than that in the ER [19]. Sphingolipids are also low in the mitochondrial membranes [19, 20].

It should be noted that the methodology of organelle purification has not changed much over the past decades. To gain a better understanding of lipid distribution in different organellar membranes, organelle isolation needs to be improved to match the superior capabilities of mass spectrometry (MS)-based analysis. The astonishing sensitivity of present mass spectrometry technology will stimulate novel approaches to organelle purification [21–23]. Furthermore, previous studies often focused only on a limited set of lipids and employed less informative methods like thin-layer chromatography (TLC) and gas chromatography (GC), which did not provide knowledge of the molecular species. However, this information is essential for the understanding of the mechanistic details of the way lipids exert their function within the membrane. Time is now ripe for an inventory of lipids in organelles in different cell types and the beauty of mass spectrometric lipidomics is that quantitative readouts can be provided.

2.3 Lipid Distribution by Nonvesicular Routes

So far we have only discussed lipid sorting by membrane trafficking, however, it is well known that many lipids such as cholesterol and ceramide use other means of moving from one organelle to another. There are a number of so-called lipid-transfer proteins that were shown to facilitate transport between membranes *in vitro* [24]. These proteins can transfer ceramide, phospholipids, sterols, or sphingolipids. The lipid transfer between organelles was already discovered in 1969 [25]. In fact, they were first called exchange proteins because that is what they exactly do. These proteins exchange lipids when they bump into a membrane and they do this mostly passively down the concentration gradient [26]. Therefore, it is very difficult to envisage how these proteins could move lipids over the cytosol between organelles. For instance, the bulk of newly synthesized cholesterol is moved out of the ER to the PM by a route that does not involve the membrane transport route over the Golgi complex [27]. How could lipid transfer proteins transfer cholesterol from the ER across the cytosol – up the concentration gradient to the PM? The same problem applies for most lipid transport processes between organelles, the efficiency would in all likelihood be too low to play a significant role. Also to be noted is that the stoichiometry of the process would also limit the dynamics of lipid transfer across the cytosol.

Therefore, other mechanisms have to be involved. Membrane contact sites have been implicated in facilitating lipid transfer between organelles [24]. These have been identified between the ER membrane and those of the mitochondria (outer membrane), of the Golgi complex, of plasma membranes, of peroxisomes, of lipid droplets, and of late endosomes and lysosomes [28]. Structurally they are mostly based on electron-microscopic images, but the molecular machinery localizing to the contact sites has also been identified. Also, lipid transfer proteins have been implicated to be active in transport across the contact sites [24]. The most convincing evidence for lipid transfer protein-mediated transport has come from studies of ceramide transport from the ER to the Golgi [29]. The ER cisternal network extends to most parts of the cell and can therefore potentially form contacts with all organelles within a cell. But so far the molecular machinery that operates to move lipids from one organelle to another remains poorly understood. It is well known that certain lipids are made in organelles such as mitochondria or peroxisomes and have to be delivered to other organelles for membrane use. One such example is PE. PS is synthesized in the ER and subsequently imported into the mitochondria, where the enzyme Psd1p in the inner mitochondrial membrane decarboxylates PS to generate PE, which is then transferred back to the ER from where it is distributed to other membranes in the cell [28]. PE plasmalogens are synthesized in the peroxisomes from where they have to be transported to other organelles, probably the ER, for further transport elsewhere [30].

In summary, cells are using a wide variety of means to generate organelle-specific lipid compositions, like localized biosynthesis, vesicular transport, lipid-specific transport protein, and membrane–membrane contact sites.

2.4 Lipids in Different Cell Types

The lipid composition of different cell types follows as far as is known the basic outline described above. However, from the little we know, it is obvious that different cell types have different lipid compositions. The greatest variation concerns the glycosphingolipids [31]. Glycosphingolipids are known to carry hundreds of different glycan chains, of which only a small selection is present in each cell type. The variety is in contrast to the glycerophospholipids that present mostly the same head groups but in different proportions in the cell membranes of each cell type.

One example of lipid specialization is offered by the disks of retinal rod photoreceptor cells [32] and the ocular lens membrane [33]. The disk membrane is in the outer segment of the photoreceptor epithelial cells and has a unique phospholipid composition, characterized by a remarkable 60% of the omega-3 docosahexaenoic acid (22:6) (DHA) as fatty acid moiety, as well as containing very long-chain fatty acids (up to 32–36 carbon atoms long) [32]. DHA is mostly derived from the diet and is also highly enriched in neurons and the sperm tails [34]. The function of rhodopsin, the major membrane protein in the disk, has been shown to depend on these polyunsaturated fatty acids [35]. Interestingly, of all cell membranes, the photoreceptor disk with its unique polyunsaturated lipid content is the most rapidly diffusing membrane known [36].

An example of a cell membrane that has evolved in the opposite direction with respect to fatty acid composition and membrane order is the ocular lens membrane [33]. The lens is made up of fibers formed from the plasma membrane of lens epithelium. The lipids of this membrane are characterized by an unusually high content of saturated fatty acids [37]. The major lipid species are cholesterol and sphingomyelin with dihydrosphingomyelin, making up 77% of the total sphingomyelins [38]. This lipid composition makes the overall membrane organization highly ordered and rigid.

Oligodendrocytes produce another remarkable specialization of the PM, myelin [39, 40]. Myelin membranes wrap around axons in the central nervous system. The main function of myelin is to insulate the axon and to cluster sodium channels into the nodes of Ranvier. This organization enables the action potential to travel with remarkable speed from one node to the other in the direction of the synaptic terminal. Myelin has an unusually high lipid content, in which glycosphingolipids, galactosylceramide, and sulfatide as well as PE plasmalogens are strikingly enriched. The process of myelin formation can be followed in tissue culture, during which oligodendrocytes generate large membrane sheets that differentiate from the surrounding PM [41].

Secretory organelles that secrete proteins are well studied in exocrine and endocrine cells. However, there are also remarkable examples of secretory organelles that secrete lipids. The skin is an informative case [42]. The keratinocytes in stratum corneum differentiate into lipid-secreting cells, in which lamellar bodies, also called Odland bodies, are generated from the Golgi complex [43]. These organelles contain lipid lamellae, composed of phospholipids, glucosylceramide, sphingomyelin, and cholesterol. When the lamellar bodies fuse with the PM, the lamellae are externalized and the lipids become modified. Phospholipids are hydrolyzed into glycerol and fatty acids, while glucosylceramide and sphingomyelin are hydrolyzed into ceramides. These breakdown products form the matrix of the skin [43]. The stratum corneum that forms the hydrophobic permeability barrier of the skin has been suggested to be built like a wall where the keratinocytes (corneocytes) are the "bricks" and the extracellular lipids are the "mortar" [44].

Another lipid-secreting cell is the alveolar epithelial cell in the lung. These cells also produce lamellar bodies that probably form from the Golgi apparatus [45]. They contain lamellae of lipids of a completely different lipid composition compared to those of skin keratinocytes [46]. Of the total lipids, 85–90% are phospholipids, of which 40% is dipalmitoyl PC and around 5% is cholesterol. Important constituents of the lipid lamellae are pulmonary surfactant proteins that are essential for the functioning of the lung alveolae [47]. After secretion of the lipids, dipalmitoyl PC together with surfactants generate a liquid surface film covering the alveolar network of the lung. If the formation of this liquid film is impaired, the alveolae can collapse with respiratory dysfunction as the outcome [47].

The alveolar epithelial cells and the skin keratinocytes are specialized to secrete lipids that play an important role in the tissues where these cells are present. Most other cells do secrete lipids as well, but in the form of exosomes [48]. These membrane vesicles are derived in the endocytic trafficking system from multivesicular bodies that instead of turning into late endosomes and lysosomes are promoted to fuse with the PM and thereby release their content into the extracellular medium. The lipid composition of exosomes is enriched in sphingolipids (sphingomyelin and hexosylceramide) and cholesterol [49]. There is furthermore an increase in saturated PC species at the expense of unsaturated PC. Also, ceramide is unusually abundant compared to total cellular membrane lipid.

An interesting example of lipid changes during cellular differentiation is epithelium formation. A comprehensive lipidomic analysis was performed on epithelial MDCK cells to follow the changes occurring during polarization from the contact-naive (unpolarized state) to the final epithelial sheet [50]. In the fully polarized state, the sphingolipids were longer, more hydroxylated, and more glycosylated than their counterparts in the unpolarized state. Conversely, the glycerolipids acquired generally longer and more saturated fatty acids. Most interestingly, the Forssman antigen, which is a pentasaccharide glycosphingolipid, practically absent in unpolarized MDCK cells, becomes the major sphingolipid in the polarized epithelial state. When the MDCK cells are forced to depolarize toward the mesenchymal state, the lipids change back to that of the contact-naive cells. Most of these changes could be traced back to the fact that during polarization, an apical membrane is introduced into the PM to generate the asymmetric epithelial architecture. Apical membranes have long been known to be enriched in glycosphingolipids in comparison to the basolateral PM domain, which forms the pole of the cell directed toward the basement membrane and the interior milieu [51]. The purified apical membrane of MDCK cells is enriched in the lipids that characterize the fully polarized state. Therefore, it is apparent that cellular morphology is reflected by its lipid composition.

This brief summary of lipid compositions and distribution demonstrates the remarkable capability of different cell types to generate membranes with different lipidomes and functions. So far we know little of this aspect of tissue organization. Until now, the new capabilities of lipidomic analysis by mass spectrometry have not been fully employed, except in a few studies [50, 52–59].

2.5
Functional Implications of Membrane Lipid Composition

It seems obvious that the reason why cells synthesize hundreds of different lipids is that these complex lipid mixtures are required for cellular function. One important role will be to regulate membrane protein activities. This can be accomplished in different ways [4, 60]. One is to allosterically regulate membrane protein conformation and function [61]. Thus, cell membranes with special lipid composition potentially provide interaction partners with proteins specific for that membrane [4]. But also the general properties of the lipid bilayer play a role [61].

The ER is capable of integrating a great variety of transmembrane proteins into its membrane despite the fact that these proteins that are destined for exit from the ER are often designed to function in membranes with different properties. For instance, the plasma membrane is thicker than that of the ER and PM proteins have been shown to have longer transmembrane domains (TMDs) than those of the ER and the Golgi complex [62]. Thus, the ER lipid bilayer must adapt to the different TMDs to avoid hydrophobic mismatch. However, the situation changes when newly synthesized proteins leave the ER and they encounter membranes with increasing cholesterol content. Cholesterol serves to both thicken and rigidify these membranes, accentuating potential hydrophobic mismatching between the hydrophobic protein TMDs and the lipid bilayer core. Theoretical studies have shown that this effect of cholesterol potentiates the intrinsic sorting capability of mismatched systems [63]. In the ER, the bilayer adapts locally to newly synthesized proteins having different TMD lengths because the ER membrane is cholesterol-poor and therefore more plastic. This prediction has been confirmed in recent experiments showing that shorter Golgi TMD proteins segregate from longer PM TMD proteins when cholesterol concentration is increased in model bilayers [64].

Thus, the cholesterol gradient from the ER to the cell surface can regulate sorting of membrane proteins to their correct membrane site, while allowing broad-spectrum incorporation into the ER. Consistent with this idea, protein translocation has been shown to be inhibited by elevated levels of cholesterol in the ER membrane [65]. In the Golgi apparatus, cholesterol concentration increases toward the *trans*-side, promoting sorting of shorter Golgi proteins from PM proteins with longer TMDs.

Another addition to the eukaryotic lipid repertoire takes place in the Golgi complex: ceramide-based sphingolipids are synthesized, either with phosphocholine as a head group for sphingomyelins or with oligosaccharide units for glycosphingolipids. Together with cholesterol, these are routed to the PM [51, 66]. Sphingolipids introduce another sorting principle for proteins destined to the PM, based on preferential association of sphingolipids with cholesterol [67]. Sphingolipid–cholesterol assemblies associate with specific PM proteins to form dynamic nanoscale rafts, having the capacity to coalesce to larger platforms [68, 69].

In the biosynthetic pathway, sorting of not only proteins but also of lipids has to occur to generate the high concentrations of sterols and sphingolipids at the PM. The increasing concentration of these lipids toward the *trans*-side of the Golgi complex is enhanced by retrograde COPI – mediated transport from the Golgi to the ER [70]. The COPI vesicles have been shown to be depleted in cholesterol and sphingomyelin, also explaining why the ER membrane is so low in these lipids.

Direct evidence for lipid sorting in the TGN has been demonstrated in yeast. First of all, yeast mutants implicating sterols and sphingolipids were identified in a genome-wide screen for post-Golgi transport to the cell surface [71]. These mutants led to impaired exit of a raft transmembrane protein from the TGN. Employing an immunoisolation protocol with a raft transmembrane protein as bait, post-Golgi transport vesicles carrying this cargo were isolated [52]. Lipidomic analysis of the purified vesicles showed that sterol and sphingolipids were enriched compared to isolated donor organelle. Further studies have established that such sterol and sphingolipid sorting is a generic feature in plasma membrane-destined transport vesicles deriving from the Golgi apparatus [72]. Moreover, the comparison of the lipid composition of the transport carriers with that of the isolated yeast plasma membrane, showed that sphingolipid species with a total chain length of 46 and 42 carbon atoms were enriched in the vesicles but depleted in the PM. This observation implies that the PM lipidome is further modified after the transport vesicles deliver their load of newly synthesized lipids and proteins to the cell surface. This modification could be accomplished through delivery from other biosynthetic routes [73] and by endocytic trafficking [74]; however, *in situ* lipid remodeling cannot be excluded.

The route from the TGN to the apical surface in epithelial cells must also involve lipid sorting to generate a glycolipid-rich apical membrane. However, the direct evidence is still missing here. Interestingly, several studies have implicated a lectin, galectin-9, in apical membrane biogenesis [75]. When galectin-9 expression is knocked down by RNAi, the MDCK cells fail to polarize and to establish apical–basolateral polarity. The transport of apical raft proteins is impaired, while transport of the basolateral cargo from the TGN to the surface is even enhanced compared to polarized control cells. Strikingly, when exogenous recombinant galectin-9 was added to unpolarized MDCK cells, depleted of endogenous galectin-9, the cells polarized again and formed an asymmetric cell layer. Galectin-9 is secreted by a mechanism that bypasses the ER and the Golgi apparatus to the apical side of the epithelial cell layers [76]. It was also found that galectin-9 binds the Forssman glycolipid and since the binding of the galectin to its ligands is pH sensitive, it is probable that the galectin-9 partially dissociates from the glycolipid and other galactose-containing ligands in the acid endosomes after internalization [75]. Thus, after reaching the *trans*-Golgi network, the recycling galectin is potentially capable of scaffolding Forssman glycolipid and other glycosylated raft cargo. This scaffolding could induce a coalescence of raft lipids and proteins destined for the apical part of plasma membrane into a raft domain that could generate the apical raft carrier. A similar galectin–glycolipid circuit has been identified in epithelial HT-29 cells, where galectin-4 binds to glycolipid sulfatide and becomes part of the apical sorting machinery [77].

Thus, lectin–glycolipid interactions in general might play a major role in the sorting of lipids in both biosynthetic and endocytic trafficking. The potential mechanism of generating the raft-enriched carriers has been discussed in recent reviews [78, 79]. Shortly, the process would be propelled by a phase separation of a lipid raft domain from the surrounding membrane, and the membrane bending and vesicular carrier formation would be driven by domain-induced budding and by the action of auxiliary proteins that, for example, wedge into the bilayer [80, 81].

An irritating issue in the field has been the lack of genetic evidence for the lipid raft sorting model in the generation and maintenance of the apical membrane in epithelial cells. Until now, the detailed studies on different model organisms failed to identify lipid raft elements in their genetic screens of epithelial polarity. Since the power of genetics is undisputed, this has been annoying for those who maintained that glycolipids are important in apical biogenesis. However, this gap has been closed recently. By a combination of genetic screens, lipid analysis, and imaging methods, it was established that glycosphingolipids indeed play a role in mediating apical sorting in the gut of *Caenorhabditis elegans* (embryogenesis) [82].

Phase separation could also be driving the formation of the myelin membrane produced in oligodendrocytes. In this case, the oligodendrocytes produce massive amounts of myelin lipids and proteins and when they reach a critical concentration in the PM, the PM could start to phase separate into a myelin phase from the surrounding membrane. Similar principles might be operating in the production of the lipid lamellae in lamellar bodies in alveolar epithelial cells and skin keratinocytes. Also, the ocular membrane is potentially the result of a phase separation process in the PM of the ocular cells.

We postulate that a common denominator for all these membrane differentiations would be coalescence, depending on raft lipids and proteins. Important to note is that the sphingolipids involved in the process are cell type- and membrane-specific. The important lipid for apical raft carriers in MDCK cells would be the Forssman glycolipid, in HT29 cells sulfatide, in HeLa cell raft endocytosis Gb3, in myelin galactosylceramide and sulfatide, in alveolar lamellae saturated dipalmitoyl PC, and in skin keratinocytes glucosylceramide and dihydrosphingomyelin. This demonstrates the astonishing flexibility that such a phase separation mechanism potentially displays. The challenge will now be to test this concept *in vitro*. By reconstituting the essential constituents, it will be possible to analyze whether phase separation can indeed account for the biogenesis of this diverse set of membranes.

So far a little explored issue is how lipids contribute to cellular architecture. For instance, the ER bilayer has a distinct morphology forming cisternae and tubular networks. This intricate tubular membrane system has the propensity to fold into interesting curved structures that are characterized by cubic membrane morphologies that could potentially form multiple spaces within the continuous organelle [83]. Interestingly, virus infections seem to lead the formation of cubic ER [84]. Also, the overexpression of chimeric membrane proteins that can be artificially scaffolded into clusters has been found to induce folding of the ER into cubic membrane structures [85].

The capability of membrane lipids to form morphologically different structures has been extensively studied and mapped by Luzzati and Husson in the 1960s [86]. So far functional implications of this morphological potential have received little attention. However, this astounding plasticity is bound to be part of the toolkit that cells use to generate their architecture [87].

2.6 Outlook: Collectives and Phase Separation

The astonishing developments in lipid mass spectrometry now make it possible to analyze the full complement of lipids in membranes. This work is only beginning. Most of it lies ahead of us, where a whole new field of research is to be explored. What makes this area so attractive is that not only is the cell biology of lipids poorly understood, but rather we also have to combine biology with physics to come to grips with the diversity of lipids in order to understand their biological function. Considering that there are thousands of different lipid species, this seems like a tantalizing task, which indeed it is!

However, there are features that characterize cell membranes that will simplify the endeavor. Lipids and proteins in membranes form collectives. Hammond *et al.* have demonstrated that three-component lipid membranes containing one disordered membrane phase of DOPC, sphingomyelin, and cholesterol as well as small amounts of GM1, can be induced to separate into two phases, one liquid-ordered and another liquid-disordered by the clustering action of cholera toxin [88]. In contrast to the simplicity of model bilayers, no one would have imagined that a plasma membrane, which is built from some thousand lipid and protein constituents, could be induced to separate into *two* phases. However, by a swelling procedure, 431 cells were found to blow up its plasma membrane into "balloons." When cholera toxin, a pentavalent lectin that binds to the ganglioside GM1, was added, it resulted in a cholesterol-dependent phase separation at 37 °C [89]. Clustering of

GM1 gave rise to a GM1 raft phase separating from the surrounding PM. During this phase separation, the PM lipids and proteins reorganized laterally according to their predicted affinity for raft domains [89]. The most obvious explanation of these observations is that the raft phase in membranes contains a collective of lipids and proteins, showing similar physical properties and propensities [68]. The raft phase is the cohesive phase, coming together by lipid–lipid–protein interactions, separating from the other lipids and proteins that are left in a more disordered phase not held together by such cohesive interactions. It implies that the physicochemical properties of the molecules underlying cohesiveness must be a product of selection during evolution. The features responsible for this collective behavior would not have survived otherwise. A corollary of this hypothesis is that the physicochemical language characterizing the raft collective should be decipherable. We now have to unravel the structural features that bring raft lipids and proteins together. This insight is reassuring because if membrane constituents were mostly to behave independent of each other, the hope to unravel underlying principles of membrane organization would be moot. Obviously, the assembly of large microdomain–raft assemblies is prevented in living cells. Rafts are usually present as dynamic nanoscale platforms that can be specifically induced to form larger and more stable platforms in a multitude of different ways, each type of platform within the membrane having a specific function.

We propose that this is a general behavior of cell membranes where sphingolipid–sterol–protein rafts can generate larger raft domains by scaffolding raft constituents that bring the raft mixture over the phase boundary into two phases. Obviously, the right concentration and mix of lipids will be essential for this to occur. Most likely, the composition has to be close to a phase boundary. Otherwise, the separation cannot be easily induced. Being close to a phase separation boundary is perhaps a property that characterizes many biological processes in cells [90, 91]. This would allow easily regulatable transition from one state to the other. We predict that the concept of collectives and phase transitions emerging from cell membrane research will fundamentally transform studies on biological organization.

References

1 Gorter, E. and Grendel, F. (1925) On bimolecular layers of lipoids on the chromocytes of the blood. *J. Exp. Med.*, **41**, 439–443.
2 Danielli, J.F. and Davson, H. (1935) A contribution to the theory of permeability of thin films. *J. Cell Comp. Physiol.*, **5**, 495–508.
3 Bretscher, M.S. (1971) A major protein which spans the human erythrocyte membrane. *J. Mol. Biol.*, **59**, 351–357.
4 Coskun, U. and Simons, K. (2011) Cell membranes: the lipid perspective. *Structure*, **19**, 1543–1548.
5 van Meer, G., Voelker, D.R., and Feigenson, G.W. (2008) Membrane lipids: where they are and how they behave. *Nat. Rev. Mol. Cell Biol.*, **9**, 112–124.
6 van Meer, G. (1989) Lipid traffic in animal cells. *Annu. Rev. Cell Biol.*, **5**, 247–275.
7 Futerman, A.H. and Riezman, H. (2005) The ins and outs of sphingolipid

synthesis. *Trends Cell Biol.*, **15**, 312–318.
8 Hannun, Y.A. and Obeid, L.M. (2011) Many ceramides. *J. Biol. Chem.*, **286**, 27855–27862.
9 Kalvodova, L., Sampaio, J.L., Cordo, S. *et al.* (2009) The lipidomes of vesicular stomatitis virus, semliki forest virus, and the host plasma membrane analyzed by quantitative shotgun mass spectrometry. *J. Virol.*, **83**, 7996–8003.
10 Warnock, D.E., Roberts, C., Lutz, M.S. *et al.* (1993) Determination of plasma membrane lipid mass and composition in cultured Chinese hamster ovary cells using high gradient magnetic affinity chromatography. *J. Biol. Chem.*, **268**, 10145–10153.
11 Gerl, M.J., Sampaio, J.L., Urban, S. *et al.* (2012) Quantitative analysis of the lipidomes of the influenza virus envelope and MDCK cell apical membrane. *J. Cell Biol.*, **196** (2), 213.
12 Coskun, U. and Simons, K. (2010) Membrane rafting: from apical sorting to phase segregation. *FEBS Lett.*, **584**, 1685–1693.
13 Kobayashi, T., Stang, E., Fang, K.S. *et al.* (1998) A lipid associated with the antiphospholipid syndrome regulates endosome structure and function. *Nature*, **392**, 193–197.
14 Liscum, L. and Munn, N.J. (1999) Intracellular cholesterol transport. *Biochim. Biophys. Acta*, **1438**, 19–37.
15 Simons, K. and Gruenberg, J. (2000) Jamming the endosomal system: lipid rafts and lysosomal storage diseases. *Trends Cell Biol.*, **10**, 459–462.
16 De Matteis, M.A. and Godi, A. (2004) PI-loting membrane traffic. *Nat. Cell Biol.*, **6**, 487–492.
17 di Paolo, G. and de Camilli, P. (2006) Phosphoinositides in cell regulation and membrane dynamics. *Nature*, **443**, 651–657.
18 Colbeau, A., Nachbaur, J., and Vignais, P.M. (1971) Enzymic characterization and lipid composition of rat liver subcellular membranes. *Biochim. Biophys. Acta*, **249**, 462–492.
19 Comte, J., Maisterrena, B., and Gautheron, D.C. (1976) Lipid composition and protein profiles of outer and inner membranes from pig heart mitochondria: comparison with microsomes. *Biochim. Biophys. Acta*, **419**, 271–284.
20 Futerman, A.H. (2006) Intracellular trafficking of sphingolipids: relationship to biosynthesis. *Biochim. Biophys. Acta*, **1758**, 1885–1892.
21 Han, X. and Gross, R.W. (2005) Shotgun lipidomics: electrospray ionization mass spectrometric analysis and quantitation of cellular lipidomes directly from crude extracts of biological samples. *Mass Spectrom. Rev.*, **24**, 367–412.
22 Shevchenko, A. and Simons, K. (2010) Lipidomics: coming to grips with lipid diversity. *Nat. Rev. Mol. Cell Biol.*, **11**, 593–598.
23 Wenk, M.R. (2010) Lipidomics: new tools and applications. *Cell*, **143**, 888–895.
24 Lev, S. (2010) Non-vesicular lipid transport by lipid-transfer proteins and beyond. *Nat. Rev. Mol. Cell Biol.*, **11**, 739–750.
25 Wirtz, K.W. and Zilversmit, D.B. (1969) Participation of soluble liver proteins in the exchange of membrane phospholipids. *Biochim. Biophys. Acta*, **193**, 105–116.
26 Bloj, B. and Zilversmit, D.B. (1981) Lipid transfer proteins in the study of artificial and natural membranes. *Mol. Cell Biochem.*, **40**, 163–172.
27 Ikonen, E. (2008) Cellular cholesterol trafficking and compartmentalization. *Nat. Rev. Mol. Cell Biol.*, **9**, 125–138.
28 Elbaz, Y. and Schuldiner, M. (2011) Staying in touch: the molecular era of organelle contact sites. *Trends Biochem. Sci.*, **36**, 616–623.
29 Hanada, K., Kumagai, K., Tomishige, N., and Kawano, M. (2007) CERT and intracellular trafficking of ceramide. *Biochim. Biophys. Acta*, **1771**, 644–653.
30 Rodemer, C., Thai, T.P., Brugger, B. *et al.* (2003) Inactivation of ether lipid biosynthesis causes male infertility, defects in eye development and optic nerve hypoplasia in mice. *Hum. Mol. Genet.*, **12**, 1881–1895.
31 Hakomori, S.I. (2008) Structure and function of glycosphingolipids and sphingolipids: recollections and future trends. *Biochim. Biophys. Acta*, **1780**, 325–346.

32 Jastrzebska, B., Debinski, A., Filipek, S., and Palczewski, K. (2011) Role of membrane integrity on G protein-coupled receptors: rhodopsin stability and function. *Prog. Lipid Res.*, **50**, 267–277.
33 Borchman, D. and Yappert, M.C. (2010) Lipids and the ocular lens. *J. Lipid Res.*, **51**, 2473–2488.
34 Valentine, R.C. and Valentine, D.L. (2004) Omega-3 fatty acids in cellular membranes: a unified concept. *Prog. Lipid Res.*, **43**, 383–402.
35 Soubias, O. and Gawrisch, K. (2012) The role of the lipid matrix for structure and function of the GPCR rhodopsin. *Biochim. Biophys. Acta*, **1818**, 234–240.
36 Poo, M. and Cone, R.A. (1974) Lateral diffusion of rhodopsin in the photoreceptor membrane. *Nature*, **247**, 438–441.
37 Epand, R.M. (2003) Cholesterol in bilayers of sphingomyelin or dihydrosphingomyelin at concentrations found in ocular lens membranes. *Biophys. J.*, **84**, 3102–3110.
38 Byrdwell, W.C. and Borchman, D. (1997) Liquid chromatography/mass-spectrometric characterization of sphingomyelin and dihydrosphingomyelin of human lens membranes. *Ophthalmic Res.*, **29**, 191–206.
39 Aggarwal, S., Yurlova, L., and Simons, M. (2011) Central nervous system myelin: structure, synthesis and assembly. *Trends Cell Biol.*, **21**, 585–593.
40 Nave, K.A. (2010) Myelination and support of axonal integrity by glia. *Nature*, **468**, 244–252.
41 Aggarwal, S., Yurlova, L., Snaidero, N. *et al.* (2011) A size barrier limits protein diffusion at the cell surface to generate lipid-rich myelin-membrane sheets. *Dev. Cell*, **21**, 445–456.
42 Proksch, E., Brandner, J.M., and Jensen, J. M. (2008) The skin: an indispensable barrier. *Exp. Dermatol.*, **17**, 1063–1072.
43 Feingold, K.R. (2007) Thematic review series: skin lipids. The role of epidermal lipids in cutaneous permeability barrier homeostasis. *J. Lipid Res.*, **48**, 2531–2546.
44 Addor, F.A. and Aoki, V. (2010) Skin barrier in atopic dermatitis. *An. Bras. Dermatol.*, **85**, 184–194.
45 Andreeva, A.V., Kutuzov, M.A., and Voyno-Yasenetskaya, T.A. (2007) Regulation of surfactant secretion in alveolar type II cells. *Am. J. Physiol. Lung Cell Mol. Physiol.*, **293**, L259–271.
46 Perez-Gil, J. (2008) Structure of pulmonary surfactant membranes and films: the role of proteins and lipid–protein interactions. *Biochim. Biophys. Acta*, **1778**, 1676–1695.
47 Perez-Gil, J. and Weaver, T.E. (2010) Pulmonary surfactant pathophysiology: current models and open questions. *Physiology (Bethesda)*, **25**, 132–141.
48 Simons, M. and Raposo, G. (2009) Exosomes–vesicular carriers for intercellular communication. *Curr. Opin. Cell Biol.*, **21**, 575–581.
49 Trajkovic, K., Hsu, C., Chiantia, S. *et al.* (2008) Ceramide triggers budding of exosome vesicles into multivesicular endosomes. *Science*, **319**, 1244–1247.
50 Sampaio, J.L., Gerl, M.J., Klose, C. *et al.* (2011) Membrane lipidome of an epithelial cell line. *Proc. Natl. Acad. Sci. USA*, **108**, 1903–1907.
51 Simons, K. and van Meer, G. (1988) Lipid sorting in epithelial cells. *Biochemistry*, **27**, 6197–6202.
52 Klemm, R.W., Ejsing, C.S., Surma, M.A. *et al.* (2009) Segregation of sphingolipids and sterols during formation of secretory vesicles at the trans-Golgi network. *J. Cell Biol.*, **185**, 601–612.
53 Takamori, S., Holt, M., Stenius, K. *et al.* (2006) Molecular anatomy of a trafficking organelle. *Cell*, **127**, 831–846.
54 Camera, E., Ludovici, M., Galante, M. *et al.* (2010) Comprehensive analysis of the major lipid classes in sebum by rapid resolution high-performance liquid chromatography and electrospray mass spectrometry. *J. Lipid Res.*, **51**, 3377–3388.
55 Andreyev, A.Y., Fahy, E., Guan, Z. *et al.* (2010) Subcellular organelle lipidomics in TLR-4-activated macrophages. *J. Lipid Res.*, **51**, 2785–2797.
56 Kiebish, M.A., Bell, R., Yang, K. *et al.* (2010) Dynamic simulation of cardiolipin remodeling: greasing the wheels for an interpretative approach to lipidomics. *J. Lipid Res.*, **51**, 2153–2170.

57 Sandhoff, R. (2010) Very long chain sphingolipids: tissue expression, function and synthesis. *FEBS Lett.*, **584**, 1907–1913.

58 Bird, S.S., Marur, V.R., Sniatynski, M.J. *et al.* (2011) Lipidomics profiling by high-resolution LC-MS and high-energy collisional dissociation fragmentation: focus on characterization of mitochondrial cardiolipins and monolysocardiolipins. *Anal. Chem.*, **83**, 940–949.

59 Bartz, R., Li, W.H., Venables, B. *et al.* (2007) Lipidomics reveals that adiposomes store ether lipids and mediate phospholipid traffic. *J. Lipid Res.*, **48**, 837–847.

60 Hunte, C. and Richers, S. (2008) Lipids and membrane protein structures. *Curr. Opin. Struct. Biol.*, **18**, 406–411.

61 Coskun, U., Grzybek, M., Drechsel, D., and Simons, K. (2011) Regulation of human EGF receptor by lipids. *Proc. Natl. Acad. Sci. USA*, **108**, 9044–9048.

62 Bretscher, M.S. and Munro, S. (1993) Cholesterol and the Golgi apparatus. *Science*, **261**, 1280–1281.

63 Lundbaek, J.A., Andersen, O.S., Werge, T., and Nielsen, C. (2003) Cholesterol-induced protein sorting: an analysis of energetic feasibility. *Biophys. J.*, **84**, 2080–2089.

64 Kaiser, H., Orlowski, A., Rog, T. *et al.* (2011) Lateral sorting in model membranes by cholesterol-mediated hydrophobic matching. *Proc. Natl. Acad. Sci. USA*, **108** (40), 16628–16633.

65 Nilsson, I., Ohvo-Rekila, H., Slotte, J.P. *et al.* (2001) Inhibition of protein translocation across the endoplasmic reticulum membrane by sterols. *J. Biol. Chem.*, **276**, 41748–41754.

66 van Meer, G., Stelzer, E.H., Wijnaendts-van-Resandt, R.W., and Simons, K. (1987) Sorting of sphingolipids in epithelial (Madin–Darby canine kidney) cells. *J. Cell Biol.*, **105**, 1623–1635.

67 Simons, K. and Ikonen, E. (1997) Functional rafts in cell membranes. *Nature*, **387**, 569–572.

68 Lingwood, D. and Simons, K. (2010) Lipid rafts as a membrane-organizing principle. *Science*, **327**, 46–50.

69 Simons, K. and Gerl, M.J. (2010) Revitalizing membrane rafts: new tools and insights. *Nat. Rev. Mol. Cell Biol.*, **11**, 688–699.

70 Brugger, B., Sandhoff, R., Wegehingel, S. *et al.* (2000) Evidence for segregation of sphingomyelin and cholesterol during formation of COPI-coated vesicles. *J. Cell Biol.*, **151**, 507–518.

71 Proszynski, T.J., Klemm, R.W., Gravert, M. *et al.* (2005) A genome-wide visual screen reveals a role for sphingolipids and ergosterol in cell surface delivery in yeast. *Proc. Natl. Acad. Sci. USA*, **102**, 17981–17986.

72 Surma, M.A., Klose, C., Klemm, R.W. *et al.* (2011) Generic sorting of raft lipids into secretory vesicles in yeast. *Traffic*, **12**, 1139–1147.

73 Harsay, E. and Schekman, R. (2002) A subset of yeast vacuolar protein sorting mutants is blocked in one branch of the exocytic pathway. *J. Cell Biol.*, **156**, 271–285.

74 Weinberg, J. and Drubin, D.G. (2012) Clathrin-mediated endocytosis in budding yeast. *Trends Cell Biol.*, **22**, 1–13.

75 Mishra, R., Grzybek, M., Niki, T. *et al.* (2010) Galectin-9 trafficking regulates apical–basal polarity in Madin–Darby canine kidney epithelial cells. *Proc. Natl. Acad. Sci. USA*, **107**, 17633–17638.

76 Friedrichs, J., Torkko, J.M., Helenius, J. *et al.* (2007) Contributions of galectin-3 and -9 to epithelial cell adhesion analyzed by single cell force spectroscopy. *J. Biol. Chem.*, **282**, 29375–29383.

77 Delacour, D., Gouyer, V., Zanetta, J.P. *et al.* (2005) Galectin-4 and sulfatides in apical membrane trafficking in enterocyte-like cells. *J. Cell Biol.*, **169**, 491–501.

78 Simons, K. and Gerl, M.J. (2010) Revitalizing membrane rafts: new tools and insights. *Nat. Rev. Mol. Cell Biol.*, **11**, 688–699.

79 Surma, M.A., Klose, C., and Simons, K. (2011) Lipid-dependent protein sorting at the trans-Golgi network. *Biochim. Biophys. Acta.*, **1821**, 1059–1067.

80 Cao, X., Coskun, U., Rossle, M. *et al.* (2009) Golgi protein FAPP2 tubulates membranes. *Proc. Natl. Acad. Sci. USA*, **106**, 21121–21125.

81 Lenoir, M., Coskun, U., Grzybek, M. *et al.* (2010) Structural basis of wedging the

Golgi membrane by FAPP pleckstrin homology domains. *EMBO Rep.*, **11**, 279–284.

82 Zhang, H., Abraham, N., Khan, L.A. *et al.* (2011) Apicobasal domain identities of expanding tubular membranes depend on glycosphingolipid biosynthesis. *Nat. Cell Biol.*, **13**, 1189–1201.

83 Landh, T. (1996) *Cubic Cell Membranes Architectures: Taking Another Look at Membrane Bound Cell Spaces*, Lund University, Lund, Sweden, p. 188.

84 Deng, Y., Almsherqi, Z.A., Ng, M.M., and Kohlwein, S.D. (2010) Do viruses subvert cholesterol homeostasis to induce host cubic membranes? *Trends Cell Biol.*, **20**, 371–379.

85 Lingwood, D., Schuck, S., Ferguson, C. *et al.* (2009) Generation of cubic membranes by controlled homotypic interaction of membrane proteins in the endoplasmic reticulum. *J. Biol. Chem.*, **284**, 12041–12048.

86 Luzzati, V. and Husson, F. (1962) The structure of the liquid-crystalline phases of lipid–water systems. *J. Cell Biol.*, **12**, 207–219.

87 Almsherqi, Z.A., Kohlwein, S.D., and Deng, Y. (2006) Cubic membranes: a legend beyond the Flatland∗ of cell membrane organization. *J. Cell Biol.*, **173**, 839–844.

88 Hammond, A.T., Heberle, F.A., Baumgart, T. *et al.* (2005) Crosslinking a lipid raft component triggers liquid ordered–liquid disordered phase separation in model plasma membranes. *Proc. Natl. Acad. Sci. USA*, **102**, 6320–6325.

89 Lingwood, D., Ries, J., Schwille, P., and Simons, K. (2008) Plasma membranes are poised for activation of raft phase coalescence at physiological temperature. *Proc. Natl. Acad. Sci. USA*, **105**, 10005–10010.

90 Brangwynne, C.P., Eckmann, C.R., Courson, D.S. *et al.* (2009) Germline P granules are liquid droplets that localize by controlled dissolution/condensation. *Science*, **324**, 1729–1732.

91 Brangwynne, C.P., Mitchison, T.J., and Hyman, A.A. (2011) Active liquid-like behavior of nucleoli determines their size and shape in *Xenopus laevis* oocytes. *Proc. Natl. Acad. Sci. USA*, **108**, 4334–4339.

3
High-Throughput Molecular Lipidomics

Marcus Ståhlman, Jan Borén, and Kim Ekroos

3.1
Introduction

Once considered passive bystanders, with the only functions as providers of energy and a basic structure of membranes, lipids are now considered multifunctional and it is difficult to find an area of cell biology where lipids do not play important roles as signaling or regulatory molecules. Lipidomics has the ambitious task of identifying and quantifying all the lipid entities within a biological system. If taken a step further, lipidomics can be seen as a tool for determining the biological function of each individual molecular lipid and for elucidating how lipids, as single entities or collectively, contribute to the function (or dysfunction) of cellular processes and effect the system as a whole. Since the number of different molecular lipids is very high (probably hundreds of thousands) [1, 2] and the concentration levels of different lipids might vary thousand- or even million-fold, it has become increasingly clear that this presents a formidable challenge when considering the analytical tools needed. Furthermore, as the relatively young field of lipidomics starts to move into new fields of science, new requirements will be put upon its applications. Both in the pharmaceutical industry and in the clinical chemistry, lipidomics will become a potential resource. However, in these environments, there is a strong demand for high throughput.

In this chapter, we will discuss how with the aid of modern equipment we can use a high-throughput workflow to achieve comprehensive characterization of lipids at molecular level.

3.2
Lipid Diversity

Originally lipids were defined as "naturally occurring substances that are hydrophobic in nature and soluble in organic solvents" [3]. This very broad definition, although true in many cases, has in recent years been refined to "hydrophobic or amphipathic small molecules that may originate entirely or in part by carbanion-based condensation of thioesters or isoprene units" [4, 5]. Even using this

Lipidomics, First Edition. Edited by Kim Ekroos.

definition, lipids are still a very diverse group containing molecules with a wide range of polarity, size, and structure. In conjunction with the proposed definitions, a new scheme for classification and nomenclature has also been suggested [4, 5]. In this scheme, the lipids are divided into eight groups based on their chemical structure and biosynthetic perspective. These groups are fatty acyls (FAs), glycerolipids (GLs), glycerophospholipids (GPs), sphingolipids (SPs), sterol lipids (STs), prenol lipids (PRs), saccharolipids (SLs), and polyketides (PKs). In addition, each of these groups is further divided into several subgroups (Table 3.1).

The FA group contains the simplest and one of the most important groups of lipids, namely, the fatty acids and their oxidized forms. The fatty acids are saturated or unsaturated hydrocarbon chains that normally vary in length between 12 and 24 carbons. Functionally, the fatty acids are used as energy as they are substrates for beta-oxidation. The fatty acids are also the precursor of very potent signaling molecules such as the leukotrienes and prostaglandins (see Chapter 11 for more details). Moreover, they are also essential building blocks for many of the other lipid groups (e.g., GLs and GPs).

GLs have a relatively simple structure with monoacylglycerols (MGs), diacylglycerols (DGs), and triacylglycerols (TGs) being common lipid classes. TGs are often used as an example of the structural diversity present within the lipid family. The TGs contain three fatty acids. If we assume the number of naturally occurring fatty acids to be around 30, there are tens of thousands of possible combinations. Even though only a few molecular TG species represent the majority of the total TG mass, the number of unique molecular species is probably still enormous. Functionally, the TGs are mainly used as energy stores, while both DGs and MGs have been shown to be involved in signaling [6, 7].

Common with GLs, GPs are also constituted of a glycerol backbone. However, due to their high abundance and importance as membrane constituents and signaling molecules, they have been awarded a separate group. The amphipathic GPs contain both hydrophobic fatty acids and a polar head group attached to the glycerol backbone. This property makes them the major building block in vesicles and membranes in virtually all biological systems. Examples of head groups that further divide this lipid class into smaller subgroups are choline, ethanolamine, serine, inositol, and glycerol. As will be discussed further on, the nature of the head group is a determinant of the methodology used for their analysis.

SPs are a very complex group of lipids that have received increased interest and recognition in recent years as important signaling molecules [8]. All SLs share a common structural feature, a sphingoid base backbone that is synthesized *de novo* from the condensation of serine and fatty acyl-CoA. In humans, the fatty acyl-CoA is almost exclusively palmitate, which results in the formation of a C18-sphingoid backbone upon which many of the different SPs are based. In other species, the backbone might have different chain lengths and branching. In *Caenorhabditis elegans*, for example, the most common sphingoid base consists of iso-branched C17 carbon chain [9]. Many SPs are further modified by the addition of one or several sugar residues. The glucosylceramides and galactosylceramides, for example, contain one sugar, while the gangliosides contain several charged sugar residues. The SPs are described in greater detail in Chapter 5.

Table 3.1 Lipid classification scheme according to Fahy *et al.* [4].

Lipid class	Examples of subclasses	Example of molecular lipids
Fatty acyls	Fatty acids and conjugates Eicosanoids Prostaglandins Leukotrienes Tromboxanes	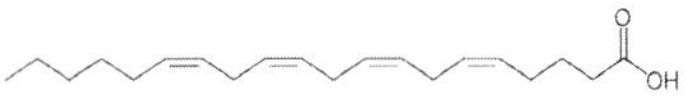 Arachidonic acid
Glycerolipids	Monoradylglycerols Monoacylglycerols Diradylglycerols Diacylglycerols Triradylglycerols Triacylglycerols	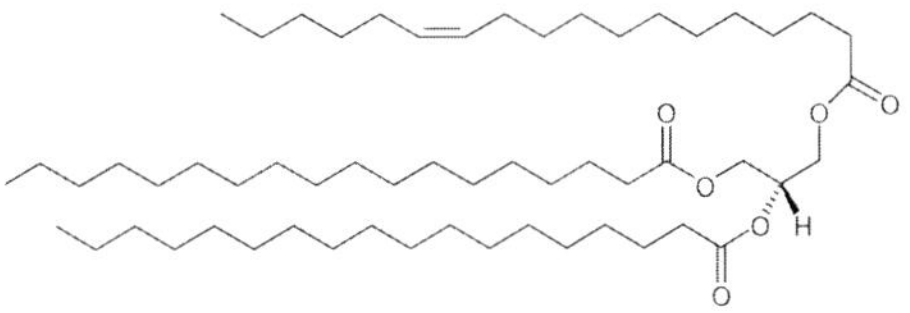 Triacylglycerol (18:1/18:0/18:0)
Glycerophos-pholipids	Phosphatidylcholines Phosphatidylethanolamines Phosphatidylserines Phosphatidylglycerols Phosphatidylinositols	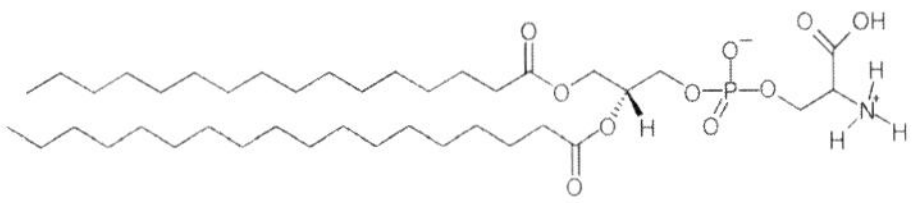 Phosphatidylserine (16:0/18:0)
Sphingolipids	Sphingoid bases Ceramides Neutral glycosphingolipids Glucosylceramides Acidic glycosphingolipids Gangliosides	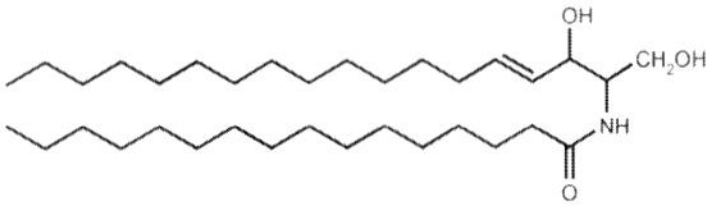 Ceramide (d18:1/16:0)
Sterol lipids	Cholesterol and derivates Cholesteryl esters Steroids Bile acids and derivates	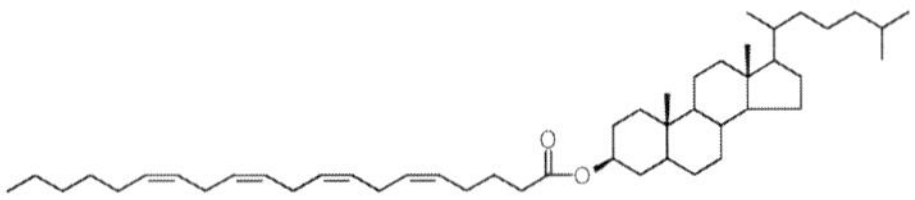 Cholesteryl ester 20:4
Prenol lipids	Isoprenoids Polyprenols	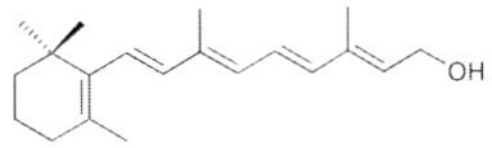 Vitamin A
Saccharolipids	Acylaminosugars Acylaminosugar glycans	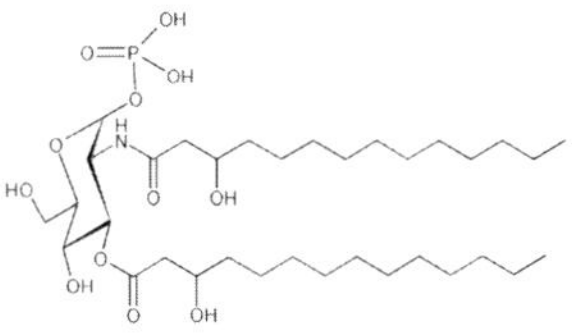 Lipid X
Polyketides	Macrolide polyketides Aromatic polyketides	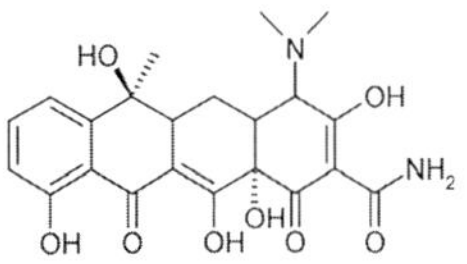 Tetracycline

3.3 Function of Molecular Lipids

It is not surprising, considering the major structural diversity of lipids, that lipids also show a broad spectrum of functionality. As mentioned briefly in the previous section, lipids play important roles as membrane constituents, signaling molecules, and energy storage. In some cases, the function of the lipid is unaffected by the composition and structural properties of the lipid. For example, all triglycerides, independent of their fatty acid composition, can be used as energy source. However, it is becoming increasingly clear that for several, perhaps the majority, of lipids, the function depends on the structure and can therefore be very different for the different molecular lipids within the sample lipid class. An example of this was shown a few years back when Shinzawa-Itoh and colleagues showed that a specific phosphatidylglycerol molecule containing palmitic acid (on the *sn*-1 position) and vaccenic acid (on the *sn*-2 position) was needed for normal function of cytochrome c oxidase [10]. Another example has been shown for the survival of *C. elegans* exposed to anoxia. Specific molecular ceramides containing C24 and C26 carbons were associated with death, while molecular lipids with shorter chains within the same class were involved in survival [11]. Also, we recently showed that the fatty acid composition of VLDL TG had a major effect on the inflammatory response when the VLDL was incubated on human smooth muscle cells [12]. These examples highlight the fact that molecular lipids need to be identified and quantified in order to elucidate the underlying biological function.

Even though traditional analyses such as thin-layer chromatography (TLC) and normal-phase high-performance liquid chromatography (HPLC) are valuable tools, they lack the required sensitivity and selectivity. To attain the molecular details of individual lipid species, we therefore have to turn to other analytical tools. Here, lipid analysis driven by mass spectrometry technologies has evolved as the primary solution for analyzing lipids at the greatest detail (Figure 3.1).

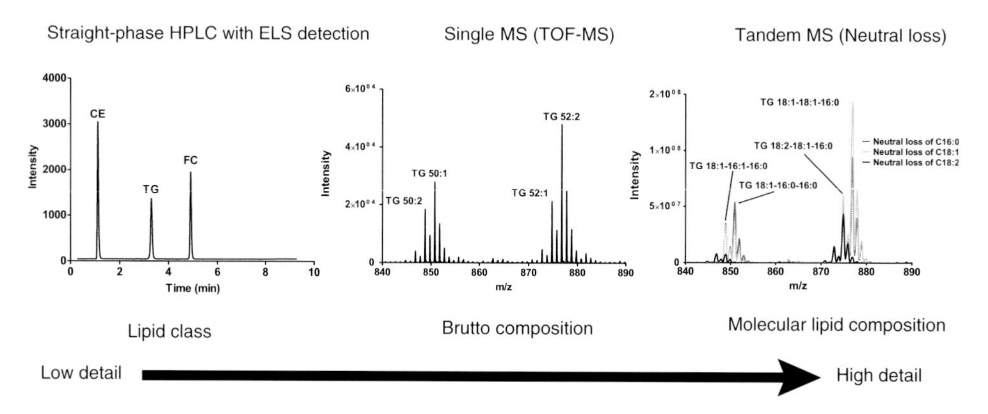

Figure 3.1 Lipid analysis can be performed using different approaches and at different levels of complexity. However, to achieve a comprehensive characterization of molecular lipids, we need to employ tandem mass spectrometry techniques.

3.4 Automated Sample Preparation

In lipidomics, the major purpose of the sample preparation is to purify the lipids (including extracting them from the tissue) while reducing interfering substances that might compromise the analysis. The sample preparation procedures can be performed in a number of different ways depending on which tissue the lipids are to be extracted from and also on the physical properties of the lipid(s) of interest. While a polar lipid classes such as lyso-lipids can be recovered accurately from plasma with a simple methanol precipitation [13], a more hydrophobic lipid class located within a tissue often needs a harder extraction involving tissue homogenization and the addition of a more hydrophobic solvent such as chloroform.

Tissue homogenization was previously, and in certain cases still is, a rather laborious task involving manual use of a rotor–stator homogenizer, which is capable of processing only one sample at a time. Today in our laboratory, automated sample homogenization is performed with a combination of Precellys (Bertin Technologies) and Mixer Mill instruments (RETSCH). Tissues (20–100 mg) are placed in polypropylene tubes together with ceramic beads and a homogenization buffer is added. After a 30 s burst in the Precellys, the samples are further homogenized for 10 min in the Mixer Mill instrument. Custom-made aluminum blocks that are kept at −20 °C, together with the use of liquid nitrogen in the Precellys, ensure that the sample temperature is always kept below 0 °C. By using this approach, 48 samples can be homogenized simultaneously and the procedure takes about 20 min.

After the homogenization, the lipids can be extracted. In terms of total lipid extraction, where lipids within a wide range of polarity are being extracted, there are two golden standard procedures commonly used: the Folch [14] and the Bligh & Dyer [15] protocols. Both of these are liquid–liquid extraction procedures using chloroform and methanol at different ratios. Similar to the tissue homogenization, lipid extraction is often performed manually. However, manual lipid extraction is very labor intensive and not compatible with large-scale studies containing hundreds or thousands of samples. Furthermore, manual lipid extraction has also been shown to be less accurate than robot-assisted extraction [16]. Recently the automation of the Folch protocol using a Hamilton Robotics Microlab Star system (Hamilton Robotics) was described [17]. This system has several technical features such as rapid high- and low-volume pipetting, antidroplet control, and pressure sensing, which all ensure accurate and reliable extraction. Using this robotic setup, the extraction efficiency was shown to be comparable to the manually performed protocol, and, importantly, it resulted in an improved analytical reproducibility.

Alternative methods to perform automated total lipid extraction are emerging. Recently, Matyash *et al.* [18] published a method where chloroform has been replaced with methyl-*tert*-butyl ether (MTBE). Due to the low density of MTBE, this resulted in the lipids being recovered in the upper organic phase of the two-phase system. Another alternative is to use a combination of butanol and methanol (BUME-method) to generate an initial one-phase extraction system [19]. After addition of heptane, ethylactete, and water (1% acetic acid), a two-phase system is formed with the lipids again contained in the upper organic phase. Both these methods benefit from having the organic phase at the top. Since the pipette tip of the robot does not have to penetrate into the water phase and interlayer containing the debris and protein precipitation, this might facilitate lipid recovery and reduce contamination. The BUME method further benefits from having a relatively high sample to solvent ratio (1:3) during the initial one-phase extraction and has been shown to have a low extraction yield of polar substances (e.g., sucrose). The automated BUME extraction is performed in 1.2 ml glass inserts contained in custom-made aluminum blocks. The procedure, which has been evaluated on a Velocity 11 Bravo robot (Agilent, Santa Clara, CA), is based on the 96-well format and can be performed in about 60 min. Since the BUME method shows almost identical extraction recoveries compared to the gold standard procedure, we believe that this method could be an attractive alternative to the automated Folch procedure.

For a comprehensive lipidomic analysis where lipids with a wide range of polarity are to be analyzed, a total lipid extraction as described above might not be sufficient. Instead, several extractions are often needed to recover all lipids of interest with sufficient extraction yields. For this, the samples are typically divided for separate extractions. The other alternative strategy is to make several extractions in series from the same sample. An elegant example of the latter was recently published by Ejsing *et al.* [20]. Here, the authors performed an initial extraction using chloroform/methanol (17:1, v/v). After removing the lower fraction containing 80–99% of the neutral lipids such as TG, DG, phosphatidylcholine (PC), phosphatidylethanolamine (PE), and lyso-phosphatidylcholine (LPC), subsequent re-extraction of the aqueous phase was performed. For this, chloroform/methanol (2:1, v/v) was added thereby enabling a high-yield (74–95%) extraction of more polar lipids such as phosphatidic acid (PA), cardiolipin (CL), phosphatidylserine (PS), phosphatidylinositol (PI) together with several inositol-containing sphingolipids. This procedure, when performed using a robotic setup, could potentially increase the comprehensiveness of a high-throughput approach.

In addition to performing regular lipid extraction protocols, the robot system might also be used to perform postextraction protocols. We have, for example, used the robot for separation of a total lipid extract into a nonpolar and polar fraction. By adding heptane and methanol (containing a small amount of alkaline water) to the dried total lipid extract, a two-phase system is formed with the nonpolar lipids (cholesteryl ester (CE), TG, DG, and free cholesterol) in the upper phase and the polar lipids (phospholipids and lyso-species) retained in the lower methanol–water phase. The procedure takes about 30 min and can be used when a more targeted analysis will be made or to remove high abundant lipids that are not included in the analysis. For example, the analysis of polar phospholipids in adipose tissue would certainly be facilitated once the neutral TG are removed using the procedure above.

A robot system can also be used for performing an alkaline hydrolysis of abundant phospholipids [21]. This will facilitate subsequent analysis of SM using shotgun lipidomics as described below. The robot can then be used further to transform the fatty acids, formed by the alkaline hydrolysis, into methyl esters, which can then be analyzed using gas chromatography.

For some applications, the use of solid-phase extraction (SPE) protocols are common. This is especially true for the extraction of low abundant arachidonic acid derivatives such as prostaglandins and leukotrienes. Since these molecules are relatively polar, they will not be accurately recovered during a total lipid extraction such as the Folch procedure. Therefore, these lipids are often recovered by an initial methanol or ethanol precipitation followed by a SPE purification step [22]. Also, for this purpose, automated protocols can be developed. Several manufacturers are now providing 96-well plates containing SPE material for high-throughput purification.

Even though much focus in this section has been on total lipid extraction, it is also important to realize that even though a highly sophisticated lipid extraction with several wash steps might produce an extremely pure extract, it might not be the optimal goal since a more "quick-and-dirty" extraction might produce a sufficiently pure extract and greatly increase sample throughput. For example, even though plasma LPC are extracted in high yield using the Folch or the newly developed BUME protocol, a simple protein precipitation using methanol might be sufficient for high recovery [13]. It is also important to remember that the choice of sample preparation procedure also depends on the methodology used for sample analysis. An LC-MS method is generally better in handling a more complex matrix compared to a shotgun approach.

In conclusion, sample preparation can be automated to a very high degree. With the aid of robot-assisted extraction procedures, it is possible to extract lipids from hundreds of samples per day in a high-throughput fashion.

3.5 Different Approaches to Molecular Lipidomics

Today, almost all lipidomic approaches are mass spectrometry based. Due to the recent technological advances in mass spectrometry, there are now several different types of instruments in the arsenal. The most commonly used are the triple quadrupole (QQQ), the quadrupole–linear ion trap (QTRAP), and also high-resolution instruments such as the time-of-flight (TOF) and orbitrap, which all have their own advantages and disadvantages. As all of these instruments can be combined with several types of sample introduction systems, such as reversed-phase HPLC, straight-phase HPLC, and direct infusion, there is a large number of different methodological platforms available. However, due to the enormous variation in lipid diversity, there is currently no single methodological platform that can be used for a full characterization of the lipidome. Instead, several lipid-selective platforms are necessary. As this will often require several sample preparation strategies, it becomes even more important to use automated sample preparation in order to maintain a high-throughput focus.

3.5.1 Untargeted versus Targeted Approaches

Untargeted (sometimes called global lipidomics) approach uses an unbiased approach for detecting lipid species. This means that no lipid class-specific scans are used, instead the mass spectrometer scans are used over a wide mass range for the detection of lipids. The benefit of using an untargeted approach for lipidomics is that the search is made with an "open mind," meaning that no prerequisites are defined but instead everything that is being ionized and present at sufficient levels will be detected. This allows detection of unexpected or novel lipids.

Although untargeted lipidomics can be used in combination with shotgun approach, most platforms use chromatographic separation prior to MS detection. Furthermore, in order to efficiently identify all the lipids in the sample, these analyses are often performed using high-resolution instruments such as TOF, orbitrap, or Fourier transform ion cyclotron resonance mass spectrometry (FT-ICR-MS) that can reach sub ppm mass accuracy. Despite this high accuracy, important information might be missed if the MS is used in scanning mode only. This is so because no distinction can be made between structurally different molecular lipids with the same mass (isobars). Therefore, the MS scans are often combined with fragmentation experiments (MS/MS) for identification of molecular lipids.

A drawback of using the untargeted approach is that the methods are often associated with relatively long runtimes (up to 90 min per analysis for sufficient separation) and are therefore not suited for high-throughput analysis. Therefore, ultraperformance liquid chromatography (UPLC) has been employed [23], which ideally can reduce the analytical runtime without compromising chromatographic resolution. However, as this leads to narrowing of the eluting peak, it becomes highly important to have MS instrumentation that is fast enough to ensure accurate detection. Finally, the complex sample matrix with accompanying ion suppression and the lack of sufficient internal standards might compromise accurate quantification.

A targeted approach focuses on specific lipids or lipids classes. This means that selective MS methods such as multiple reaction monitoring (MRM), precursor ion scanning (PIS), and neutral loss scanning (NLS) are typically used, which increases the sensitivity of the analysis. Using this definition, a targeted approach means everything from a simple MRM method quantifying one single analyte to profiling of several hundreds of TGs using NLS [24]. Using this approach, which is more common than a global approach, quantitative data for hundreds of lipids can be attained with high accuracy and precision. A targeted approach can be based on either HPLC or shotgun method.

3.5.2
Shotgun Lipidomics

The term shotgun lipidomics was coined almost a decade ago by Han and Gross [25] and refers to a situation where the sample mixture is infused directly into the mass spectrometer without any online separation. A prerequisite for this technique has been the recent advances in MS technology that has led to improvements in, for example, sensitivity, acquisition speed, mass resolution, and mass accuracy. For many applications, the sample is introduced into the MS using a syringe pump. However, since this approach has been difficult to automate, the establishment of a reliable syringe-based high-throughput methodology has been hampered. Another alternative is to perform direct flow injection using an autosampler, such as the CTC (CTC analytics) [26]. Both of these approaches infuse samples at μl/min flow rates. Recently, a new system based on chip technology and static sample infusions at nanoflow rates was introduced [27]. In addition to its high-throughput capabilities, the major advantages of this system are the reduced sample requirements (10 μl is sufficient), reduced ion suppression resulting in increased sensitivity, and finally, practically no restriction in sample analysis time due to the extended and stable infusions. We have previously demonstrated that using this system a very stable spray over a long period of time can be produced, which is the prerequisite for a robust and reproducible analysis [16]. Furthermore, the system is automated and easily incorporated into a high-throughput workflow.

In order to achieve optimal ionization of specific lipid classes, the sample matrix and instrumental settings can be optimized. This is sometimes called intrasource separation and is based on creating an environment in the ion source that results in maximal ionization of the lipids of interest, while reducing this ionization of other lipids [25]. Examples of parameters that can be adjusted, in order to attain optimal ionization, are polarity of the electrospray, adduct formation, pH, and declustering potentials.

As mentioned above, the shotgun approach is mainly used for a targeted analysis and often the lipid class-specific scans, such as NLS or PIS, are employed. These scans take advantage of the structural similarities of molecular lipids within a specific lipid class. The CEs, for example, which are composed of a cholesterol and a fatty acid part, all give rise to the cholestadiene fragment at m/z 369.3 when exposed to collision-induced dissociation (CID) in positive mode [26]. Therefore, by using precursor ion scanning of m/z 369.3, there is possibility to selectively and sensitively detect CE species even though present at low amounts and analyzed directly from total lipid extracts. By monitoring the fragment of m/z 184.1 in positive ion mode, molecular species of PC and SM molecular species can be readily detected and quantified [28]. However, since PC contains two fatty acids, only the brutto composition will be attained (the total number of carbons and double bonds in the fatty acids) and no information about which fatty acids that are present will be revealed. For this purpose, applications using multiple PIS (MPIS) in negative mode have been developed [29, 30]. During CID in negative mode, acyl anions of respective fatty acid will be formed. By monitoring the most abundant and interesting fatty acids (up to 40 fragments), information on the molecular species can therefore be acquired. MPIS using multiple fatty acids were originally performed using a hybrid quadrupole TOF instrument [29]. This instrument is a very attractive alternative for performing MPIS since it allows simultaneous recording of a theoretically unlimited number of ions. Furthermore, the high mass accuracy of the TOF allows the detection of ions within a small mass range (typically 0.1 Da), which minimizes the chance of detecting a false positive and biased quantification. The latest MS hardware developments enable to perform similar analyses on QQQ and QTRAP instruments. Although these instruments still lack the high mass accuracy, they offer superior sensitivity and speed. In our lipidomic workflow, we routinely use QTRAP 5500 (AB Sciex) type of instrumentations to acquire a total of 70–80 PIS and 20–30 NL scans to cover both fatty acids and lipid head group fragments. This approach is mainly used for medium to highly abundant lipid classes such as glycerophospholipids, glycerolipids, and CEs, and produces quantitative data on molecular lipid species [17].

3.5.3
Analytical Validation of the Shotgun Approach

In addition to the requirements of high throughput, it is also important that the methods used for lipidomics are validated and preferably GLP compliant. This becomes even more important if the methods are to be used in a clinical setting or during pharmaceutical trials. Compared to LC-MS methods, which have been implemented routinely within GLP environments, shotgun lipidomics still needs to be accepted in such standards. Therefore, it is important to validate and show that these methods are robust and capable of producing reproducible results.

Due to the complexity and relatively high concentration of lipids in the sample mixture, the shotgun approach will be associated with some degree of ion suppression (charge competition in the ion source). For the critics of the shotgun approach, this is the major argument since it is suggested to hamper accurate quantification [31]. However, using the shotgun approach, the ion suppression is expected to be more constant during the analysis of each sample and, furthermore, several investigators have shown that ion suppression is lipid class dependent and only moderately affected by the fatty acid composition [32, 33]. This means that these effects will be corrected for by the use of one or several lipid class-specific internal standards, which should be endogenously nonexisting [16]. In our lab we add several internal standards with heptadecanoyl fatty acid during the extraction procedure. The standards are supplied from several companies such as Avanti Polar Lipids (Alabaster, AL) or Larodan Fine Chemicals (Malmö, Sweden). We also use deuterated standards from C/D/N isotopes (Pointe-Claire, QC).

In some aspects, the ion suppression might be a bigger problem during LC-MS analysis. The reason is that the ion suppression will vary during the chromatographic run due to the difference in the eluting mobile-phase composition (for gradient LC-MS methods) and sample matrix. In other words, the signal intensity of different molecular lipids from the same lipid class will depend not only on the inherent physicochemical properties of the molecule but also on the retention time. Data, supporting the idea that ion suppression might be more complex during LC-MS analysis, were recently published [17].

In addition, to guarantee adequate quantification, it is equally important to ensure analytical stability and reproducibility. Recently, we published data showing that by including 29 sample reference samples and standard curves during the sample analysis of 500 human serum samples, the analytical variation could be controlled and monitored [17]. The results showed that under the conditions used for the analysis, the methodology was robust and the variation in the reference samples was low (around 10% on average).

Despite being a fast and attractive tool for performing lipidomic analysis, the shotgun approach has certain disadvantages. One is a potential isotopic overlap where the third isotope of one molecular lipid might overlap with the first isotope of another. This requires an accurate isotope correction algorithm to avoid incorrect lipid identification and quantification. Other challenges might be the abundance of isobars and isomers interfering with accurate identification and quantification.

3.5.4 Targeted LC-MS Lipidomics

Even though shotgun lipidomics is an attractive alternative for automated and high-throughput analysis of several lipid classes, it is not suitable for low abundant molecular lipids. Due to high ion density in the ions source during shotgun analysis, the signal from low abundant molecular lipids will be suppressed and hard to detect. If a shotgun approach is to be used for low abundant lipids, prior purification using SPE, TLC, or straight-phase HPLC is often required. Instead, low abundant lipids are normally analyzed using targeted LC-MS approaches. Here, the lipid extract is separated using HPLC in conjunction with MS analysis. The rationale for using this approach is that the chromatographic system can be adjusted for optimal retention of the lipids of interest while the sample matrix will be removed. Therefore, the lipids will be eluted in a very clean surrounding allowing maximal ionization in the ion source. This will in turn lead to a very sensitive analysis, especially when performed in MRM mode. In this analysis, which is performed almost exclusively on QQQ instruments, both the Q1 and Q3 are locked at specific masses allowing very selective and sensitive detection. Quantification using this approach is often performed using combinations of internal and external standards. Although the lack of appropriate standards is still a concern, new standards are constantly being developed, to a large extent due to the LIPID MAPS initiative.

Although the use of HPLC has been a common analytical technique for several decades, the recent development of ultra-HPLC (UHPLC) has made several new high-throughput applications possible. UHPLC employs columns with small particle size and operates at high pressures at normal flow rates. The benefit is reduced analysis time due to the high-resolution chromatograms with peak widths reduced to a few seconds. Due to the narrow peak widths, the scan speed of the MS instrument can suddenly become an apparent limitation. If too many lipids are monitored simultaneously using MRM, there will not be enough data points generated over the peak for accurate quantification. However, solutions to this problem are not only development of faster instruments but also through software solutions offering more optimal analyses. For example, "scheduled MRM" enables the user to limit the individual MRM transitions to certain time windows. This means that the MS is not required to monitor all MRM transitions simultaneously.

3.6
Data Processing and Evaluation

A comprehensive lipidomic analysis can generate an enormous amount of data. Especially if the analysis is based on an untargeted approach or on a shotgun approach employing MPIS or NL. Using these approaches, a single sample can easily give rise to more than 100 different spectra, which in a large study, involving hundreds of people, amounts to thousands of spectra that need to be processed and evaluated. In a sense, this part of the workflow is the most important because this is where the data have to be turned into knowledge and understanding of the biological processes being studied. Since the lipidomic bioinformatics is described in more detail in Chapter 8, it is described only briefly below.

The bioinformatics workflow consists of several steps with the first one being the processing of lipidomic data. Here, the raw data generated by the MS are being turned into data sets with, preferably, quantitative information on identified lipids. For an MRM analysis, this is not a very complex process as the monitored lipids are known. The current MS instruments typically offer built-in data processing tools facilitating MRM data processing and quantification. It becomes more complicated to extract data from shotgun lipidomics or LC-MS-based untargeted lipidomic approaches. Some important steps in the data processing workflow are peak detection, peak alignment (untargeted approach), lipid identification, isotope correction, and normalization. Several solutions exist as downloadable, open-source software packages. Examples of tools used for LC-MS-based data are XCMS [34] and the recently described LDA [35], while the new AMDMS-SL [36] and LipidXplorer [37] are comprehensive freeware packages designed for shotgun lipidomics. Perhaps the most comprehensive lipid software currently available is the LipidView software. However, this commercial software, which is built on the previous LipidProfiler [27], is developed by ABSciex and currently only compatible with data generated by their instruments.

Once the data have been processed and made available in some sort of table or matrix format, a number of different statistical tools can be used. The choice depends heavily on the experimental design and can involve both univariate and multivariate methods. Common tools are simple univariate analyses such as ANOVA and correlation analysis. However, multivariate approaches such as principal component analysis (PCA) and partial least-squares discriminant analysis (PLS/DA) are also valuable tools.

Besides performing the statistical evaluation of the lipidomic data, it is also important to put the generated result in the context of biological networks such as metabolic pathways. For example, the KEGG (http://www.genome.jp/kegg) database is a resource commonly used for these purposes and includes a collection of manually drawn pathway maps representing current knowledge about interaction and reaction networks. However, due to the detailed information on today's lipidomic analyses, novel tools are still lacking for construction of metabolic pathways at molecular level.

In the context of high throughput, it is crucial to have tools for data processing, evaluation, and visualization integrated in the lipidomic workflow. Over the last years, we have established an integrated bioinformatics solution for processing high-throughput lipidomic data in a semiautomated fashion. This bioinformatics pipeline is not only restricted to shotgun lipidomic data but can also handle LC-MS data.

In conclusion, the time it takes from the start of the processing of the data generated by the MS instruments to the final report and conclusions totally depends on the scientific question. However, in cases where the sample sets are large and the questions are complex, this part can easily become the bottleneck in the lipidomic workflow. Therefore, it is important to develop bioinformatics tools that can integrate several parts needed for lipidomic data evaluation. This should preferentially occur in a user-friendly environment.

3.7 Lipidomic Workflows

In this chapter, we have discussed the available analytical capabilities and how they can be used for quantitative analysis of molecular lipids. We have focused on how we can achieve high throughput and where efforts should be made in order to reduce bottlenecks associated with the analysis of large sample sets. In a lipidomic workflow as outlined in Figure 3.2, there are several parts that need to be optimized. Today it is possible, with the aid of robot-assisted sample preparation, to extract lipids from hundreds of samples within hours. The samples will then be

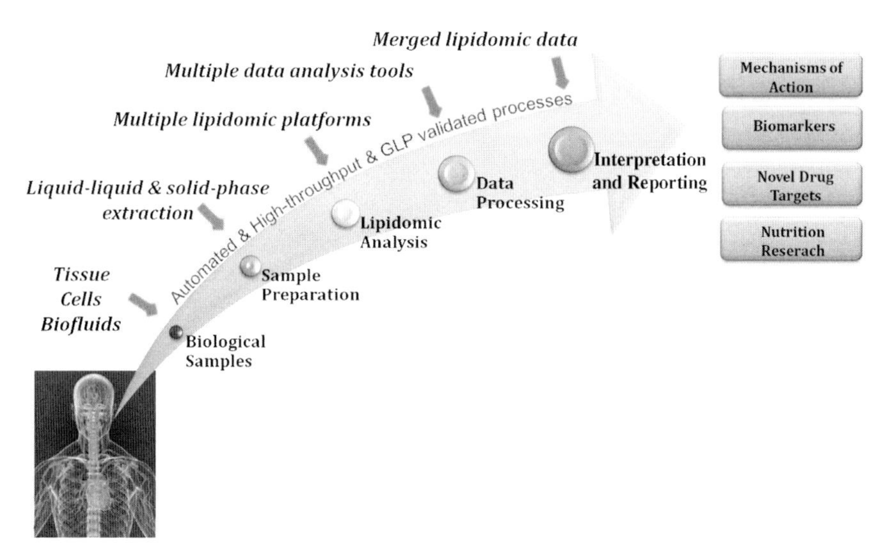

Figure 3.2 A schematic picture showing the workflow from sample preparation to the final evaluation.

run on several different platforms in order to attain a comprehensive profile of different molecular lipids. Although this will require a substantial amount of time, technical developments such as UHPLC and sensitive instruments with high scanning speeds have helped reducing the analytical runtimes. Instead, much focus should be put on the data processing and evaluation. One of the challenges is to standardize the data generated from several platforms and integrate all this into one database. In our lab, much efforts and resources have been invested in achieving a high-throughput approach and today a plasma lipidome characterization, from initial sample preparation to a final concentration table comprising 50 000 molecular lipid entries from 96 samples (e.g., a 96-well plate) can be achieved within 5 days.

3.8 Conclusions and Future Perspectives

Although being an intense area of research since the 1960s, the field of lipid biochemistry is currently experiencing a renaissance. One of the explanations for this is the recent development within the field of advanced mass spectrometry, which has empowered us with better tools for the characterization of lipids. In this chapter, we have discussed the different analytical capabilities available and shown that despite being a relatively young field, there are a large number of applications and approaches available. Moreover, these will increase further as the technology continues to develop. Even though we in this chapter claim to be able to attain quantitative information about molecular lipids, there is, at least, two more levels of complexity that we need to decipher. The first one is double bond position. Even though this information can partly be attained by performing a lipid hydrolysis followed by subsequent fatty acid analysis using GC, applications for comprehensive characterization of intact molecular lipids species are still lacking. However, the newly developed OzID holds great potential [38]. Using this technique, intact lipids are exposed to ozone vapor within an ion-trap mass spectrometer. This induces ozonolysis at the position of the double bond, producing fragment ions with mass-to-charge ratios specific to the position of the double bond. The OzID is performed directly on crude extracts using the shotgun approach, which is beneficial when including it in a lipidomic workflow. The second challenge is that of stereochemistry. Several lipids contain chiral centers and double bonds can have *cis*- and *trans*- configuration. Both of these questions are further discussed in Chapter 6.

As lipidomics is still a young field, it will continue to grow and hopefully be found in new areas of research. Within the pharmaceutical industry today, lipidomics is mainly used in the discovery phase. However, it is clear that it could be highly relevant in later stages also and even during clinical trials. Lipidomic screens of patients involved in these trials could not only provide information on and understanding of the beneficial effects of the drug but could also be used for identifying biomarkers and unraveling the pathophysiological pathways responsible for the lack of effect in the patients that are not responding to the drug, the so-called nonresponders. However, several requirements have to be met in order for lipidomics to become a functional tool during clinical trials. One is to be able to perform the analysis with validated methods within a GLP environment. Another requirement is to become more high throughput oriented since these trials can consist of several thousand samples. As described in this chapter, with the help of robotics, advanced sample preparation can today be performed relatively quickly. This means that hundreds of samples can be prepared for multiple platform analysis each day. By using a combination of shotgun and targeted UPLC-MS approaches, the analytical runtimes have also been reduced, which means that data from hundreds of molecular lipids within several lipid classes can be attained relatively quickly. Instead, the major bottleneck in the high-throughput workflow is the data processing and perhaps most of all, the data evaluation and interpretation in which major efforts have still to be invested.

References

1 Brown, H.A. and Murphy, R.C. (2009) Working towards an exegesis for lipids in biology. *Nat. Chem. Biol.*, **5**, 602–606.

2 Griffiths, W.J. and Wang, Y. (2009) Mass spectrometry: from proteomics to metabolomics and lipidomics. *Chem. Soc. Rev.*, **38**, 1882–1896.

3 Smith, A.D., Datta, S.P., Smith, G.H., Campbell, P.N., Bentley, R., and McKenzie, H.A. (eds) (2000) *Oxford Dictionary of Biochemistry and Molecular Biology*, Oxford University Press, Oxford.

4 Fahy, E., Subramaniam, S., Brown, H.A., Glass, C.K., Merrill, A.H., Jr., Murphy, R.C., Raetz, C.R., Russell, D.W., Seyama, Y., Shaw, W., Shimizu, T., Spener, F., van Meer, G., VanNieuwenhze, M.S., White, S.H., Witztum, J.L., and Dennis, E.A. (2005) A comprehensive classification system for lipids. *J. Lipid Res.*, **46**, 839–861.

5 Fahy, E., Subramaniam, S., Murphy, R.C., Nishijima, M., Raetz, C.R., Shimizu, T., Spener, F., van Meer, G., Wakelam, M.J., and Dennis, E.A. (2009) Update of the LIPID MAPS comprehensive classification system for lipids. *J. Lipid Res*, (50 Suppl), S9–S14.

6 Sugiura, T., Kondo, S., Sukagawa, A., Nakane, S., Shinoda, A., Itoh, K., Yamashita, A., and Waku, K. (1995) 2-Arachidonoylglycerol: a possible endogenous cannabinoid receptor ligand in brain. *Biochem. Biophys. Res. Commun.*, **215**, 89–97.

7 Carrasco, S. and Merida, I. (2007) Diacylglycerol, when simplicity becomes complex. *Trends Biochem. Sci.*, **32**, 27–36.

8 Hannun, Y.A. and Obeid, L.M. (2008) Principles of bioactive lipid signalling: lessons from sphingolipids. *Nat. Rev. Mol. Cell Biol.*, **9**, 139–150.

9 Chitwood, D.J., Lusby, W.R., Thompson, M.J., Kochansky, J.P., and Howarth, O.W. (1995) The glycosylceramides of the nematode *Caenorhabditis elegans* contain an unusual, branched-chain sphingoid base. *Lipids*, **30**, 567–573.

10 Shinzawa-Itoh, K., Aoyama, H., Muramoto, K., Terada, H., Kurauchi, T., Tadehara, Y., Yamasaki, A., Sugimura, T., Kurono, S., Tsujimoto, K., Mizushima, T., Yamashita, E., Tsukihara, T., and Yoshikawa, S. (2007) Structures and physiological roles of 13 integral lipids of bovine heart cytochrome c oxidase. *EMBO J.*, **26**, 1713–1725.

11 Menuz, V., Howell, K.S., Gentina, S., Epstein, S., Riezman, I., Fornallaz-Mulhauser, M., Hengartner, M.O., Gomez, M., Riezman, H., and Martinou, J.C. (2009) Protection of *C. elegans* from anoxia by HYL-2 ceramide synthase. *Science*, **324**, 381–384.

12 Stahlman, M., Pham, H.T., Adiels, M., Mitchell, T.W., Blanksby, S.J.,

Fagerberg, B., Ekroos, K., and Boren, J. (2012) Clinical dyslipidaemia is associated with changes in the lipid composition and inflammatory properties of apolipoprotein-B-containing lipoproteins from women with type 2 diabetes. *Diabetologia*, **55** (4), 1156–1166.

13 Zhao, Z. and Xu, Y. (2009) An extremely simple method for extraction of lysophospholipids and phospholipids from blood samples. *J. Lipid Res.*, **51** (3), 652–659.

14 Folch, J., Lees, M., and Sloane Stanley, G.H. (1957) A simple method for the isolation and purification of total lipides from animal tissues. *J. Biol. Chem.*, **226**, 497–509.

15 Bligh, E.G. and Dyer, W.J. (1959) A rapid method of total lipid extraction and purification. *Can. J. Biochem. Physiol.*, **37**, 911–917.

16 Stahlman, M., Ejsing, C.S., Tarasov, K., Perman, J., Boren, J., and Ekroos, K. (2009) High-throughput shotgun lipidomics by quadrupole time-of-flight mass spectrometry. *J. Chromatogr. B Analyt. Technol. Biomed. Life Sci.*, **877**, 2664–2672.

17 Jung, H.R., Sylvanne, T., Koistinen, K.M., Tarasov, K., Kauhanen, D., and Ekroos, K. (2011) High throughput quantitative molecular lipidomics. *Biochim. Biophys. Acta*, **1811**, 925–934.

18 Matyash, V., Liebisch, G., Kurzchalia, T.V., Shevchenko, A., and Schwudke, D. (2008) Lipid extraction by methyl-*tert*-butyl ether for high-throughput lipidomics. *J. Lipid Res.*, **49**, 1137–1146.

19 Lofgren, L., Stahlman, M., Forsberg, G. B., Saarinen, S., Nilsson, R., and Hansson, G. I. (2012) The BUME method: a novel automated chloroform-free 96-well total lipid extraction method for blood plasma. *J. Lipid Res.* (accepted for publication).

20 Ejsing, C.S., Sampaio, J.L., Surendranath, V., Duchoslav, E., Ekroos, K., Klemm, R.W., Simons, K., and Shevchenko, A. (2009) Global analysis of the yeast lipidome by quantitative shotgun mass spectrometry. *Proc. Natl. Acad. Sci. USA*, **106**, 2136–2141.

21 Jiang, X., Cheng, H., Yang, K., Gross, R.W., and Han, X. (2007) Alkaline methanolysis of lipid extracts extends shotgun lipidomics analyses to the low-abundance regime of cellular sphingolipids. *Anal. Biochem.*, **371**, 135–145.

22 Deems, R., Buczynski, M.W., Bowers-Gentry, R., Harkewicz, R., and Dennis, E.A. (2007) Detection and quantitation of eicosanoids via high performance liquid chromatography–electrospray ionization–mass spectrometry. *Methods Enzymol.*, **432**, 59–82.

23 Nygren, H., Seppanen-Laakso, T., Castillo, S., Hyotylainen, T., and Oresic, M. (2011) Liquid chromatography–mass spectrometry (LC-MS)-based lipidomics for studies of body fluids and tissues. *Methods Mol. Biol.*, **708**, 247–257.

24 Murphy, R.C., James, P.F., McAnoy, A.M., Krank, J., Duchoslav, E., and Barkley, R.M. (2007) Detection of the abundance of diacylglycerol and triacylglycerol molecular species in cells using neutral loss mass spectrometry. *Anal. Biochem.*, **366**, 59–70.

25 Han, X. and Gross, R.W. (2005) Shotgun lipidomics: electrospray ionization mass spectrometric analysis and quantitation of cellular lipidomes directly from crude extracts of biological samples. *Mass Spectrom. Rev.*, **24**, 367–412.

26 Liebisch, G., Binder, M., Schifferer, R., Langmann, T., Schulz, B., and Schmitz, G. (2006) High throughput quantification of cholesterol and cholesteryl ester by electrospray ionization tandem mass spectrometry (ESI-MS/MS). *Biochim. Biophys. Acta*, **1761**, 121–128.

27 Ejsing, C.S., Duchoslav, E., Sampaio, J., Simons, K., Bonner, R., Thiele, C., Ekroos, K., and Shevchenko, A. (2006) Automated identification and quantification of glycerophospholipid molecular species by multiple precursor ion scanning. *Anal. Chem.*, **78**, 6202–6214.

28 Brugger, B., Erben, G., Sandhoff, R., Wieland, F.T., and Lehmann, W.D. (1997) Quantitative analysis of biological membrane lipids at the low picomole level by nano-electrospray ionization tandem

mass spectrometry. *Proc. Natl. Acad. Sci. USA*, **94**, 2339–2344.

29 Ekroos, K., Chernushevich, I.V., Simons, K., and Shevchenko, A. (2002) Quantitative profiling of phospholipids by multiple precursor ion scanning on a hybrid quadrupole time-of-flight mass spectrometer. *Anal. Chem.*, **74**, 941–949.

30 Ekroos, K., Ejsing, C.S., Bahr, U., Karas, M., Simons, K., and Shevchenko, A. (2003) Charting molecular composition of phosphatidylcholines by fatty acid scanning and ion trap MS3 fragmentation. *J. Lipid Res.*, **44**, 2181–2192.

31 Ivanova, P.T., Milne, S.B., Myers, D.S., and Brown, H.A. (2009) Lipidomics: a mass spectrometry based systems level analysis of cellular lipids. *Curr. Opin. Chem. Biol.*, **13**, 526–531.

32 Han, X. (2002) Characterization and direct quantitation of ceramide molecular species from lipid extracts of biological samples by electrospray ionization tandem mass spectrometry. *Anal. Biochem.*, **302**, 199–212.

33 Yang, K., Zhao, Z., Gross, R.W., and Han, X. (2009) Systematic analysis of choline-containing phospholipids using multi-dimensional mass spectrometry-based shotgun lipidomics. *J. Chromatogr. B Analyt. Technol. Biomed. Life Sci.*, **877**, 2924–2936.

34 Smith, C.A., Want, E.J., O'Maille, G., Abagyan, R., and Siuzdak, G. (2006) XCMS: processing mass spectrometry data for metabolite profiling using nonlinear peak alignment, matching, and identification. *Anal. Chem.*, **78**, 779–787.

35 Hartler, J., Trotzmuller, M., Chitraju, C., Spener, F., Kofeler, H.C., and Thallinger, G.G. (2011) Lipid data analyzer: unattended identification and quantitation of lipids in LC-MS data. *Bioinformatics*, **27**, 572–577.

36 Yang, K., Cheng, H., Gross, R.W., and Han, X. (2009) Automated lipid identification and quantification by multidimensional mass spectrometry-based shotgun lipidomics. *Anal. Chem.*, **81**, 4356–4368.

37 Herzog, R., Schwudke, D., Schuhmann, K., Sampaio, J.L., Bornstein, S.R., Schroeder, M., and Shevchenko, A. (2011) A novel informatics concept for high-throughput shotgun lipidomics based on the molecular fragmentation query language. *Genome. Biol.*, **12**, R8.

38 Thomas, M.C., Mitchell, T.W., Harman, D.G., Deeley, J.M., Nealon, J.R., and Blanksby, S.J. (2008) Ozone-induced dissociation: elucidation of double bond position within mass-selected lipid ions. *Anal. Chem.*, **80**, 303–311.

4 Multidimensional Mass Spectrometry-Based Shotgun Lipidomics

Hui Jiang, Michael A. Kiebish, Daniel A. Kirschner, and Xianlin Han

4.1 Introduction

Development of lipidomic research has been greatly accelerated with technological advancements in mass spectrometry (MS), particularly the instruments associated with soft ionization techniques, for example, electrospray ionization (ESI) and matrix-assisted laser desorption/ionization (MALDI). The approaches based on ESI-MS can be largely classified into two categories: liquid chromatography (LC)-MS and shotgun lipidomics, depending on whether or not an LC is coupled to a mass spectrometer. Shotgun lipidomics broadly refers to those methodologies with direct infusion. Most approaches in shotgun lipidomics maximally exploit the chemical and physical properties of lipids to facilitate the high-throughput analysis of a cellular lipidome on a large scale [1] instead of separation science in the LC-MS approaches. A unique feature of shotgun lipidomics is that a mass spectrum that displays a lipid class can be acquired at a "constant" concentration of the solution. This feature allows researchers to perform analysis of a lipid class with numerous scans in the precursor ion scanning (PIS) and/or neutral loss scanning (NLS) modes without the time constraints that are typically encountered through "on the fly" analysis during chromatographic elution. There exist three main platforms of shotgun lipidomics in practice, that is, tandem MS-based shotgun lipidomics [2, 3], high mass accuracy-based shotgun lipidomics [4–7], and multidimensional mass spectrometry (MDMS)-based shotgun lipidomics [1, 8–12]. This chapter focuses on MDMS-based shotgun lipidomics.

4.2 Multidimensional Mass Spectrometry-Based Shotgun Lipidomics

MDMS-based shotgun lipidomics [8, 11–13] is a well-recognized platform in the current practice of lipidomics. This technology maximally exploits the unique chemistry inherent in discrete lipid classes. In this platform, diluted lipid extracts of biological samples are directly infused into a mass spectrometer without prior separation of

Lipidomics, First Edition. Edited by Kim Ekroos.

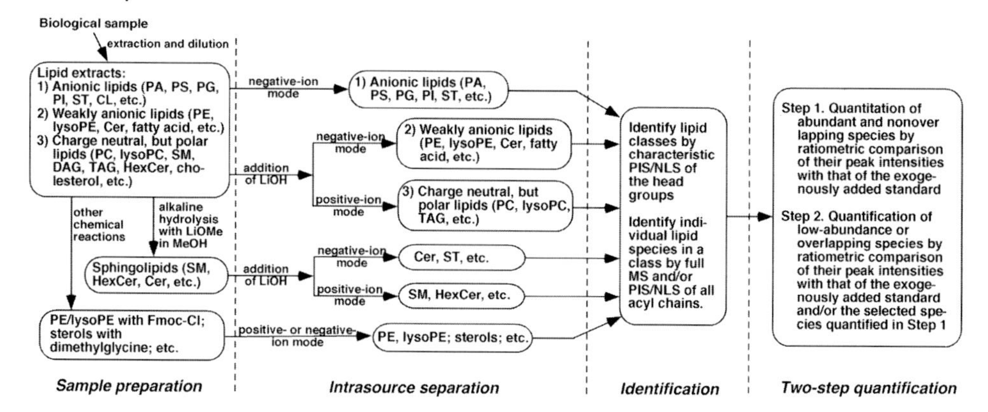

Figure 4.1 A schematic workflow used in multidimensional mass spectrometry-based shotgun lipidomics for identification and quantitation of individual molecular species in a lipid extract of a biological sample. The entire platform is comprised of four components (including sample preparation, intrasource separation, multidimensional mass spectrometry identification, and two-step quantification) in addition to data processing and bioinformatics, which is not included in the workflow. The abbreviations for lipid classes including PA, PS, PG, PI, ST, CL, PE, Cer, PC, SM, DAG, TAG, and HexCer stand for phosphatidic acid, phosphatidylserine, phosphatidylglycerol, phosphatidylinositol, sulfatide, cardiolipin, ethanolamine glycerophospholipid, ceramide, choline glycerophospholipid, sphingomyelin, diacylglycerol, triacylglyceride, and hexosylceramide, respectively. PIS and NLS denote precursor ion scanning and neutral loss scanning, respectively.

individual lipid classes in most cases (Figure 4.1). In some occasions, a chemical reaction may be conducted for analysis of a particular lipid class or a category of lipid classes. For example, alkaline treatment of lipid extracts is used for the analysis of sphingolipidome [14] and Fmoc derivative (i.e., addition of Fmoc chloride) is employed for the analysis of ethanolamine-containing lipid classes (e.g., ethanolamine glycerophospholipid (PE) and lysoPE) [15]. Next, intrasource separation of the lipid classes is realized by exploiting different charge properties of lipid classes to selectively ionize a certain category of lipid classes under multiplexed experimental conditions (Figure 4.1). After collision-induced dissociation, the fragments that correspond to the head groups and fatty acyl chains are identified with MDMS analysis. Each of these fragments can be monitored with at least one PIS and/or NLS (Figure 4.1). These diagnostic ions are also exploited for quantification of individual lipid species through a two-step procedure (Figure 4.1) [1, 16]. In addition, variation of each of the conditions related to extraction, infusion, ionization, collision, and so on provides an additional dimension of analysis that is integrated to encapsulate MDMS analysis.

4.2.1 Intrasource Separation

Based on charge properties, lipids can be reclassified into three categories: anionic lipids, weakly anionic lipids, and charge neutral but polar lipids.

Anionic lipids (e.g., cardiolipin (CL), phosphatidylglycerol (PG), phosphatidylinositol (PI), phosphatidylserine (PS), phosphatidic acid (PA), sulfatide (ST), their lysolipids, and acyl CoA) carry at least one net negative charge under weakly acidic conditions. Weakly anionic lipids (e.g., PE, lysoPE, free fatty acids and their derivatives, bile acids, and ceramide (Cer)) are charge neutral under weakly acidic conditions, but negatively charged under alkaline conditions. Charge neutral but polar lipids (e.g., phosphatidylcholine (PC), lysoPC, sphingomyelin (SM), hexosylceramide (HexCer), acylcarnitine, monoacylglycerol (MAG), diacylglycerol (DAG), triacylglycerol (TAG), and cholesterol and its esters) carry no separable charges. Three categories of lipids behave differentially in different infusion solvents, which can be acidic, basic, or neutral depending on the modifier [17]. An acidic modifier is favorable for the ionization of charge neutral but polar lipids and weakly acidic lipids as the protonated ions in the positive ion mode, while it disfavors the ionization of acidic lipids in the negative ion mode (Figure 4.2). A neutral modifier promotes the ionization of charge neutral but polar lipids as either the protonated ions or the cationic adducts of the modifier in the positive ion mode (Figure 4.2) [5, 18]. In the negative ion mode, acidic lipids plus PE species are selectively ionized as the deprotonated ions, while PC species can be ionized as anionic adducts with a neutral modifier (Figure 4.2). With addition of a basic modifier, anionic lipids and weakly anionic lipids are selectively ionized in the deprotonated form in the negative ion mode, while PC species are selectively ionized as cationic adducts or the protonated ions (when the modifier is NH_4OH) in the positive ion mode. Without a modifier, the ionization of anionic lipids as deprotonated ions is dominant in the negative ion mode. These types of selective ionization form the basis of intrasource separation [1, 9, 17].

Although the presence of different modifiers could substantially affect the molecular ion profiles of different lipid classes, the profile of individual molecular species of each class is not influenced. For example, the consistent ratios can be observed between individual PE species (e.g., di16:1 PE versus 18:0–22:6 PE at m/z 688.4 and 792.5 in Figure 4.2a and at m/z 686.4 and 790.5 in Figure 4.2b, d, and f) and individual PC species (e.g., di14:0 PC versus 16:0–22:6 PC as protonated at m/z 674.4 and 806.5 in Figure 4.2a and c, as lithiated at m/z 680.4 and 812.5 in Figure 4.2e and as acetate adducts at m/z 732.4 and 864.5 in Figure 4.2b). These observations indicate that individual molecular species of a lipid class can be quantified by ratiometric comparison with the selected internal standard of the class.

4.2.2 The Principle of Multidimensional Mass Spectrometry

There exist many variables in mass spectrometry, including conditions of lipid extraction, infusion, ionization, and tandem MS analysis. Varying each of these variables adds a dimension to MS, and integratively leads to MDMS [1]. Specifically, MDMS is defined as the comprehensive mass spectrometry analyses

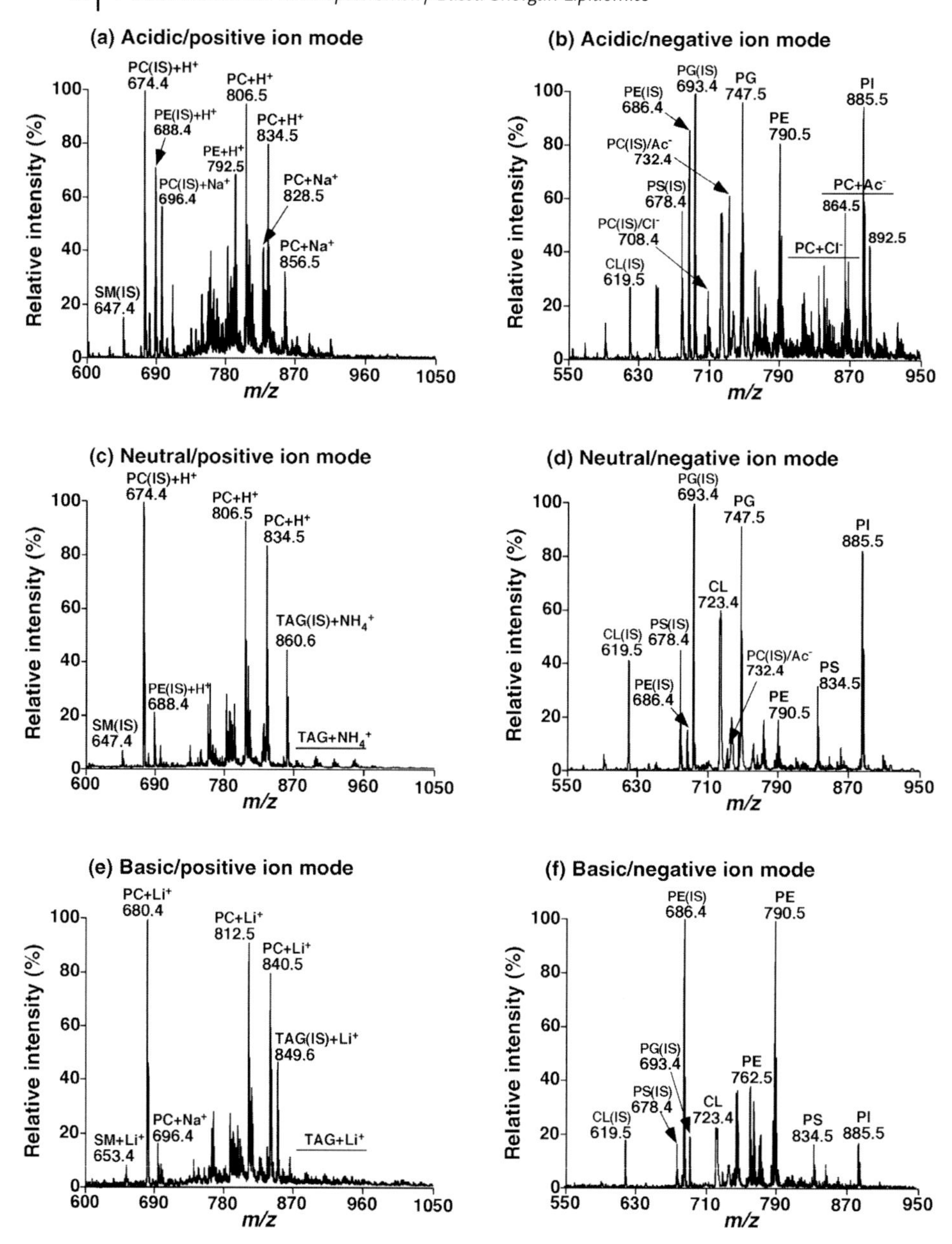

Figure 4.2 Representative positive and negative ion ESI mass spectra acquired under weak acidic, neutral, and weak basic conditions. A lipid extract of mouse myocardium was prepared and mass spectrometry analysis was performed as described previously [44]. Positive and negative ion ESI mass spectra as indicated were acquired after direct infusion in the presence of 10% acetic acid (a and b), 5 mM ammonium acetate (c and d), and 10 μM lithium hydroxide (e and f) in the infused solution. IS and Ac stand for "internal standard" and "acetate," respectively. The abbreviations for lipid classes are listed in the legend of Figure 4.1.

conducted under a variety of instrumental variables. At the current stage, MDMS is decomposed into multiple 2D MS for the ease of use, and is displayed by varying only one of the variables at a time. It is anticipated that advanced computational approaches will facilitate the direct use of 3D MS or MDMS and provide an additional level of information directly obtained from the mass spectrometry analysis in the next generation of MDMS-based shotgun lipidomics.

4.2.3 Variables in Multidimensional Mass Spectrometry

4.2.3.1 Variables in Fragment Monitoring by Tandem MS Scans

The variables include *m/z* values of fragment ions in PIS, *m/z* values of molecular ions in product ion scan, and mass values of the neutral fragments in NLS. In theory, these variables should vary in a range of interest unit by unit. In practice, one can learn the fragments of all species in a lipid class from the analysis of a few representative lipid species in the product ion mode. Hence, the variable of *m/z* (fragment ions) in PIS can vary only those potential product ions. The neutral fragment mass in NLS equals the difference between a molecular ion and a product ion, so this variable can vary only those masses corresponding to the product ions of interest. In most cases, the identity of a lipid species can be fully determined with only two or three characteristic product ions that carry the information about the head group and fatty acyl chains of the species in addition to the molecular ion *m/z*.

4.2.3.2 Variables Related to the Infusion Conditions

These variables include the solvent(s) of the lipid solution, the modifier(s) of the lipid solution, and the lipid concentration in the solution. Infusion solvent(s) largely affect the solubility of lipids and the ionization efficiency. In this regard, lipids in polar solvents form micelles that are not readily ionized. In the current lipidomic analysis by ESI-MS, the most commonly used solvent system is comprised of chloroform (or dichloromethane) and methanol or to a lesser extent, isopropanol. The modifier of a solvent system substantially affects the ionization sensitivity and efficiency. The modifiers can be acidic (e.g., acetic acid), basic (e.g., LiOH and NH_4OH), and neutral (e.g., ammonium acetate). These three types of modifiers have distinct impacts on different lipid classes. Electrospray ionization of lipids is a concentration-dependent process [19–21]. As lipid concentration increases, lipids tend to form aggregates that are not readily ionized and the formation of aggregates depends on lipid molecular species. We found that lipids with different acyl chain lengths and unsaturation degrees show apparent differences in ionization efficiency at a concentration of total lipids higher than 0.1 nmol/μl in chloroform–methanol [1:1, v/v) [21, 22].

4.2.3.3 Variables under Ionization Conditions

The variables in this category include capillary temperature, spray voltage, and flow rate. These variables are critical for ionization efficiency and/or intrasource separation [17]. Although they are usually predetermined prior to the analysis, the predetermined values may not be optimal for a particular sample or a specific lipid class. Thus, varying these variables is indispensable for providing the best condition for the analysis of each lipid class in each sample.

4.2.3.4 **Variables under Collision Conditions**

Collision conditions can be varied with collision energy, collision gas pressure, and collision gas type. Collision energy provides kinetic energy to molecular ions in a collision cell; collision gas pressure determines collision frequency and pathway; and different collision gas types possess different sizes and intrinsic energies. Accordingly, variation of these variables affects the fragmentation patterns of lipids, which has recently been utilized to identify the double bond locations of fatty acid isomers [10, 23].

4.2.3.5 **Variables Related to the Sample Preparations**

The conditions related to the lipid sample preparations can be varied with pH condition and solvent polarity during lipid extraction and during chemical reactions with the extracted lipids. The polarity and/or the charge properties of many lipid classes are pH dependent. For example, PE species are positively charged under acidic conditions, whereas they are negatively charged under basic conditions. Accordingly, variation of pH conditions of an extraction solution could lead to different extraction efficiency. In practice, an acidic condition is favorable for the extraction of anionic lipid species, whereas a neutral condition is preferred for the extraction of PE species. A weakly acidic condition may be used for "total" lipid extraction. Extraction efficiency of lipid species also largely depends on the solvent(s) used in the extraction. Nonpolar solvents (e.g., hexane and ethyl ether) can be used to extract nonpolar lipids such as TAG, cholesterol, cholesterol esters, and free fatty acids. Most lipids can be effectively extracted with a method based on the use of chloroform (or dichloromethane) such as the Folch method or a modified Bligh–Dyer method [24]. Very polar lipids (e.g., acyl CoAs, acylcarnitines, lysophospholipids, and gangliosides) can be recovered from the aqueous phase of a solvent extraction method with reversed-phase or affinity cartridge columns or special solvents [25, 26]. In addition, the intrinsic chemical stability and/or reactivity of lipids can be used as a variable in lipidomic analysis. For example, the chemical stability of a sphingoid backbone to alkaline hydrolysis has been used to enrich sphingolipids [14, 27, 28]; the lability of vinyl ether moiety under acidic conditions has been used to identify the presence of plasmalogen species [29, 30]; special derivative procedures have been developed for the enhanced analyses of a few lipid classes (e.g., PE and lysoPE with Fmoc chloride [15], *N*-methylpiperazine acetic acid *N*-hydroxysuccinimide ester [31], or 4-(dimethylamino)benzoic acid [32], cholesterol and oxysterols with dimethylglycine [33] or with Girard P hydrazones [34]; DAG with *N*-chlorobetainyl chloride [35]; and eicosanoids with *N*-(4-aminomethylphenyl) pyridinium [36]).

4.3
Application of Multidimensional Mass Spectrometry-Based Shotgun Lipidomics for Lipidomic Analysis

4.3.1
Identification of Lipid Molecular Species by 2D Mass Spectrometry

Although accurate mass in combination with an isotopologue pattern can be used to identify individual lipid molecular species, the detailed identity of individual lipid species including a variety of isomers can be identified only through product ion MS analysis of individual species or through the analysis of building blocks including head groups and aliphatic chains by 2D MS [12]. Table 4.1 summarizes the representative building blocks for identification of lipid species in each class by 2D MS.

4.3.1.1 Identification of Anionic Lipids

After a particular sample preparation (with a final concentration of total lipids at ~50 pmol/μl) [1, 24], PIS of all acyl chains potentially present in a biological lipid extract are acquired sequentially with no modifier in the infusion solution and in the negative ion mode. For example, the ion peaks detected from the spectra of PIS255.2 and 281.2 support the presence of 16:0 and 18:1 fatty acyl chains, respectively. PIS and/or NLS of head groups or their derivatives are acquired to identify individual lipid classes in the category of anionic lipids. For example, the ion peaks in PIS of glycerol phosphate derivative (PIS153.0 at 35 eV) show the presence of all the anionic glycerophospholipids; NLS of serine (NLS87.0 at 24 eV) shows the presence of PS species.

4.3.1.2 Identification of Weakly Anionic Lipids

Next, a small amount of LiOH (10–50 pmol/μl) is added to the infusion solution used in the last section, and this basified sample is analyzed in the negative ion mode again. PIS of all potential acyl chains and PIS and/or NLS of head groups identify the acyl chain and the head group building blocks of the species in the category of weakly anionic lipids. Alternatively, PE and lysoPE species can be identified through Fmoc derivatization and NLS222.1 [15].

4.3.1.3 Identification of Charge Neutral but Polar Lipids

Finally, in the positive ion mode with this basified solution, NLS of all potential fatty acids identify the acyl chain building blocks for charge neutral but polar lipids. Alternatively, NLS of fatty acid plus 59 (i.e., trimethylamine) provides increasing specificity and sensitivity for PC species [37]. Again, PIS and/or NLS of head groups are acquired to classify individual lipid classes in this category.

4.3.1.4 Identification of Sphingolipids

Sphingolipids in most biological samples are present in low abundance. Detailed analysis of sphingolipids can be performed after alkaline hydrolysis (Figure 4.1) [13, 38]. Ceramide content in biological samples is quite low, so assessment of ceramide content heavily depends on tandem MS spectra.

Table 4.1 Summary of the representative building blocks used to identify individual lipid species by MDMS.

Lipid class	Ion format	Scans for class-specific prescreen (building block derivative)	Scans for identification of acyl chain and/or regioisomers
PC [37]	$[M + Li]^+$	NLS189 (lithium cholinephosphate), NLS183 (phosphocholine), −35 eV	NLS(59 + FA), −40 eV
LysoPC [37]	$[M + Na]^+$	NLS59 (trimethylamine), −22 eV	PIS104, −34 eV
		NLS205 (sodium cholinephosphate), −34 eV	PIS147, −34 eV
PE, lysoPE [15]	$[M - H]^-$	PIS196 (glycerol phosphoethanolamine derivative), 50 eV	PIS(FA-H), 30 eV
	$[M - H + Fmoc]^-$	NLS222 (Fmoc), 30 eV	
PI, lysoPI [13]	$[M - H]^-$	PIS241 (inositol derivative), 45 eV	PIS(FA-H), 47 eV
PS, lysoPS [13]	$[M - H]^-$	NLS87 (serine), 24 eV	PIS(FA-H), 30 eV
PG, PA, lysoPG, lysoPA [13]	$[M - H]^-$	PIS153 (glycerol phosphate), 35 eV	PIS(FA-H), 30 eV
Cardiolipin (CL), monolysoCL [45]	$[M - 2H]^{2-}$	Full MS at high resolution	PIS(FA-H) at high resolution, 25 eV; NLS(FA-H_2O) at high resolution, 22 eV
Triacylglycerol [8]	$[M + Li]^+$		NLS(FA), −35 eV
Sphingomyelin [37]	$[M + Li]^+$	NLS213 (lithium cholinephosphate plus methyl aldehyde), −52 eV	NLS (neutral fragments from sphingoid backbone)
Ceramide [38]	$[M - H]^-$	NLS256, NLS327, and NLS240 (derivatives of d18:1 backbone); NLS258, NLS329, and NLS242 (derivatives of d18:0 backbone); NLS284, NLS355, and NLS268 (derivatives of d20:1 backbone); 32 eV	NLS (neutral fragments from sphingoid backbone), (e.g., NLS256 (d18:1 nonhydroxy species)), 32 eV
Hexosylceramide [49, 50]	$[M + Li]^+$	NLS162 (monohexose), −50 eV	NLS (neutral fragments from sphingoid backbone)
Sulfatide [51]	$[M - H]^-$	PIS 97 (sulfate), 65 eV	NLS (neutral fragments from sphingoid backbone)
Sphingoid-1-phosphate [52]	$[M - H]^-$	PIS79, 24 eV	

(*continued*)

Table 4.1 (Continued)

Lipid class	Ion format	Scans for class-specific prescreen (building block derivative)	Scans for identification of acyl chain and/or regioisomers
Sphingoid base [14]	$[M + H]^+$	NLS48, −18 eV	
Psychosine [53]	$[M + H]^+$	NLS180, −24 eV	
Cholesterol [54]	[Cholesteryl methoxyacetate +MeOH + Li]$^+$	PIS97, −22 eV	
Acylcarnitine [55]	$[M + H]^+$	PIS85, −30 eV	PIS85, −30 eV for all species PIS145, −30 eV for hydroxy species
Acyl CoA [25]	$[M - H]^-$, $[M - 2H]^{2-}$ $[M\text{-}3H]^{3-}$	PIS134, 30 eV	PIS134, 30 eV

FA and (FA-H) denote free fatty acid and fatty acyl carboxylate anion, respectively. The abbreviations for other lipid classes are listed in the legend of Figure 4.1.

Specifically, NLS327.3 and NLS256.2 scans are used to identify sphingosine-based ceramide molecular species with or without a hydroxyl group at the α-position, whereas NLS240.2 is used to assess the contents of these molecular species [38]. To sphinganine or other sphingoid-based ceramide molecular species, the corresponding sets of building blocks are respectively analyzed for identification and quantification (Table 4.1) [39–41].

4.3.1.5 The Concerns of the MDMS-Based Shotgun Lipidomics for Identification of Lipid Species

The caveats of the MDMS-based shotgun lipidomics for identification of individual species of a cellular lipidome are at least in two aspects. First, it would be impossible to identify the isomers that possess identical fragmentation patterns by using this platform. Examples of this category include chiral isomers as well as GluCer and GalCer in the positive ion mode. Second, the application of this technology for discovery of any novel lipid class is limited because a precharacterization of the class of interest is always required prior to its application for identification at the current stage of MDMS platform. For this particular case, LC-MS methods offer advantages over shotgun lipidomic approaches.

4.3.2 Quantification of Lipid Molecular Species by MDMS-Based Shotgun Lipidomics

Quantification of individual lipid species is one of the most important, yet challenging, components of lipidomic analysis by MS. Thus, a two-step quantification method has been developed for MDMS-based shotgun lipidomics.

4.3.2.1 The Principle of Quantification of Individual Lipid Species by MS

Due to the high sensitivity of the mass spectrometer, the ion intensity of an analyte might be affected by even minor differences in sample preparation, ionization conditions, tuning conditions, the analyzer or detector used in the mass spectrometer, and so on. Accordingly, quantification of any compound by ESI-MS has to be made by comparison with either an internal or an external standard similar to the compounds of interest, such as their isotopologues, under identical experimental conditions.

To determine the concentration of an analyte, the following formula is generally employed:

$$c_u/c_i = I_u/I_i \tag{4.1}$$

where c_u and c_i are the contents of the unknown species and the standard, respectively, while I_u and I_i are the peak intensities of the unknown species and the standard, respectively. Note that, ^{13}C deisotoping has to be performed prior to the ratiometric comparison [1, 13]. The formula (4.1) could be derived from the linear correlation between the content (c) and the ion intensity (I):

$$c = a(I - b) \tag{4.2}$$

where a is the response factor and b is the background noise. When $I \gg b$ (e.g., $S/N > 10$), $c \approx aI$. Formula (4.1) is obtained if the response factors of different molecular species in a lipid class are essentially identical. Identical response factors of different molecular species of a lipid class hold true only for polar lipid classes in the low concentration region. Polar lipid classes guarantee that the ionization efficiency of individual molecular species of a class predominantly depends on the head group [1, 11, 20, 42, 43]. The low concentration guarantees that the linear dynamic range does not interfere with aggregation [19–21]. It should be emphasized that the response factors of different nonpolar lipid species to ESI-MS are quite different and have to be predetermined for an accurate quantification [8].

4.3.2.2 Quantification by Using a Two-Step Procedure in MDMS-Based Shotgun Lipidomics

After intrasource separation and identification of individual lipid species by MDMS (Figure 4.1), quantification of the identified species is performed in a two-step procedure [11, 16]. In the first step, the abundant and nonoverlapping lipid species are first quantified with ratiometric comparison (i.e., formula (4.1)) with the internal standard of the class using survey scan spectra. The determined contents of these nonoverlapping and abundant species plus the exogenously added internal standard become the candidate standards for the second step of quantification. Next, the second step is performed to quantify the overlapping and/or low-abundance species by using the standards selected after the first step. An algorithm is generated based on two variables (i.e., the differences in the number of total carbon atoms and the number of total double bonds in fatty acyl chains of each species from that of the standards) to determine the correction factors for each individual species [12]. The corrected peak intensities of the overlapping and/or low-abundance species from the class-specific PIS or NLS are used to quantify these species by ratiometric comparison with the peak intensities of the standards. This approach has been used to quantify individual species of nearly 30 lipid classes from extracts of biological samples [12]. With the second step in quantification, over 5000-fold linear dynamic range for many lipid classes can be achieved [44].

The quantitative accuracy in the two-step procedure of MDMS-based shotgun lipidomics has been validated by a series of experiments, in which various known amounts of exogenously added internal standards or lipid analytes are spiked into the lipid samples prior to or after extraction [10]. The results of the experiments indicate that the matrix effects on quantification of individual lipid species through ratiometric comparison are minimal, and the linear relationships hold very consistent for the data determined from both the full MS and tandem MS analyses. These results well validate the feasibility of the two-step quantification procedure of MDMS-based shotgun lipidomics.

The caveats of this two-step quantification methodology include that the experimental error for the species measured in the second step of quantification is propagated and is larger than that in the first step. To minimize this effect in the second step, it is very important to use the species in high abundance determined from the first step as standards for quantification of other species in the second step. In addition, the two-step quantification procedure cannot be applied to any lipid class for which a class-specific and class-sensitive PIS or NLS is not present, for example, TAG, Cer, PE, and CL. Special quantification methods for these lipid classes have been developed in MDMS-based shotgun lipidomics [13, 15, 38, 45].

4.3.2.3 Quantitative Analysis of PEX7 Mouse Brain Lipidome by MDMS-Based Shotgun Lipidomics

The application of MDMS-based shotgun lipidomics is demonstrated by a recent study on alterations in mouse brain lipidome induced by PEX7 (peroxisomal biogenesis factor 7) knockout (Jiang and Han, unpublished data). PEX7 is associated with multiple catabolic and anabolic functions, including β-oxidation of very long-chain fatty acids (VLCFA) and synthesis of bile acids and plasmalogens [46–48]. MDMS-based shotgun lipidomics revealed that dramatic changes of various lipid species are present in PEX7 knockout mouse brain samples relative to those of wild-type mice (see below) in addition to a virtual depletion of plasmalogen species (Figure 4.3).

For example, both composition and content of PS species in PEX7 knockout mouse cerebellum are altered in comparison to those present in wild-type mice (Figure 4.4). Specifically, after identification, two abundant species at m/z 788.5 and 834.6, corresponding to 18:0–18:1 and 18:0–22:6 PS are accurately determined by ratiometric comparison with the internal standard (i.e., 14:0–14:0 PS at m/z 678.4) after ^{13}C deisotoping from the survey scan mass spectra (Figure 4.4a and c). NLS87.1 (serine) is quite specific to PS species, which shows multiple additional low-abundance PS species (Figure 4.4b and d). These low abundance PS species are determined using either 18:0–18:1 or 18:0–22:6 PS or both as the selected

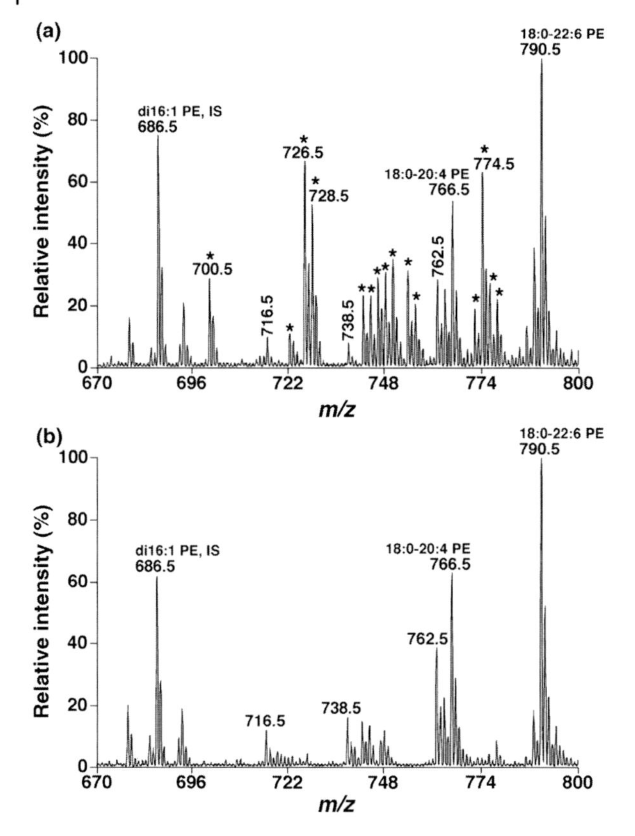

Figure 4.3 Demonstration of virtual depletion of plasmalogen species in lipid extracts of PEX7 mouse brain samples. The lipid extracts from cerebella of PEX7 mice (b) and their littermates (a) were prepared by a modified Bligh–Dyer procedure [24] and properly diluted prior to direct infusion into a mass spectrometer by using a Nanomate device. Negative ion electrospray ionization mass spectrometry analysis of ethanolamine glycerophospholipid (PE) species were performed as previously described [44]. IS stands for internal standard. The asterisks indicate the apparent plasmalogen species that are present in wild-type mouse samples (a), but virtually disappear in PEX7 mouse samples (b).

standard(s) after compensating the influence from different acyl carbon numbers and double bond numbers.

Ceramide species are usually low abundant in biological samples and can be barely seen in the survey scan. Thus, alkaline hydrolysis is frequently performed prior to the analysis of ceramide species [13, 38]. Alternatively, quantification of this lipid class can be conducted with tandem MS analysis as described previously [38]. Specifically, NLS256.2 can be used to identify and quantify sphingosine-based ceramide species not containing α-hydroxy group in the fatty acyl amide chain, while NLS240.2 can be employed to quantify all sphingosine-based ceramide species with ratiometric comparison with a selected internal standard (i.e., d18:1/N17:0 Cer at *m/z* 550.5). With this approach, we demonstrated significant

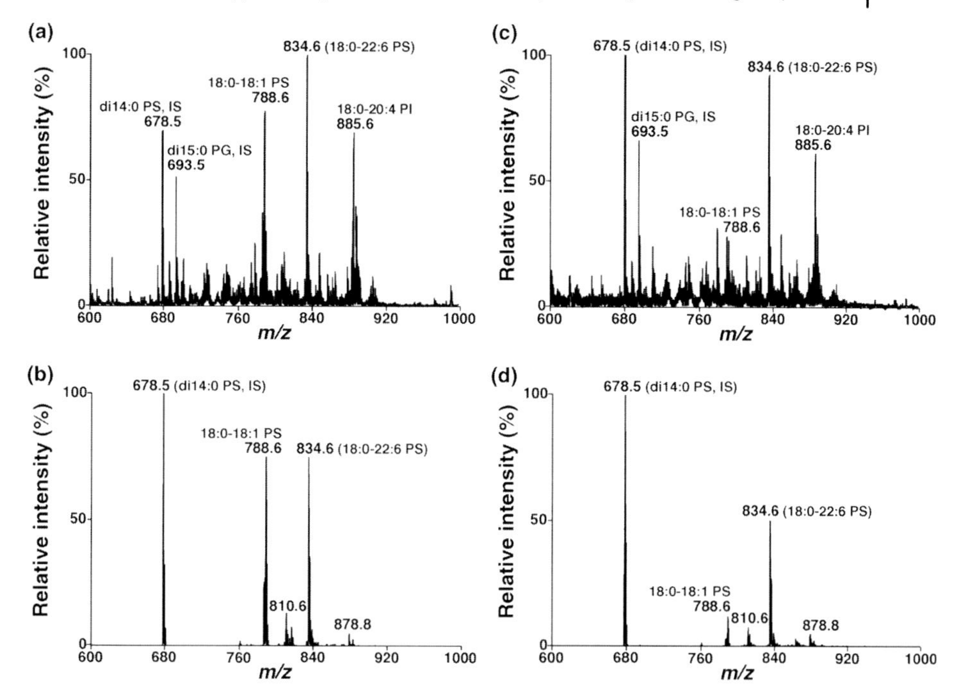

Figure 4.4 Two-step quantitative analysis of phosphatidylserine species in lipid extracts of mouse brain cerebellum. The lipid extract was prepared by a modified Bligh–Dyer procedure [24] and properly diluted prior to direct infusion into a mass spectrometer by using a Nanomate device. The survey scans (a and c) and scans of neutral loss of 87.0 (b and d) of lipid extracts of brain cerebellum from wild-type (a and b) and PEX7 knockout (c and d) mice were acquired in the negative ion mode by sequentially programmed customized scans operating utilizing Xcalibur software as previously described [44]. IS stands for internal standard.

increases of Cer species levels in brain tissues of PEX7 knockout mice. For example, Figure 4.5 displays the representative mass spectra of NLS256.2 of brainstem lipid extracts from PEX7 knockout mice and their littermates. These spectra clearly show the markedly increased mass levels of Cer species in PEX7 knockout brainstem (Figure 4.5b) relative to those in wild-type mice (Figure 4.5a).

Similar to Cer analysis, alkaline hydrolysis is frequently performed prior to the analysis of SM species [13, 38]. Survey scan mass spectral analysis of the alkaline-treated lipid solution with addition of a small amount of LiOH (~10 pmol/μl) in the positive ion mode showed some abundant SM species (Figure 4.6a and c) (e.g., lithiated N18:0 and N24:1 SM at *m/z* 737.6 and 819.6, respectively), which were identified and can be quantified with ratiometric comparison with the selected internal standard (i.e., lithiated N12:0 SM at *m/z* 653.5) after ^{13}C deisotoping. Other low-abundance SM species can then be quantified with NLS213.2 (which is specific to SM class) (Figure 4.6b and d) with lithiated N12:0, N18:0, and N24:1 SM species as standards. Moreover, the NLS213.2 mass spectra clearly show that PEX7 deficiency leads to substantial reduction of the mass levels of SM species (Figure 4.6).

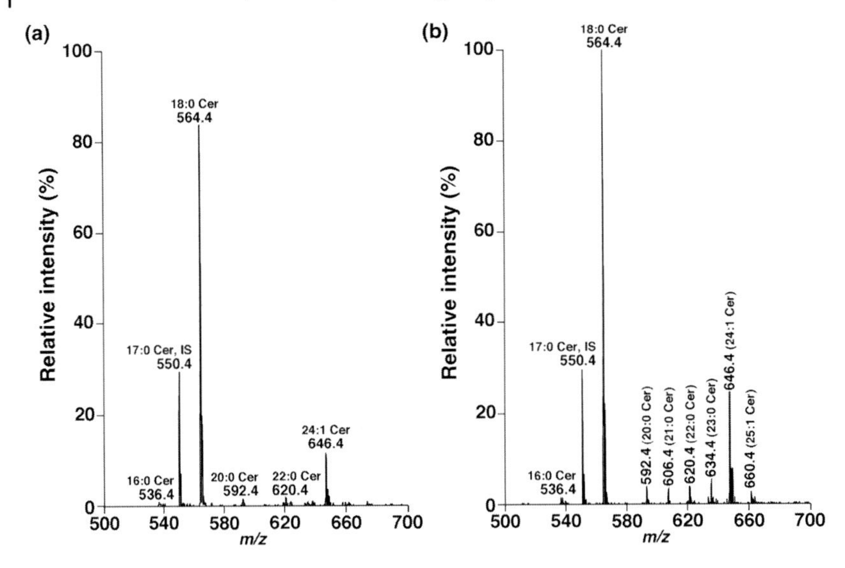

Figure 4.5 Tandem mass spectrometry analysis of ceramide species in lipid extracts of mouse brainstem. Lipid samples from brainstem of wild-type (a) and PEX7 knockout (b) mice were extracted by a modified Bligh–Dyer procedure [24] and analyzed by tandem mass spectrometry with a neutral loss of 256.2 as previously described [38]. IS stands for internal standard. The mass spectrum in (a) is displayed after normalization to the internal standard peak in (b) (i.e., the peaks corresponding to "IS" are equally intense in the paired spectra) for direct comparison.

Collectively, MDMS-based shotgun lipidomic analysis of PEX7 mouse brain lipidome not only confirmed the depletion of plasmalogen species but also uncovered significant mass level changes of numerous lipid classes and individual molecular species. Elucidation of the linkage between the PEX7 gene and the altered lipids as well as interrogation of the consequence resulting from the altered lipids are beyond the scope of this report. However, this example clearly indicates that global lipidomic analysis is a very powerful approach to reveal novel phenotypes of a gene, a physiological stimulus, or a pathological condition and enable us to uncover the biochemical mechanisms underpinning the (patho)physiological changes.

4.4
Conclusions

Lipidomics as a new research field has emerged for nearly 10 years [9]. Substantial progresses on development of this new field, particularly in methodology toward automation and high throughput by means of bioinformatics, have been made over the course of this period. We believe that many biochemical mechanisms underlying lipid metabolism critical to the metabolic diseases associated with lipid abnormalities will be increasingly identified as lipidomics penetrates into the lower abundance regions of individual lipid molecular species. Through application of a

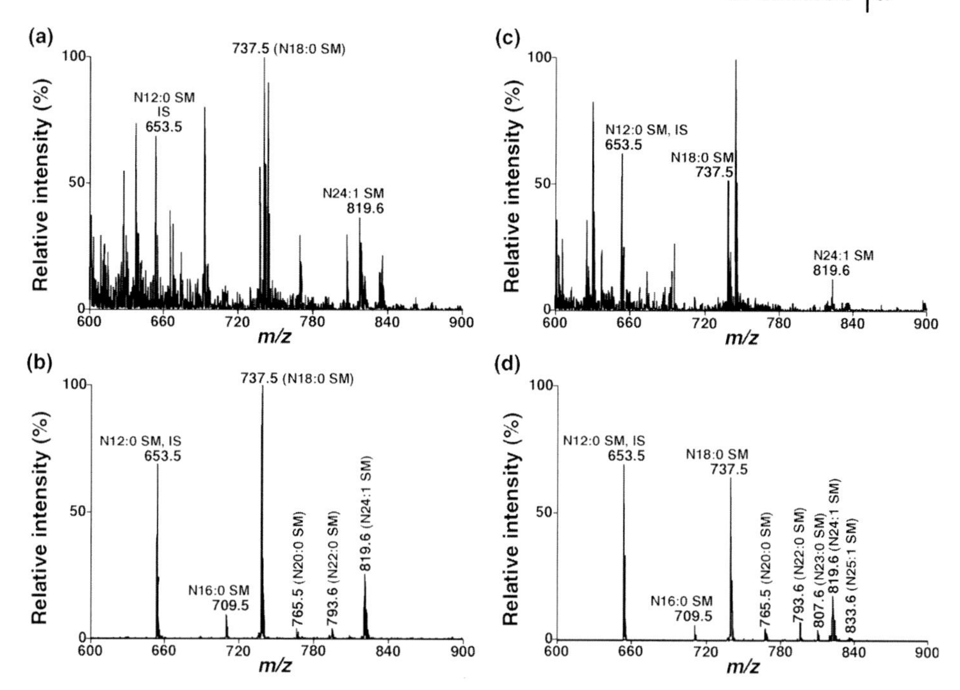

Figure 4.6 Two-step quantitative analysis of sphingomyelin molecular species in the alkaline-treated lipid extracts of mouse cerebellum in the positive ion mode in the presence of LiOH. Mouse cerebellum lipid extracts were prepared by using a Bligh–Dyer extraction procedure [24]. A part of each lipid extract was treated with LiOMe as described previously [14]. Positive ion ESI mass spectrum in the full MS scan mode (a and c) and scans for neutral loss of 213.2 (b and d) of lipid extracts of brain cerebellum from wild-type (a and b) and PEX7 knockout (c and d) mice were acquired in the positive ion mode by sequentially programmed customized scans operating utilizing Xcalibur software as previously described [44]. IS stands for internal standard. The mass spectrum in (d) is displayed after normalization to the internal standard peak in (b) (i.e., the peaks corresponding to "IS" are equally intense in the paired spectra) for direct comparison.

systems biology approach, the markers for lipid mediated/associated diseases, which are diagnostic of disease onset, progression, or severity, as well as evaluation of drug efficacy and safety will be revealed. We speculate that lipidomics integrated with genomic and proteomic studies will significantly enhance our understanding of the role of lipids in biological systems and lead to the diagnosis of lipid-related disease states at earlier time to enhance therapeutic efficacy and to tailor pharmacological therapy.

Acknowledgments

This work was supported by National Institute on Aging/National Institute of Diabetes and Digestive and Kidney Diseases Grant R01 AG31675 (XH). The research described here on the PEX7 transgenic mouse was supported by the European Leukodystrophy Association (ELA 2008-009C4) (DAK).

References

1 Han, X. and Gross, R.W. (2005) Shotgun lipidomics: electrospray ionization mass spectrometric analysis and quantitation of the cellular lipidomes directly from crude extracts of biological samples. *Mass Spectrom. Rev.*, **24**, 367–412.

2 Brugger, B., Erben, G., Sandhoff, R., Wieland, F.T., and Lehmann, W.D. (1997) Quantitative analysis of biological membrane lipids at the low picomole level by nano-electrospray ionization tandem mass spectrometry. *Proc. Natl. Acad. Sci. USA*, **94**, 2339–2344.

3 Welti, R., Li, W., Li, M., Sang, Y., Biesiada, H., Zhou, H.-E., Rajashekar, C.B., Williams, T.D., and Wang, X. (2002) Profiling membrane lipids in plant stress responses: role of phospholipase Da in freezing-induced lipid changes in *Arabidopsis*. *J. Biol. Chem.*, **277**, 31994–32002.

4 Ekroos, K., Chernushevich, I.V., Simons, K., and Shevchenko, A. (2002) Quantitative profiling of phospholipids by multiple precursor ion scanning on a hybrid quadrupole time-of-flight mass spectrometer. *Anal. Chem.*, **74**, 941–949.

5 Ejsing, C.S., Duchoslav, E., Sampaio, J., Simons, K., Bonner, R., Thiele, C., Ekroos, K., and Shevchenko, A. (2006) Automated identification and quantification of glycerophospholipid molecular species by multiple precursor ion scanning. *Anal. Chem.*, **78**, 6202–6214.

6 Schwudke, D., Oegema, J., Burton, L., Entchev, E., Hannich, J.T., Ejsing, C.S., Kurzchalia, T., and Shevchenko, A. (2006) Lipid profiling by multiple precursor and neutral loss scanning driven by the data-dependent acquisition. *Anal. Chem.*, **78**, 585–595.

7 Schwudke, D., Liebisch, G., Herzog, R., Schmitz, G., and Shevchenko, A. (2007) Shotgun lipidomics by tandem mass spectrometry under data-dependent acquisition control. *Methods Enzymol.*, **433**, 175–191.

8 Han, X. and Gross, R.W. (2001) Quantitative analysis and molecular species fingerprinting of triacylglyceride molecular species directly from lipid extracts of biological samples by electrospray ionization tandem mass spectrometry. *Anal. Biochem.*, **295**, 88–100.

9 Han, X. and Gross, R.W. (2003) Global analyses of cellular lipidomes directly from crude extracts of biological samples by ESI mass spectrometry: a bridge to lipidomics. *J. Lipid Res.*, **44**, 1071–1079.

10 Han, X., Yang, K., and Gross, R.W. (2012) Multi-dimensional mass spectrometry-based shotgun lipidomics and novel strategies for lipidomic analyses. *Mass Spectrom. Rev.*, **31**, 134–178.

11 Han, X. and Gross, R.W. (2005) Shotgun lipidomics: multi-dimensional mass spectrometric analysis of cellular lipidomes. *Expert Rev. Proteomics*, **2**, 253–264.

12 Yang, K., Cheng, H., Gross, R.W., and Han, X. (2009) Automated lipid identification and quantification by multi-dimensional mass spectrometry-based shotgun lipidomics. *Anal. Chem.*, **81**, 4356–4368.

13 Han, X., Yang, J., Cheng, H., Ye, H., and Gross, R.W. (2004) Towards fingerprinting cellular lipidomes directly from biological samples by two-dimensional electrospray ionization mass spectrometry. *Anal. Biochem.*, **330**, 317–331.

14 Jiang, X., Cheng, H., Yang, K., Gross, R. W., and Han, X. (2007) Alkaline methanolysis of lipid extracts extends shotgun lipidomics analyses to the low abundance regime of cellular sphingolipids. *Anal. Biochem.*, **371**, 135–145.

15 Han, X., Yang, K., Cheng, H., Fikes, K.N., and Gross, R.W. (2005) Shotgun lipidomics of phosphoethanolamine-containing lipids in biological samples

after one-step *in situ* derivatization. *J. Lipid Res.*, **46**, 1548–1560.

16 Han, X., Cheng, H., Mancuso, D.J., and Gross, R.W. (2004) Caloric restriction results in phospholipid depletion, membrane remodeling and triacylglycerol accumulation in murine myocardium. *Biochemistry*, **43**, 15584–15594.

17 Han, X., Yang, K., Yang, J., Fikes, K.N., Cheng, H., and Gross, R.W. (2006) Factors influencing the electrospray intrasource separation and selective ionization of glycerophospholipids. *J. Am. Soc. Mass Spectrom.*, **17**, 264–274.

18 Han, X., Abendschein, D.R., Kelley, J.G., and Gross, R.W. (2000) Diabetes-induced changes in specific lipid molecular species in rat myocardium. *Biochem. J.*, **352**, 79–89.

19 DeLong, C.J., Baker, P.R.S., Samuel, M., Cui, Z., and Thomas, M.J. (2001) Molecular species composition of rat liver phospholipids by ESI-MS/MS: the effect of chromatography. *J. Lipid Res.*, **42**, 1959–1968.

20 Han, X. and Gross, R.W. (1994) Electrospray ionization mass spectroscopic analysis of human erythrocyte plasma membrane phospholipids. *Proc. Natl. Acad. Sci. USA*, **91**, 10635–10639.

21 Koivusalo, M., Haimi, P., Heikinheimo, L., Kostiainen, R., and Somerharju, P. (2001) Quantitative determination of phospholipid compositions by ESI-MS: effects of acyl chain length, unsaturation, and lipid concentration on instrument response. *J. Lipid Res.*, **42**, 663–672.

22 Zacarias, A., Bolanowski, D., and Bhatnagar, A. (2002) Comparative measurements of multicomponent phospholipid mixtures by electrospray mass spectroscopy: relating ion intensity to concentration. *Anal. Biochem.*, **308**, 152–159.

23 Yang, K., Zhao, Z., Gross, R.W., and Han, X. (2011) Identification and quantitation of unsaturated fatty acid isomers by electrospray ionization tandem mass spectrometry: a shotgun lipidomics approach. *Anal. Chem.*, **83**, 4243–4250.

24 Christie, W.W. and Han, X. (2010) *Lipid Analysis: Isolation, Separation, Identification and Lipidomic Analysis*, The Oily Press, Bridgwater, UK.

25 Kalderon, B., Sheena, V., Shachrur, S., Hertz, R., and Bar-Tana, J. (2002) Modulation by nutrients and drugs of liver acyl-CoAs analyzed by mass spectrometry. *J. Lipid Res.*, **43**, 1125–1132.

26 Tsui, Z.C., Chen, Q.R., Thomas, M.J., Samuel, M., and Cui, Z. (2005) A method for profiling gangliosides in animal tissues using electrospray ionization–tandem mass spectrometry. *Anal. Biochem.*, **341**, 251–258.

27 Merrill, A.H., Jr., Sullards, M.C., Allegood, J.C., Kelly, S., and Wang, E. (2005) Sphingolipidomics: high-throughput, structure-specific, and quantitative analysis of sphingolipids by liquid chromatography tandem mass spectrometry. *Methods*, **36**, 207–224.

28 Bielawski, J., Szulc, Z.M., Hannun, Y.A., and Bielawska, A. (2006) Simultaneous quantitative analysis of bioactive sphingolipids by high-performance liquid chromatography–tandem mass spectrometry. *Methods*, **39**, 82–91.

29 Kayganich, K.A. and Murphy, R.C. (1992) Fast atom bombardment tandem mass spectrometric identification of diacyl, alkylacyl, and alk-1-enylacyl molecular species of glycerophosphoethanolamine in human polymorphonuclear leukocytes. *Anal. Chem.*, **64**, 2965–2971.

30 Yang, K., Zhao, Z., Gross, R.W., and Han, X. (2007) Shotgun lipidomics identifies a paired rule for the presence of isomeric ether phospholipid molecular species. *PLoS ONE*, **2**, e1368.

31 Berry, K.A. and Murphy, R.C. (2005) Analysis of cell membrane aminophospholipids as isotope-tagged derivatives. *J. Lipid Res.*, **46**, 1038–1046.

32 Zemski Berry, K.A., Turner, W.W., VanNieuwenhze, M.S., and Murphy, R.C. (2009) Stable isotope labeled 4-(dimethylamino)benzoic acid derivatives of glycerophosphoethanolamine lipids. *Anal. Chem.*, **81**, 6633–6640.

33 Jiang, X., Ory, D.S., and Han, X. (2007) Characterization of oxysterols by electrospray ionization tandem mass spectrometry after one-step derivatization with dimethylglycine. *Rapid Commun. Mass Spectrom.*, **21**, 141–152.

34 Griffiths, W.J., Liu, S., Alvelius, G., and Sjovall, J. (2003) Derivatisation for the characterisation of neutral oxosteroids by electrospray and matrix-assisted laser desorption/ionisation tandem mass spectrometry: the Girard P derivative. *Rapid Commun. Mass Spectrom.*, **17**, 924–935.

35 Li, Y.L., Su, X., Stahl, P.D., and Gross, M. L. (2007) Quantification of diacylglycerol molecular species in biological samples by electrospray ionization mass spectrometry after one-step derivatization. *Anal. Chem.*, **79**, 1569–1574.

36 Bollinger, J.G., Thompson, W., Lai, Y., Oslund, R.C., Hallstrand, T.S., Sadilek, M., Turecek, F., and Gelb, M.H. (2010) Improved sensitivity mass spectrometric detection of eicosanoids by charge reversal derivatization. *Anal. Chem.*, **82**, 6790–6796.

37 Yang, K., Zhao, Z., Gross, R.W., and Han, X. (2009) Systematic analysis of choline-containing phospholipids using multi-dimensional mass spectrometry-based shotgun lipidomics. *J. Chromatogr. B*, **877**, 2924–2936.

38 Han, X. (2002) Characterization and direct quantitation of ceramide molecular species from lipid extracts of biological samples by electrospray ionization tandem mass spectrometry. *Anal. Biochem.*, **302**, 199–212.

39 Han, X. (2007) Neurolipidomics: challenges and developments. *Front. Biosci.*, **12**, 2601–2615.

40 Han, X. and Jiang, X. (2009) A review of lipidomic technologies applicable to sphingolipidomics and their relevant applications. *Eur. J. Lipid Sci. Technol.*, **111**, 39–52.

41 Han, X., Cheng, H., Jiang, X., and Zeng, Y. (2010) Mass spectrometry methods for the analysis of lipid molecular species: a shotgun lipidomics approach, in *Lipid-Mediated Signaling* (eds. E.J. Murphy and T.A. Rosenberger), CRC Press, Boca Raton, FL, pp. 149–173.

42 Han F X., Gubitosi-Klug, R.A., Collins, B. J., and Gross, R.W. (1996) Alterations in individual molecular species of human platelet phospholipids during thrombin stimulation: electrospray ionization mass spectrometry-facilitated identification of the boundary conditions for the magnitude and selectivity of thrombin-induced platelet phospholipid hydrolysis. *Biochemistry*, **35**, 5822–5832.

43 Han, X. (2005) Lipid alterations in the earliest clinically recognizable stage of Alzheimer's disease: implication of the role of lipids in the pathogenesis of Alzheimer's disease. *Curr. Alzheimer Res.*, **2**, 65–77.

44 Han, X., Yang, K., and Gross, R.W. (2008) Microfluidics-based electrospray ionization enhances intrasource separation of lipid classes and extends identification of individual molecular species through multi-dimensional mass spectrometry: development of an automated high throughput platform for shotgun lipidomics. *Rapid Commun. Mass Spectrom.*, **22**, 2115–2124.

45 Han, X., Yang, K., Yang, J., Cheng, H., and Gross, R.W. (2006) Shotgun lipidomics of cardiolipin molecular species in lipid extracts of biological samples. *J. Lipid Res.*, **47**, 864–879.

46 Braverman, N., Steel, G., Obie, C., Moser, A., Moser, H., Gould, S.J., and Valle, D. (1997) Human PEX7 encodes the peroxisomal PTS2 receptor and is responsible for rhizomelic chondrodysplasia punctata. *Nat. Genet.*, **15**, 369–376.

47 Motley, A.M., Brites, P., Gerez, L., Hogenhout, E., Haasjes, J., Benne, R., Tabak, H.F., Wanders, R.J., and Waterham, H.R. (2002) Mutational spectrum in the PEX7 gene and functional analysis of mutant alleles in 78 patients with rhizomelic chondrodysplasia punctata type 1. *Am. J. Hum. Genet.*, **70**, 612–624.

48 van den Brink, D.M., Brites, P., Haasjes, J., Wierzbicki, A.S., Mitchell, J., Lambert-Hamill, M., de Belleroche, J., Jansen, G. A., Waterham, H.R., and Wanders, R.J. (2003) Identification of PEX7 as the second gene involved in Refsum disease. *Am. J. Hum. Genet.*, **72**, 471–477.

49 Han, X. and Cheng, H. (2005) Characterization and direct quantitation

of cerebroside molecular species from lipid extracts by shotgun lipidomics. *J. Lipid Res.*, **46**, 163–175.

50 Hsu, F.F. and Turk, J. (2001) Structural determination of glycosphingolipids as lithiated adducts by electrospray ionization mass spectrometry using low-energy collisional-activated dissociation on a triple stage quadrupole instrument. *J. Am. Soc. Mass Spectrom.*, **12**, 61–79.

51 Hsu, F.-F., Bohrer, A., and Turk, J. (1998) Electrospray ionization tandem mass spectrometric analysis of sulfatide: determination of fragmentation patterns and characterization of molecular species expressed in brain and in pancreatic islets. *Biochim. Biophys. Acta*, **1392**, 202–216.

52 Jiang, X. and Han, X. (2006) Characterization and direct quantitation of sphingoid base-1-phosphates from lipid extracts: a shotgun lipidomics approach. *J. Lipid Res.*, **47**, 1865–1873.

53 Jiang, X., Yang, K., and Han, X. (2009) Direct quantitation of psychosine from alkaline-treated lipid extracts with a semi-synthetic internal standard. *J. Lipid Res.*, **50**, 162–172.

54 Cheng, H., Jiang, X., and Han, X. (2007) Alterations in lipid homeostasis of mouse dorsal root ganglia induced by apolipoprotein E deficiency: a shotgun lipidomics study. *J. Neurochem.*, **101**, 57–76.

55 Su, X., Han, X., Mancuso, D.J., Abendschein, D.R., and Gross, R.W. (2005) Accumulation of long-chain acylcarnitine and 3-hydroxy acylcarnitine molecular species in diabetic myocardium: identification of alterations in mitochondrial fatty acid processing in diabetic myocardium by shotgun lipidomics. *Biochemistry*, **44**, 5234–5245.

5
Targeted Lipidomics: Sphingolipidomics

Ying Liu, Yanfeng Chen, and M. Cameron Sullards

5.1
Introduction

Lipids constitute a broad category of molecules that serve not only as components of biological structures but also as energy stores, signaling molecules, and regulators of numerous cell functions [1–4]. Lipids may be generally subdivided into eight categories on the basis of well-defined chemical and biochemical principles: fatty acyls, glycerolipids, glycerophospholipids, sphingolipids (SL), sterol lipids, prenol lipids, saccharolipids, and polyketides [5, 6]. Therefore, the large-scale investigation focusing on complete identification and quantitation of all the lipid molecules just categorized in a biological system at different time points under various states of stimulation versus control samples is known as "lipidomics." The resulting lipidomic data contain global changes in different lipid species, which may then be identified within a system-integrated context, and the interactive roles of lipids in metabolic pathways may be determined.

Recent advances in analytical technologies such as mass spectrometry (MS) and liquid chromatography (LC) have greatly facilitated advances in lipidomic research. These tools and techniques have provided new discoveries regarding changes in lipid metabolism and in the regulation of these pathways under physiological and pathological conditions at the system level. Analytical strategies regarding the collection of lipidomic data can be broadly divided into two complimentary approaches: global and targeted workflows.

In a global lipidomic methodology, both the lipid extraction and the subsequent analyses are performed in a manner in which conditions are not optimized for recovery of specific lipid species and detection methods do not specifically focus on particular individual molecular species. Typically, global approaches perform lipid analyses via only mass spectrometry techniques in several manners, two of the more prominent being "shotgun" lipidomics [7] or "top-down" lipidomics [8]. In the former method, extracted samples are directly infused and ionized by electrospray ionization (ESI). Tandem mass spectrometry (MS/MS) is then subsequently used to identify and determine the structures of numerous lipid species. Here, multiple precursor ion scans (Pre or PIS) and constant neutral loss scans (CNL or

Lipidomics, First Edition. Edited by Kim Ekroos.

NLS) of structurally unique fragment ions reveal the various head group/acyl chain combinations of lipids present. This method of analysis is considered focused because tandem mass spectrometry scanning techniques are used; therefore, broad m/z ranges may be examined for specific lipid structures. This may result in the discovery of unusual or unexpected lipid species. However, it could be argued that specifying particular PIS or NLS is itself a form of targeting, since any results will arise from only those fragments that are queried. The latter technique directly addresses this by utilizing chip-based nanospray ionization (nESI) in conjunction with single-stage MS analyses at ultrahigh resolution and accurate mass. Here, the resolving power and mass accuracy are sufficient to determine the empirical formula of a lipid by mass in most cases. MS/MS can be performed subsequently to elucidate the structures of lipids of interest. As with the previous methodology, a scanning technique is employed in which wide m/z ranges may be examined for lipids; however, in this case, it is without bias with regard to structure. This "top-down" technique is essentially a modified "shotgun" method that further addresses limitations of "shotgun" lipidomics with regard to isobaric or isotopic interferences as a result of the high resolution and accurate mass determination. It should also be noted that both methods utilize direct infusion and that both ESI and nESI may be used with either technique.

Global methods have several positive attributes. The analyses tend to be easily performed, relatively fast, and result in a relatively high throughput. These approaches work well for highly abundant or easily ionized lipids [7, 9–12]. Furthermore, these approaches are greatly enhanced by using new tools and techniques such as the orbitrap mass analyzer and nESI. However, they do have limitations with regard to comprehensive lipid extraction, ionization suppression, detection of low-abundance species, and the ability to differentiate isomeric species.

A targeted lipidomic approach directly addresses the limitations of global methodologies. Here, both the extraction and the analytical analysis protocols are optimized for a specific lipid class, such as sphingolipids. Liquid chromatography is then mainly used to separate the complex class of lipids (SL) into subclasses: ceramide (Cer), monohexosylceramide (HexCer), lactosylceramide (LacCer), and sphingomyelin (SM), as well as individual components: sphingosine (So), sphinganine (Sa), sphingosine-1-phosphate (So1P), and sphinganine-1-phosphate (Sa1P). This serves to reduce ionization suppression (via separation of species having different ioniziabilities), enhance ionization (due to desalting and chromatographically focused samples), mitigate interferences from isotopic and isobaric species, and differentiate isomeric species such as glucosylceramide (GlcCer) and galactosylceramide (GalCer). Finally, highly specific tandem mass spectrometry protocols known as multiple reaction monitoring (MRM) are used for the detection of individual molecular species. Here, both the intact precursor ion and a structurally specific fragment ion are detected as a pair, and no scanning is performed. Furthermore, both ionization and fragmentation parameters are optimized for the individual molecular species, so sensitivity is enhanced and optimized. Each MRM pair may be monitored for only a few milliseconds, and it is possible to monitor a given pair at a specific retention time window, known as scheduled MRM. Therefore, it is possible to monitor hundreds of individual molecular species of SL in a single liquid chromatography–tandem mass spectrometry (LC-MS/MS) run.

The resulting targeted sphingolipid analyses require multiple stages of method development and validation to ensure analytical rigor and reproducibility. The foundation upon which all measurements rest is the establishment of high quality and purity internal standards (IS). Following this, protocols for the extraction of sphingolipids as completely and comprehensively as possible must be developed. Next, highly specific methods must be optimized for the separation, ionization, fragmentation, and detection of sphingolipids via LC-MS/MS. Quantitation may then be achieved by using the ratio of the signal arising from the unknown to that of the internal standard. Finally, the methods may then be validated for comprehensiveness and reproducibility by comparison with other structural analogues at different concentrations over multiple injections.

This chapter primarily focuses on methods used for a targeted extraction and analysis of one class of lipids, sphingolipids. Here, a set of expanding and evolving protocols have been developed and rigorously validated for the extraction, identification, separation, and quantitation of sphingolipids by LC-MS/MS. This template has been adapted to and used for the analysis of other categories of lipids, and when all are combined, it forms a comprehensive "lipidomic" platform as developed by the LIPID MAPS™ consortium (http://www.lipidmaps.org/).

5.2 Sphingolipids Description and Nomenclature

The *de novo* biosynthesis of sphingolipids originates with the condensation of serine and palmitoyl-CoA [13]. The resulting 3-keto sphinganine is then reduced to form sphinganine, which serves as one of the fundamental building blocks of sphingolipids [13]. This long-chain base (LCB) is usually a linear 18-carbon alkane or alkene, which contains a 1,3 dihydroxy-2-amino functional groups. There are numerous variations on this core lipid owing to position and number of double bonds, hydroxylations, or branching [13]. The typical abbreviation for sphingoid bases is to begin by assigning a letter that is indicative of the number of hydroxylations contained in it. This can be "m," "d," or "t" for mono-, di-, or trihydroxylated. This is followed by a number representative of the number of carbon atoms present in the LCB, which is typically 18. This number is followed by a colon and a second number indicating the number of units of unsaturation (i.e., rings or double bonds). Thus, for the commonly occurring sphingosine, the abbreviation would be "d18:1."

After formation, sphinganine can undergo *N*-acylation with a wide variety of fatty acids (FA) typically of chain length C16–C24. As with the sphingoid base, the fatty acid can contain any number of double bonds, hydroxylations, or branching. This subsequently formed dihydroceramide (DHCer) now consists of two parts, the LCB and the FA. A typical DHCer that contains a sphinganine backbone esterified to palmitate would be abbreviated as "d18:0/C16:0."

DHCer species can undergo desaturation at the Δ4-position to form Cer, and together these two species form one of the main loci for the diversity of sphingolipid species. This is because any number of head groups may be attached to the 1-hydroxy position. Phosphorylation at this position yields ceramide-1-phosphate (Cer1P) and attachment of phosphoethanolamine results in the formation of ceramide phosphoethanolamine (CPE). Sphingomyelin is also formed via linkage of the phosphocholine moiety to ceramide at this position. It should be noted that Cer or Cer1P can be degraded via loss of the fatty acid to form the signaling molecules So and S1P, respectively. The nomenclature for these species is similar to that of the Cer; however, the head group is indicated first followed by the description of the ceramide core. For example, *N*-tetracosanoyl sphinganine phosphorylcholine is abbreviated as SM d18:0/C24:0 and *N*-palmitoyl sphingosine-1-phosphate as Cer1P d18:1/C16:0.

Carbohydrates represent the most complex head group(s) that may be attached at the 1-hydroxy position. The simplest of these species arise from attachment of either glucose or galactose to form GlcCer and GalCer, respectively. These two species are the foundation upon which hundreds of other diverse molecular species may be formed. Addition or modification of numerous other glycans or functional groups results in the formation of the complex glycosphingolipids (GSL). These species may be categorized as either neutral or acidic and can be subdivided into the ganglio-, globo-, isoglobo-, lacto-, neolacto-, arthro-, and mollo-species, each having a distinctive tetraglycan core structure [13].

Nomenclature for the gangliosides commonly follows that proposed by Svennerholm [14]. Here, the letter "G" is used to represent the presence of a sialic acid containing ganglioside. A second letter follows that is indicative of the number of sialic acid residues present. For example, mono-, di-, tri-, and quad- would give rise to GM, GD, GT, and GQ species, respectively. Finally, a number is assigned that is indicative of how the species are resolved from one another via thin-layer chromatography. For example, species having fewer carbohydrates move further than those having more carbohydrates giving rise to the series GM3 > GM2 > GM1. Other systems of nomenclature have been proposed and used, and these have been reviewed in more detail previously [13].

5.3 Sphingolipids Analysis via Targeted LC-MS/MS

Sphingolipids have been studied using mass spectrometry for decades. More recently, matrix-assisted laser desorption ionization (MALDI) [15–17] and electrospray ionization [18, 19] have been used to generate intact molecular ions of unmodified SL. Subsequent mass analysis of SL has been performed on many different types of mass analyzers, such as quadrupoles (Q) [20, 21], time-of-flight (TOF) [22, 23], ion traps (IT) [18, 24], and Fourier transform ion cyclotron resonance (FT ICR) [17]. These studies have established that mass spectrometry is a powerful tool for sphingolipidomic analysis.

Researchers have long recognized that MS provides a high degree of specificity with regard to identification of complex compounds via molecular mass, as well as when analyzed by MS/MS or MS^n for structure elucidation. ESI used in conjunction with MS also possesses the requisite high sensitivity for lipid detection, and can often be up to several orders of magnitude lower detection limits than classical techniques. This results in low-abundance lipids being detected even if they are present in fmol amounts per 10^6 cells (or sometimes less depending on the instrument). Finally, ESI-MS has demonstrated a wide dynamic range, which allows analysis of compounds that vary in abundance in a range of several orders of magnitude in the same biological samples (e.g., sphingomyelin versus free long-chain bases).

5.3.1 Sphingolipid Internal Standards

Well-characterized and defined internal standards are fundamental for the development and validation of extraction and analysis protocols as well as for quantitation. An appropriate internal standard will serve as control for extraction, HPLC injection, ionization variability, and identification verification. An ideal IS is generally considered to be a stable isotope-labeled analogue of each analyte of interest. Typically, IS is synthesized in a heavy isotope-labeled form and added to the matrix containing the target analyte (light form) in a known quantity. The heavy and light forms of the target compound should be coextracted, coeluted, and give identical response by MS. It is relatively a simple calculation to extrapolate the magnitude of the unknown relative to the IS when done in this manner.

It is highly impractical for isotope-labeled internal standards to be synthesized for the large numbers of compounds examined in a sphingolipidomic study when time, expertise, and cost are considered. Therefore, alternative sphingolipids that are similar in structure as well as in ionization and fragmentation characteristics may be used as internal standards. To this end, the LIPID MAPS consortium in collaboration with Avanti Polar Lipids (Alabaster, AL) developed an internal standard cocktail for sphingolipids (catalog number LM-6002). The IS mixture is provided in sealed ampoules and is certified to be >95% pure and within 10% of the specified amount (25 μM). It contains four different 17-carbon chain length sphingoid base analogues: C17-sphingosine, (2*S*,3*R*,4*E*)-2-aminoheptadec-4-ene-1,3-diol (d17:1-So); C17-sphinganine, (2*S*,3*R*)-2-aminoheptadecane-1,3-diol (d17:0-Sa); C17-sphingosine 1-phosphate, heptadecasphing-4-enine-1-phosphate (d17:1-So1P); and C17-sphinganine 1-phosphate, heptadecasphinganine-1-phosphate (d17:0-Sa1P); and five C12-fatty acid analogues of the more complex SL: C12-Cer, *N*-(dodecanoyl)-sphing-4-enine (d18:1/C12:0); C12-Cer 1-phosphate, *N*-(dodecanoyl)-sphing-4-enine-1-phosphate (d18:1/C12:0-Cer1P); C12-SM, *N*-(dodecanoyl)-sphing-4-enine-1-phosphocholine (d18:1/C12:0-SM); C12-GlcCer, *N*-(dodecanoyl)-1-β-glucosyl-sphing-4-eine (d18:1/C12:0-GlcCer); and C12-LacCer, *N*-(dodecanoyl)1-β-lactosyl-sphing-4-eine (d18:1/C12:0-LacCer). Other internal standards for quantitation of additional sphingolipids such as sulfatides (d18:1/C12:0-sulfatide, ST) and

GalCer (d18:1/C12:0-GalCer) can be obtained from Avanti and Matreya (Pleasant Gap, PA), respectively. Typically, the dihydro- (i.e., sphinganine backbone) versions of the complex sphingolipid standards are not commercially available. However, they are easily synthesized by reduction of the backbone double bond using hydrogen gas and 10% Pd on charcoal (Aldrich-Sigma, St. Louis, MO). Complete hydrogenation can be verified by LC-MS/MS analysis to prove that no starting material remains [24, 25].

Once the LC and MRM protocols have been determined, the internal standards will be used to generate standard curves for each subclass and the individual long-chain bases. The methods are then validated by comparison with several chain length analogues to ensure accurate quantitation.

5.3.2
Biological Sample Preparation and Storage

A wide variety of samples such as cultured cells, organ/tissue homogenates, blood, plasma, urine, feces, fruit flies, worms, and so on may be analyzed for their SL content. Typically, 10^6 cells are sufficient for analysis on midlevel triple quadrupole instruments. With regard to homogenized tissue, 1–10 mg at 10% wet weight per volume of phosphate buffered saline (PBS) is sufficient for SL analyses. Biological fluids such as blood, plasma, or urine, typically require 1–10 μL for effective extraction. Tissue culture medium, however, requires larger volumes (100–500 μL) to be extracted for SL analysis. When these studies are performed, the serum is reduced to 1% and the medium should be lyophilized prior to extraction to reduce the aqueous volume. It should be noted that if samples are lyophilized, they should be lyophilized in the tubes that will be used for the extraction. However, wide variations of up to an order of magnitude more or less may be required, depending on the abundance of the analyte of interest in that particular cell line or tissue.

A normalizing parameter between different samples such as microgram DNA, milligram protein, or cell count is necessary. To this end, samples should be initially prepared in aliquots or as duplicates so that one may be assayed for this purpose. Generally, analyses for the normalizing factor should be performed prior to the SL extraction. This is because any errors or variation in analysis of protein, DNA, and so on are often more likely a cause for failure at this stage than in the subsequent lipid analysis. Therefore, by doing them first, the time and effort for the extraction and MS analysis are not wasted if the normalizing parameter cannot be measured accurately enough.

Pyrex $13 \times 100\,mm^2$ borosilicate tubes with a Teflon-lined cap (catalog number 60827-453, VWR, West Chester, PA) are used to hold the samples. SL will stick to some types of glass; therefore, these specific tubes are critical to be used. It should be noted that because of this, some lipids may be left behind when samples are transferred from one container to another resulting in variability. Unextracted samples should be stored at −80 °C. Finally, if samples are to be shipped, this should be done in a container that can keep the tubes separate, in an upright orientation, and at a low temperature.

5.3.3
Sphingolipid Extraction Protocol

The extraction protocol was adopted from the standard two-phase lipid extraction method (e.g., Bligh–Dyer and Folch) (Figure 5.1) [26, 27]. There are many purposes of extraction: first, it isolates sphingolipids from samples in their native state and removes nonlipid contaminants using the organic solvent mixture chloroform/methanol or methylene chloride/methanol. Second, base treatment removes the abundant glycerolipids that can interfere with the ionization and detection of sphingolipids. Third, water that is used to wash lipid extracts removes highly water-soluble lipids such as gangliosides, residual polyphosphoinositides, lysophospholipids, acyl-carnitines, and other nonlipid contaminants. Thus, the extraction method is utilized to fractionate subcategories of sphingolipids, which are less polar, into the organic phase. Methods are under development to extract and analyze the more polar sphingolipids found in the aqueous phase.

Prior to extraction, samples are generally homogenized using sonication or another types of homogenizer to get a uniform consistency. An aliquot is then removed for normalization purposes, as mentioned previously. The remaining sample is then divided into two 13 × 100 mm^2 screw-capped glass test tubes with Teflon caps in a measured volume. One will be extracted for high recovery of free

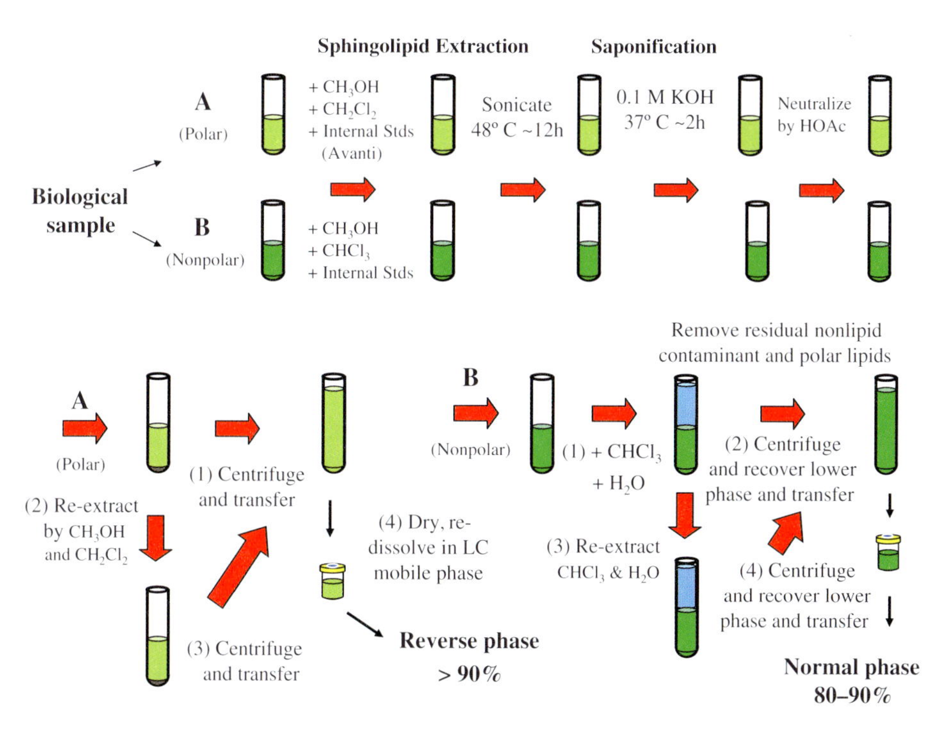

Figure 5.1 Normalization and targeted extraction protocol for sphingolipidomic analyses via LC-MS/MS.

sphingoid bases (test tube A) and the other will be extracted differently for high recovery of the complex SL (test tube B).

The extraction protocol begins by adding 1 mL of methanol (CH_3OH) to both test tubes. This is followed by addition of 0.5 mL of methylene chloride (CH_2Cl_2) to test tube A and 0.5 mL of chloroform ($CHCl_3$) to test tube B. The internal standards are then added to both test tubes A and B. Care should be taken with the amount of internal standard added so that it is well above the limit of quantitation of the instrument used for the analyses, but not so high as to suppress the ionization of the analytes. Therefore, the amount of the internal standard should be roughly estimated when performing the initial SL profiling in the presence of the internal standard. It has generally been observed that the concentration of the analytes and standards should be within 1–2 orders of magnitude of each other for accurate, reproducible results. To this end, typically 20 μL of the SL cocktail is an appropriate addition for the sample quantities mentioned previously. If additional analytes are to be quantified (e.g., sphingosylphosphocholine and sulfatides), additional standards can also be added to the sample.

Following addition of the IS, both test tubes should be capped and sonicated for ~30 s. This is then followed by incubation for overnight at 48 °C in a heating block or water bath. This step proves helpful in more effectively extracting SL because of their higher phase transition temperature [28]. Following the incubation, add 150 μL of 1 M KOH in CH_3OH to each tube after they have cooled down. Next, sonicate both test tubes for 30 s and incubate for 2 h at 37 °C. This sonication step was designed to remove interfering glycerolipids, particularly phosphatidylcholines. Allow the tubes to cool to room temperature after base hydrolysis and add an appropriate volume of glacial acetic acid (typically ~6 μL) to adjust pH to neutral.

Test tube A is now identified as the "single-phase extract" for further workup, and test tube B is designated as the "organic-phase extract." The single-phase extract should be centrifuged to pellet and insoluble material and the remaining solvent should then be transferred to a new glass test tube. The pellet may then be re-extracted by adding 1.5 mL of CH_3OH:CH_2Cl_2 in a ratio of 2:1 (v/v) to the original test tube followed by sonication, vortexing, and centrifugation. The solvent should be transferred and pooled with the first extraction. The pellet may now be discarded and the pooled extracts may be reduced to dryness via speed vac, with caution of not overheating the sample.

The organic-phase extraction from test tube B is initiated by addition of 1 mL of $CHCl_3$ and 2 mL of H_2O, followed by vortexing and then centrifugation. The lower $CHCl_3$ layer should be carefully removed via Pasteur pipette and put into a new test tube, leaving the interface. This can then be re-extracted by addition of 1 mL of $CHCl_3$ followed by vortexing and centrifugation. As done previously, remove the lower layer, leaving the interface, and pool the $CHCl_3$ layers. The recovered $CHCl_3$ may be reduced to dryness by speed vac or other reduced pressure device, being careful to not to overheat.

The dried extracts are kept refrigerated and can be analyzed when solubilized in the appropriate mobile phase. Most extracted sphingolipids are fairly stable. It has been observed that within a few weeks, there is no noticeable decomposition of LCBs (based on disappearance of the internal standards). The more complex SL (Cer, HexCer, and SM) have been reanalyzed in some samples after several years of storage on the shelf and the results were not discernibly different from the original analyses.

Both the efficiency of this extraction protocol and the appropriateness of the selected IS molecules have been determined by comparison of neat IS with IS that had undergone the full extraction process. To this end, a series of six samples with proper internal standards were extracted four different times using the single-phase extract protocol [24]. The results demonstrated that all the LCB species and their related internal standards have better than 80% recovery in the first extract, approximately 10% in the second, and virtually none in the third or fourth extract [24]. A similar experiment using the "organic-phase" extraction procedure was performed and also gave similar results [24]. This series of experiments unequivocally showed that all the chain length variants, including the internal standard of the complex SL (Cer, Cer1P, HexCer, LacCer, and SM) were 80–90% recovered in the first two extractions, thus demonstrating the efficiency of the extraction protocol and the appropriateness of the IS developed for sphingolipidomic analyses.

5.3.4 Liquid Chromatography

Liquid chromatography is an essential component for targeted sphingolipidomic analyses. It directly addresses many of the limitations of untargeted techniques such as ionization suppression, isobaric, isotopic, and isomeric interferences, as well as detection of lower abundance species. Here, complex mixtures of lipids may be separated into similar subclasses or individual molecular species. This serves to lessen the complexity of ESI at any given time reducing ionization suppression. Moreover, in the case of normal-phase separation, the IS coelutes with the analytes and thus it is ionized, fragmented, and detected under similar conditions as the analyte.

Isobaric interferences arise when two different lipid molecules have the same nominal mass and fragment to the same nominal mass product ion for detection. This situation arises with several of the longer chain length Cer1P species and the shorter chain length HexCer species. For example, the $(M+H)^+$ ions of both Cer1P d18:1/C24:1 ($C_{42}H_{83}NO_6P$) and HexCer d18:1/C18:0 ($C_{42}H_{82}NO_8$) have a nominal m/z value of 728. When these precursor ions fragment, they both yield the structure-specific m/z 264 ion, indicative of the d18:1 sphingoid base, which both core lipids contain. The normal- or reverse-phase chromatography would resolve these species by head group and lipid tail, respectively, thus enabling each to be quantitated independently. In addition, since they would be separated chromatographically, one would not suppress the ionization of the other yielding more accurate quantitative results.

Isotopic interferences arise when an isotope from one species occurs at the same nominal mass as the monoisotopic mass of another species. This situation regularly occurs in the analyses of the LCB species and their phosphate analogues. Here, the isotopologue (e.g., the M +2 ^{13}C isotopologue of d18:1) from So overlaps the monoisotopic mass of Sa at m/z 302. Furthermore, the isotopologue also fragments to yield the same fragment ion at m/z 266. Although the abundance of this isotopologue is only ~2% of So, So is often more than an order of magnitude more abundant than the Sa. This leads to a falsely high quantitative value if these species are not resolved. Fortunately, reverse-phase chromatography baseline separates these species providing the correct quantitative measurement.

Isomeric interferences arise when two distinct SL have the same mass and fragment to the same product ion. This situation occurs, for example, with the GSL such as GlcCer and GalCer. These species cannot be distinguished via MS- or MS/MS-based methods. However, they are easily separated by normal-phase HPLC.

As a final point in support for LC, it also provides additional advantages for sphingolipids: the subclasses and individual molecular species can be desalted, chromatographically focused, and selectively eluted. Desalting is critical when looking at low-abundance species such as the LCBs. The single-phase extract contains a large amount of salt relative to the LCBs and their removal primarily enhances formation of $(M+H)^+$ species, reduces formation of salt adducts, and thereby increases sensitivity. As mentioned earlier, chromatographic focusing decreases the number of compounds that are ionized and subsequently analyzed in a given volume of eluate, which serves to improve sensitivity. In addition, use of LC reduces the possibility that the electrospray droplet itself will be comprised of compounds with highly different gas-phase proton affinity, thereby minimizing ionization suppression of the analyte(s) of interest.

LC-MS/MS has been used to identify, quantify, and elucidate the structures of LCBs and their phosphates, Cer, HexCer (both GalCer and GlcCer), LacCer, SM, and other complex SL, for over a decade [25, 29]. A baseline separation of all individual molecular species within a given subclass of SL is not necessary as long as all nonchromatographically resolved species have similar chemical and physical properties and a unique precursor/product ion m/z value. This is because the similarity of the chemical and physical properties ensures that one species does not affect the ionization of others and the uniqueness of the precursor/product ion pair allows the mass spectrometer to differentiate between many components in a complex mixture (i.e., all *N*-acyl chain variants of Cer), even in the event that they coelute from the HPLC. The addition of internal standards that coelute with the analytes of interest is particularly important. This ensures the resulting signal to be normalized and any ionization suppression or enhancement that may occur during that elution window to be accounted for.

Separations of SL by HPLC have classically been performed either by reverse- or normal-phase chromatography (Figure 5.2). In reverse-phase chromatography, separations are based on the length and (un)saturation of the sphingoid base and/or *N*-acyl fatty acid [25, 30, 31]. Normal-phase separations are based on the polarity of the head group (i.e., separation of Cer, HexCer, LacCer, and SM from each other) [25, 31, 32].

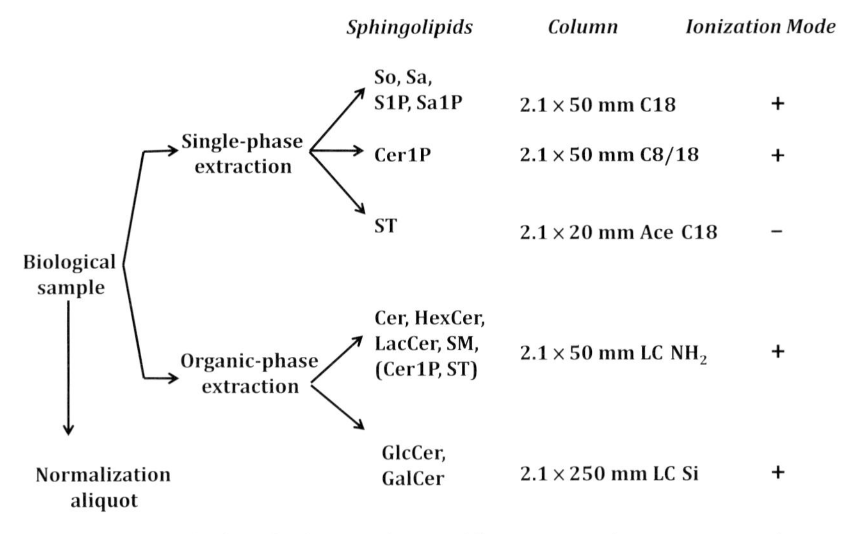

Figure 5.2 Targeted sphingolipidomic analyses workflow using specific extraction procedure, subclass, HPLC column, and ionization mode.

5.3.4.1 LCBs and Cer1P

The dried single-phase extract residue may be reconstituted in 300 μL of the initial condition mobile phase used for reverse-phase LC-MS/MS. It is sonicated for ~15 s, then transferred to separate 1.5 mL microfuge tubes (organic solvent resistant), and centrifuged for several minutes or until clear. Approximately 70 μL of the clear supernatant is transferred from each tube into 200 μL tapered glass autoinjector sample vials for LC-MS/MS analysis. The remainder may be saved for profiling and further future analyses.

Separation of LCBs is performed using reverse-phase LC on a Supelco 2.1 (i.d.) × 50 mm Discovery C18 column (Sigma, St. Louis, MO) and a binary solvent system at a flow rate of 1 mL/min (Figure 5.3a). The column is pre-equilibrated prior to injection for 0.4 min with a solvent mixture of 60% mobile phase A ($CH_3OH/H_2O/HCOOH$, 58/41/1, v/v/v, with 5 mM ammonium formate) and 40% mobile phase B ($CH_3OH/HCOOH$, 99/1, v/v, with 5 mM ammonium formate). The A/B ratio is maintained at 60/40 for 0.5 min after sample injection (50 μL), followed by a linear gradient to 100% B over 1.8 min. The run is held at 100% B for 0.8 min to flush any strongly retained species off the column. Then, the run is returned to initial conditions and a re-equilibration wash of the column with 60:40 A/B is performed for 0.5 min before the next run. The total runtime for LCB analysis is ~4 min.

Analysis of Cer1P species is performed simply by extending the 100% B hold for an additional 4.5 min, yielding a total runtime of ~9 min. Cer1P has proven to be difficult to analyze via reverse-phase chromatography. A significant amount of carryover (>1%) has been observed to elute from the column during blank runs. This occurs with reverse-phase columns regardless of vendor as well as with different lots of columns from the same vendor.

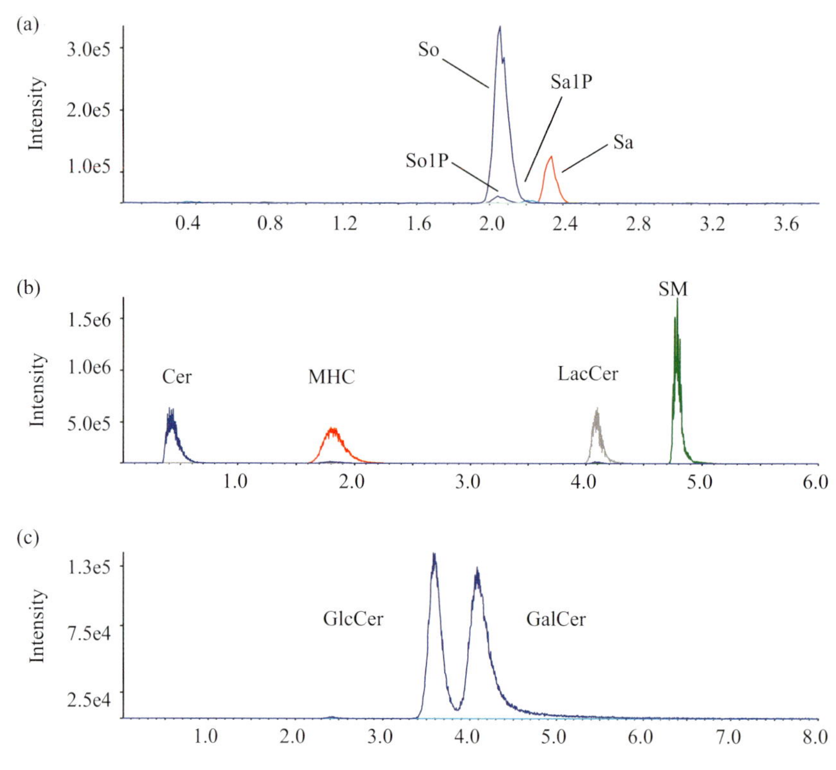

Figure 5.3 LC-MS/MS using MRM for separation and detection of (a) long-chain bases, (b) complex sphingolipids, and (c) isomeric sphingolipids.

Alternatively, Cer1P can be analyzed by using a less hydrophobic Supleco 2.1 (i.d.) × 50 mm Discovery C8 column (Sigma, St. Louis, MO). Here the column is heated to 60 °C and a binary solvent system is used at a flow rate of 0.5 mL/min. The column is equilibrated for 2 min prior to injection with a solvent mixture of 70% mobile phase A (CH_3OH/H_2O/THF/HCOOH, 68.5/28.5/2/1, v/v/v, with 5 mM ammonium formate) and 30% mobile phase B (CH_3OH/THF/HCOOH, 97/2/1, v/v/v, with 5 mM ammonium formate). The A/B ratio is maintained at 70/30 for 0.4 min after injection (30 μL). This is followed by a linear gradient to 100% B over 1.9 min. The flow is held at 100% B for 5.3 min to remove any strongly retained species, followed by a 0.5 min re-equilibration of the column with 70:30 A/B before the next run.

5.3.4.2 **Cer, HexCer, LacCer, SM, ST, and Cer1P**

The dried organic-phase extraction residue is reconstituted in 300 μL of the initial condition mobile phase used for normal-phase LC-MS/MS. A similar sonication and centrifugation processes as mentioned previously are performed. Approximately 70 μL of the clear supernatant is transferred into separate 200 μL tapered glass autoinjector sample vials for LC-MS/MS analysis.

These lipids are separated using normal-phase LC on a Supelco 2.1 (i.d.) × 50 mm LC-NH_2 column at a flow rate of 1.0 mL/min and a binary solvent system (Figure 5.3b). The column is equilibrated for 1.0 min prior to injection with 100% mobile phase A ($CH_3CN/CH_3OH/HCOOH$, 97/2/1, v/v/v, with 5 mM ammonium formate). After injection, 100% mobile phase A is continued for 3 min. A 1.0 min linear gradient to 100% mobile phase B ($CH_3OH/H_2O/HCOOH$, 89/6/5, v/v/v, with 50 mM triethylammonium acetate) is then performed. The 100% B flow is held for 3.0 min to elute any strongly retained compounds. Then, the flow is returned to 100% A by a 1.0 min linear gradient and maintained at 100% A for 1 min to re-equilibrate the column. Cer1P and ST may also be detected in the organic-phase extract, however, recoveries are low and errors increase accordingly. If it is desired to accurately quantitate these species, the samples should be spiked with C12-sulfatide (ST d18:1/C12:0) IS (Avanti Polar Lipids) since this lipid is currently not part of the SL cocktail. In addition, it should be pointed out that the "organic-phase extract" can be used as qualitative screen of whether or not sulfatides are present. ST have a higher recovery (>50%) in the "single-phase extract," therefore, it is recommended that this should be used for quantitation.

5.3.4.3 Separation of GlcCer and GalCer

Normal-phase chromatography is used to separate GlcCer and GalCer. Here, a silica-based normal-phase column (Supelco 2.1 (i.d.) × 250 mm LC-Si) and an isocratic elution protocol are required (Figure 5.3c). The initial mobile phase A is $CH_3CN/CH_3OH/H_3CCOOH$, 97/2/1, v/v/v, with 5 mM ammonium acetate (note acetate buffering versus formate buffering above) flowing at a rate of 1.5 mL/min. Prior to injection, the column is pre-equilibrated for 1.0 min, the sample (dissolved in mobile phase) injected, and then the column is eluted isocratically for 8 min. The isomeric GlcCer and GalCer elute approximately 0.5–1 min apart and should be nearly baseline separated.

5.3.5 Mass Spectrometry

5.3.5.1 Electrospray Ionization

The eluate from the LC column is directed to the ESI source of the mass spectrometer. Here, the liquid flow containing the SL is introduced via a hollow metal needle held at a high potential (either positive or negative voltage). At the tip of the needle, the solution is nebulized by a gas stream into highly charged droplets, which are drawn into the orifice of the mass spectrometer by both potential and pressure difference. Between the needle and the orifice, a heated nitrogen gas flow serves to further drive off the volatile solvents. As the droplets transition between atmospheric pressure and high vacuum, the charged droplets undergo further evaporation whereby the neutral solvent is pumped away. As the ratio of charge to surface area greatly increases, the droplet grows smaller leading to a Coulomb explosion creating many new, smaller, charged droplets [33]. This process rapidly repeats itself eventually leaving SL analytes as ions in the gas phase.

ESI is considered a soft ionization technique whereby primarily intact molecular ions of SL are formed and little fragmentation is observed when performed at low temperatures and potentials. It should be noted, however, that instruments that possess heated glass or metal capillaries can thermally decompose many SL and must be carefully controlled to obtain intact molecular ions. ESI is ideal for quantitation because the signal response of SL is proportional to analyte concentration over several orders of magnitude. One drawback of this technique is the background noise created by solvent ions and their clusters. This can make detection of some of the LCBs difficult, particularly at low concentrations.

5.3.5.2 **Tandem Mass Spectrometry**

Since ESI primarily generates intact molecular ions of SL, no structural information can be derived from single-stage mass analysis. Even at ultrahigh resolution and mass accuracy, only empirical formula may be determined and no structural details may be discerned. Product ion scans, a special form of tandem mass spectrometry (MS/MS), are used to elucidate an intact ions structure (Figure 5.4). Here, the first mass analyzer (Q1 in the case of a QQQ) is set to select a precursor ion of interest, it then fragments in the collision cell (Q2), and the subsequently formed product ions are separated by the second mass analyzer (Q3) then detected. The product ions provide important clues about the structure and reactivity of the intact molecular ion.

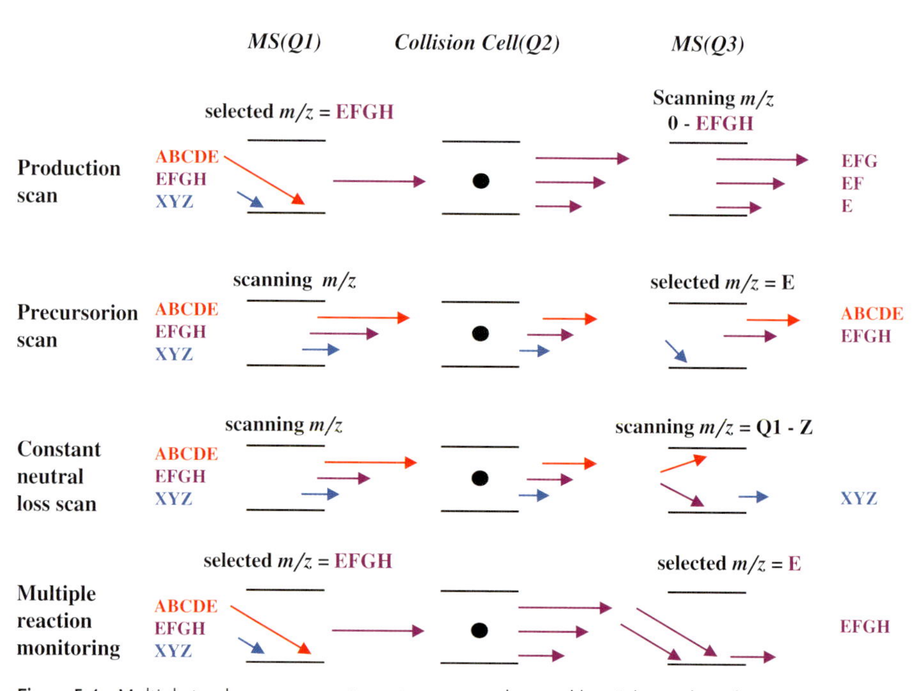

Figure 5.4 Multiple tandem mass spectrometry scan modes used by triple quadrupole instrumentation.

Product ions are usually formed by collision-induced dissociation (CID). Here, part of the precursor ions translational energy is converted into internal energy. This occurs via collisions with a neutral gas such as helium, nitrogen, or argon in the collision cell. The resulting product ion spectrum shows the fragmentation pattern of the selected precursor ions and provides structural information in the form of both product ions detected and neutral species lost. The relative ion abundances of the product ions detected reflect the kinetics of the various dissociation pathways and vary with collision energy (CE). The amount of energy deposited in fragmentation can be controlled by changing either the velocity of the precursor ions or the mass of the target gas or both.

Sphingolipids fragment under MS/MS conditions to yield product ions or neutral losses that are uniquely distinctive of either the head group, sphingoid base, or fatty acid. Sphingomyelin, for example, fragments to yield highly abundant ions indicative of the phosphorylcholine head group in the positive or the negative ion mode of m/z 184 and 168, respectively. The LCBs, Cer, HexCer, and many other SL that have a d18:1 sphingoid base fragment to yield a product ion indicative of the sphingoid base at m/z 264 in the positive ion mode. This product ion will vary according to the length or degree of unsaturation in the sphingoid base. For example, a sphingoid base of d20:1 composition will be observed at m/z 292 and a d18:0 sphingoid base will be observed at m/z 266. HexCer at low collision energies and CPE both undergo fragmentation via a neutral loss of their head groups of 162u and 141u, respectively. Given these unique fragmentations, either precursor ion or neutral loss scans may be used to profile the SL species present in samples for optimization of MRM parameters.

In a precursor ion scan (Figure 5.4), the second mass analyzer (in the case of a QQQ, Q3) is set to pass the m/z of a structure-specific product ion. Q1 is then scanned across a range of m/z values. Only product ions that fragment in Q2 to yield the product ions that have been prescribed by Q3 will reach the detector. When used in this manner, precursor ion scans reveal those ions that have the specified molecular structure in a complex mixture.

In a constant neutral loss scan (Figure 5.4), the first and second mass analyzers (Q1 and Q3) are scanned together. However, Q3 is offset from Q1 by a fixed value that corresponds to the mass of the neutral molecule(s) of interest, which is lost in the fragmentation. Thus, only those ions whose precursor loses the correct mass fragment(s) as a neutral species will be passed to the detector by Q1 and Q3. Therefore, the neutral loss scans like the precursor ion scan will reveal those ions that have the prescribed molecular structure in a complex mixture.

The results that are described here regarding SL have been obtained using two types of tandem mass spectrometers: a triple quadrupole mass spectrometer (the API 3000) and a hybrid quadrupole–linear ion trap tandem mass spectrometer (the ABI 4000 QTrap). Tandem mass spectrometry analyses such as precursor, product, and neutral loss scans, which can be combined with multiple reaction monitoring, for the analysis of both simple and complex sphingolipids in a high-throughput mode may be performed using both types of instruments.

5.3.5.3 Multiple Reaction Monitoring

Multiple reaction monitoring is a powerful tool for increasing the efficiency and accuracy of quantitative MS/MS analyses. In MRM, the first mass analyzer (Q1) is set to pass a specific precursor ion m/z, and the second mass analyzer (Q3) is set to pass a specific product ion m/z (Figure 5.4). Neither mass analyzers are scanned; rather they are targeted to single mass value windows. Therefore, only ions that meet both precursor and product ion m/z conditions simultaneously will be transmitted to the detector. The analysis is not limited to a single precursor–product pair, but can continue onto other precursor/product ion pairs and cycle between them repeatedly. Thus, multiple transitions corresponding to numerous analytes can be monitored within one LC-MS/MS run. When used in this manner, the duty cycles of both analyzers are used most efficiently and time is not wasted on the acquisition of data from m/z space that contains no SL. Moreover, this targeted approach serves to greatly enhance sensitivity by optimization of ionization and dissociation conditions for each individual molecular species m/z transition pair. In addition, it should be noted that selectivity or sensitivity can be further enhanced by decreasing or increasing the mass window each quadrupole may pass.

Optimization of fragmentation conditions is an important point regarding untargeted approaches. For example, consider the case of a small and a large ceramide species, C_{12} Cer versus C_{24} Cer. Their MRM pairs are m/z 482.6/264.4 and 650.9/264.4. Given that both are present at equimolar amounts, at low collision energies the C_{12} Cer will fragment efficiently and give a large signal at m/z 264, but the C_{24} Cer will fragment poorly and yield a weak m/z 264 signal [24, 25]. Thus, a "shotgun" analysis using a precursor ion scan will give the false result of a highly abundant short-chain species and a low-abundance long-chain species. The opposite is true at high collision energies. Here, the C_{24} Cer will fragment optimally and yield a strong signal, whereas the C_{12} Cer will fragment to increasingly smaller species and yield a weak product ion m/z 264 [24, 25]. Again a "shotgun" analysis will provide the wrong quantitative result of a highly abundant long-chain species and a low-abundance short-chain species. If, however, MRM is used where both species fragment optimally at different collision energies, the resulting signal reflects the true equimolar response [24, 25]. It is therefore clear why both precursor ion and neutral loss scans obtained at constant collision energies might yield misleading quantitative results. Thus, when used in conjunction with LC and the appropriate internal standards, MRM provides more accurate quantitative data as a result of addressing these critical issues with regard to ionization suppression, instrument duty cycle, kinetics of ion dissociation, and sensitivity.

Conditions for both maximum ionization and fragmentation efficiency may be determined for each analyte of interest as well as the internal standards. Standard curves should be generated for the internal standards across a wide range of concentrations to determine the limit of detection (LOD = 3:1 signal to noise, s/n), limit of quantitation (LOQ = 10:1 s/n), and the linear dynamic range of the mass spectrometer. Each SL standard can be dissolved at a concentration of 1–10 pmol/uL and be infused at a flow rate of 5–10 uL/min. The declustering potential (DP) and focusing potential (FP) can be ramped from their minimum to maximum value to determine their optimal settings to give the greatest signal for the $(M + H)^+$ species ensuring that no in-source fragmentation occurs under these conditions. The exact center of the mass of the precursor is then determined so that transmission of ions to Q2 for fragmentation is maximized. Following this, a structurally specific product ion, mentioned previously (i.e., *m/z* 264, 266, 184, etc.), may then be identified. Subsequently, their optimal collision energy and collision cell exit potential (CXP) may be determined. Finally, these optimized conditions may be entered into the instrument software to establish a MRM detection channel for that specific molecular species. Each channel may have a dwell time typically on the order of 20 ms with a 5 ms interchannel delay (although newer instruments may need less time). Thus, for 10 MRM pairs, it takes approximately 0.250 s to perform a complete cycle and generates 4 data points per second of HPLC elution time. Therefore, a 10 s wide LC peak would have approximately 40 data points for each transition. It should be noted that the dwell time may need to be adjusted to give the optimal number of data points depending on the number of MRM pairs, the width of the chromatographic peak (if using UPLC), and the speed with which the instrument can cycle between transitions.

The following approach is typically used to optimize these methods. First, samples representative of differing metabolic states are initially profiled via precursor ion and neutral loss scans to identify the various sphingolipids present (i.e., create a "parts list"). Second, the analytes of interest are identified and quantified by LC-MS/MS using the optimized MRM channels for each individual molecular species in the parts list. When analyzed in this manner, LC-MS/MS provides three orthogonal dimensions of separation and identification: retention time, mass, and structure. The extra dimension of analytical rigor provided by HPLC serves to minimize interferences arising from isomeric, isobaric, and isotopic species, which are problematic with other techniques.

5.3.6 Generation of Standard Curves

Standard curves may be generated once the LC and MRM protocols have been determined. The concentration of the individual components of the Avanti internal standard cocktail is 25 μM. If additional SL subspecies are intended to be quantitated, a stock solution at approximately 0.5 mg/mL concentration should be made in methanol. These may then be serially diluted into the appropriate LC solvent immediately before analysis to provide a wide concentration range for each standard per 50 μL injection. (It should be noted that every instrument will have its own unique performance characteristics, which will dictate the concentration range used for that instrument.) Each concentration may be analyzed from the lowest to the highest to generate the standard curves such as those shown previously [24] and then the linear regression lines and fit may be calculated.

5.3.7
Data Analysis

The areas under the peak for each MRM pair are used to quantify the amounts of the SL analytes of interest in biological samples. The areas generated for both analytes and internal standards are integrated via the native mass spectrometer software (i.e., Analyst, Hystar, Xcaliber, MassLynx, etc.). Identical integration parameters such as peak smooths (2–3), bunching factor (5–10), and noise threshold (1×10^4) are used to integrate both the internal standard and the analytes. The calculation of the pmol of analyte is determined using the following formula:

$$\text{pmol of analyte of interest} = K_{\text{analyte}} \times (A_{\text{analyte}}/A_{\text{IS}}) \times \text{pmol of added internal standard}$$

where K_{analyte} are the correction factors for the analyte versus the internal standard, A_{analyte} is the area under the peak of the analyte, and A_{IS} is the area under the peak of the added internal standard. The K_{analyte} factor adjusts for differences between the analyte and the internal standard with respect to ion yield per unit amount for the selected MRM pair as determined from the standard curves. This calculation also includes any correction for differences in isotopic abundance (~1.1% per carbon). These are minimal for analytes with alkyl chain lengths similar to the internal standard; however, this becomes larger when the number of carbon atoms in the analyte is much larger than the internal standards (i.e., C_{12} IS versus C_{24} or greater analytes). To determine the magnitude of this "lost signal," the ratio of $(M+H)^+$, $(M+H+1)^+$, and $(M+H+2)^+$ to the number of carbons in the internal standards and analytes are calculated. For example, consider C_{12}/C_{24} ceramide. The former has a molecular formula of $C_{30}H_{60}NO_3$ and its first three isotope ratios are 100:34.5:6.4. The latter has a formula of $C_{42}H_{84}NO_3$ with an isotope ratio of 100:48:11.9. Here, only 62% of the C_{24} Cer total mass is being measured via the monoisotopic MRM channel relative to the C12 Cer at 71%. Thus, the C_{24} Cer signal needs to be multiplied by 1.145 (or an additional 14.5%) to compensate for this "lost" signal. This can also be crudely approximated by multiplying by the difference in the number of carbon atoms by the ^{13}C isotopic abundance ($12 \times 1.1\% = 13.2\%$). Either may then be used to adjust for differences between these species.

No correction is necessary for differences in extraction recovery using these protocols. This is because the extraction conditions were optimized to achieve maximal recovery of SL species. The quantities of the SL are then expressed as pmol analyte/μg of DNA (LIPID MAPS convention). For comparison, based on our experience, 1×10^6 RAW 264.7 cells contain approximately 3 μg of DNA and approximately 0.25 mg of protein.

5.3.8
Quality Control

Internal standards serve as a control not only for extraction but also for the autosampler, HPLC column, and instrument. It is essential that each batch of samples submitted for LC-MS/MS analysis should include analysis of the internal standards alone at the beginning, middle, and end of the run. This will confirm that all parts of the analysis system are functioning properly and reproducibly. In addition, blank samples (containing only the LC solvent) may be used to assess the possibility of carryover and should be analyzed at varying intervals throughout the run. Extended batch runs should be checked periodically to assess if carryover or shifts in the LC retention times occur for any of the analytes or standards. If so, the run should be halted and the column should be back flushed and cleaned before resuming the batch run.

5.4 Applications of Sphingolipidomics in Biology and Disease

5.4.1 LC-MS/MS

Targeted sphingolipidomic analysis using LC-MS/MS is a useful technique to provide more insights into metabolic pathways and pathologies. LC-MS/MS analyses can reveal changes in the pathways of metabolic flux of sphingolipids from biosynthesis or turnover in a biological system. For example, detailed analysis of the full spectrum *N*-acyl chain lengths of ceramides and monohexosylceramdes in HEK-293T cells transiently overexpressing *LASS1* revealed an increase in *N*-acyl C18:0 chains. Furthermore, this increase also showed resistance to inhibition by fumonisin B_1 (FB_1), a known inhibitor of dihydroceramide synthase (DHCerS) [34]. The structure-specific data from LC-MS/MS analysis provided the necessary specificity to demonstrate the preference of ceramide synthase (CerS1) for stearoyl-CoA substrate [35].

In another study, sphingolipid signaling pathways triggered by the paracellular-generated versus intracellular-generated ceramides were compared. This is of interest as they induce lung endothelial cell apoptosis. Here, LC-MS/MS analysis revealed increases in both relative and absolute levels of C16:0 ceramide in response to C8:0 ceramide and TNF-α treatments. This implicates that the serine palmitoyl synthase-regulated ceramide synthesis may contribute to the amplification of pulmonary vascular injury induced by excessive ceramides [36]. The aberrant production of sphingolipid in various tissues, including skeletal muscle, pancreas, and adipocytes, promoted by elevation of plasma free fatty acids has revealed the roles of sphingolipids in many metabolic diseases, including obesity, diabetes, atherosclerosis, and metabolic syndrome [37].

Application of LC-MS/MS techniques has also facilitated the quantitative analysis of bioactive sphingolipids at their physiological and pathological levels to examine the distinct functions of these lipids in cancer pathogenesis and therapy [38]. The endogenous ceramide levels in 43 malignant breast tumors and 21 benign breast biopsies as well as those of normal tissues have been analyzed using LC-MS/MS. The total ceramide levels in malignant tumor tissue samples were statistically significantly elevated compared to that in normal tissue samples. The levels of Cer C16:0, C24:1, and C24:0 all significantly increased in malignant tumors compared to that in benign and normal tissue. The augmentation of the various ceramides could be assigned to an increase of the messenger RNA levels of ceramide synthases (CerS) LASS2 (longevity assurance gene, now designated as ceramide synthase), LASS4, and LASS6. Progression in breast cancer can be associated with increased ceramide levels due to an upregulation of specific LASS genes [39].

Targeted lipidomics using LC-MS/MS techniques are also critical in determining novel lipid biomarkers in many diseases [40–43]. A highly significant inverse relationship between the level of the LCBs So1P, Sa1P, and C24:1 Cer in the high-density lipoproteins (HDL)-containing fraction of serum and the occurrence of ischemic heart disease (IHD) has been reported. This indicates that compositional differences of sphingolipids in the HDL-containing fraction of human serum may contribute to the putative protective role of HDL in IHD [44]. Furthermore, lipids can play many different biological roles including toxicity and increased mortality [45].

The quantitation of sphingolipid metabolites by LC-MS/MS from two animal models of type 1 diabetes has also been performed to explore the role of sphingolipids and identify putative therapeutic targets and biomarkers in diabetes. So1P was found to be elevated in both diabetic models. In the meantime, plasma levels of omega-9, C24:1 (nervonic acid)-containing ceramide, sphingomyelin, and cerebrosides in diabetic animals decreased. Reduction of C24:1-esterfied SL was also observed in liver and heart, indicating the bioactive So1P and the cardio-and neuroprotective omega-9 esterified sphingolipids could be biomarkers for type 1 diabetes and represent novel therapeutic targets [46].

In a similar study of 13 obese type 2 diabetic patients and 14 lean healthy control patients, quantitative LC-MS/MS analyses of plasma demonstrated elevated ceramide levels in type 2 diabetic subjects. This suggested their effects on insulin resistance through activation of inflammatory mediators such as TNF-α [47].

5.4.2 Transcriptomic Guided Tissue Imaging Mass Spectrometry

Transcriptomic studies can reveal the details of gene expression and predict metabolite changes in biological process. Targeted metabolite profiling and imaging approaches, such as LC-MS/MS and tissue imaging mass spectrometry (TIMS), are capable of testing quantitative signatures of specific metabolites in biological processes and revealing the molecular changes at specific locations in a fast and accurate way. Imaging mass spectrometry is a powerful tool for profiling and mapping various molecules within biological systems via direct tissue analysis. Although it has been widely used for untargeted "omic" studies, it can be integrated with other transcriptomic and metabolomic approaches to improve the understanding of the mechanism(s) of diseases such as cancer [48] and prototypic neurodegenerative disease [49].

The gene expression profiles of the enzymes related to glycosphingolipid-specific biosynthesis and turnover pathways were studied in ovarian cancer [48]. Here, 12 ovarian epithelial carcinoma samples were compared with another group of 12 normal ovarian epithelial cells or 12 normal stromal tissues using microarray assay. The mRNAs for several enzymes of sulfatide (ST) biosynthesis – most notably GalCer synthase (also called ceramide galactosyltransferase, UGT8) and GalCer sulfotransferase (Gal3ST1) – were higher for the ovarian carcinoma cells versus normal stromal tissue and normal ovarian epithelial cells. This was in contrast to the mRNAs for three enzymes of ST catabolism, arylsulfatase, galactosylceramidase, and possibly the related saposins (based on the pro-saposin mRNA, PSAP) were not observed to be different. This gene expression analysis, therefore, predicted that ovarian epithelial carcinoma cells could potentially have higher ST versus normal stromal tissue. Subsequent analysis of 12 ovarian tissues graded as histologically normal or having epithelial ovarian tumors by LC-MS/MS established that most tumor-bearing tissues have higher amounts of ST. In addition, some have elevated GalCer only or both GalCer and ST.

Ovarian cancer tissues are comprised of many different cell types. Therefore, adjacent slices of tissues from tumors were analyzed by MALDI tissue imaging mass spectrometry. Three abundant ST species corresponding to d18:1/C16:0, d18:1/C24:1, and d18:1/C24:0 were observed in most of the tissue regions identified as ovarian epithelial carcinoma. This was in strong contrast to the stroma, which was essentially free of ions of *m*/*z* 778.6, 888.6, or 890.6, respectively. These data established that the ST that have been found in extracts of ovarian tumors by LC-MS/MS were derived primarily from the malignant cells rather than from other regions of the tissue [48].

MALDI TIMS analysis of a model of Tay-Sachs and Sandhoff disease yields another example of combining gene expression change and targeted lipidomic approaches. Here, profiling and localization were performed on many different lipid species, and of particular interest, ganglioside GM2, asialo-GM2 (GA2), which were predicted to be elevated due to genomic deficiency. TIMS results of the mutant Tay/Sachs-Sandhoff (hexb$^{-/-}$, knockout) mouse brain not only showed several striking features between the normal and diseased sample but also revealed the specific locations of the accumulation of disease-related molecules [49]. The knockout mouse brain slices when ionized in the negative mode showed two prominent ions of *m*/*z* 888.9 (ST d18:1/24:1) and 1383 (GM2 d18:1/18:0). The ST was localized mainly to the myelinated fiber region of the cerebellum and fairly evenly distributed in the brainstem. In contrast, the *m*/*z* 1383 ion was localized to the granular cell region in the cerebellum and was not detected in the brainstem. In the positive ion mode, the ion of *m*/*z* 772.6 (presumed HexCer) was observed primarily in the molecular layer region. This was contrasted by the *m*/*z* 1132 ion (presumed GA2) that was seen in the granular cell region, similar to that of the *m*/*z* 1383 ion in negative mode. As gene expression level predicted, the normal mouse (heterozygote (hexb$^{+/-}$) brain had neither the *m*/*z* 1383 (GM2) nor *m*/*z* 1132 (GA2) in the negative or positive mode, respectively [49].

5.5
Conclusions

Targeted lipidomics has great potential in many fields from fundamental biological research to clinical applications, because lipids are involved in so many aspects of cell function and regulation. The development of sophisticated analytical technologies such as LC-MS/MS has enabled targeted lipidomics to identify and quantify numerous individual molecular species of lipids in biological systems. The resulting data are beginning to reveal the important roles of specific lipid molecular species. In addition, the connections between lipid metabolism and pathologies are being uncovered by more and more research groups. Finally, interactive collaborations between researchers focusing on different categories of lipids will bring a more comprehensive understanding of lipids as well as provide more breakthrough achievements in related fields.

References

1 Brown, H.A. (2007) *Lipidomics and Bioactive Lipids: Specialized Analytical Methods and Lipids in Disease*, vol. 433, Methods in Enzymology, Academic Press, pp. XV–XVI.

2 Feng, L. and Prestwich, G.D. (2006) *Functional Lipidomics*, Taylor & Francis, Boca Raton.

3 Lahiri, S. and Futerman, A.H. (2007) The metabolism and function of sphingolipids and glycosphingolipids. *Cell Mol. Life Sci.*, **64**, 2270–2284.

4 Merrill, A.H., Jr., Wang, M.D., Park, M., and Sullards, M.C. (2007) (Glyco) sphingolipidology: an amazing challenge and opportunity for systems biology. *Trends Biochem. Sci.*, **32**, 457–468.

5 Fahy, E., Subramaniam, S., Brown, H.A., Glass, C.K., Merrill, A.H., Jr., Murphy, R. C., Raetz, C.R., Russell, D.W., Seyama, Y., Shaw, W., Shimizu, T., Spener, F., van Meer, G., VanNieuwenhze, M.S., White, S.H., Witztum, J.L., and Dennis, E.A. (2005) A comprehensive classification system for lipids. *J. Lipid Res.*, **46**, 839–861.

6 Fahy, E., Subramaniam, S., Murphy, R.C., Nishijima, M., Raetz, C.R., Shimizu, T., Spener, F., van Meer, G., Wakelam, M.J., and Dennis, E.A. (2009) Update of the LIPID MAPS comprehensive classification system for lipids. *J. Lipid Res*, **50** (Suppl.), S9–S14.

7 Han, X. and Gross, R.W. (2005) Shotgun lipidomics: electrospray ionization mass spectrometric analysis and quantitation of cellular lipidomes directly from crude extracts of biological samples. *Mass Spectrom. Rev.*, **24**, 367–412.

8 Schwudke, D., Hannich, J.T., Surendranath, V., Grimard, V., Moehring, T., Burton, L., Kurzchalia, T., and Shevchenko, A. (2007) Top-down lipidomic screens by multivariate analysis of high-resolution survey mass spectra. *Anal. Chem.*, **79**, 4083–4093.

9 Han, X., Yang, K., Cheng, H., Fikes, K.N., and Gross, R.W. (2005) Shotgun lipidomics of phosphoethanolamine-containing lipids in biological samples after one-step *in situ* derivatization. *J. Lipid Res.*, **46**, 1548–1560.

10 Cheng, H., Jiang, X., and Han, X. (2007) Alterations in lipid homeostasis of mouse dorsal root ganglia induced by apolipoprotein E deficiency: a shotgun lipidomics study. *J. Neurochem.*, **101**, 57–76.

11 Gross, R.W. and Han, X. (2009) Shotgun lipidomics of neutral lipids as an enabling technology for elucidation of lipid-related diseases. *Am. J. Physiol.*, **297**, E297–E303.

12 Jung, H.R., Sylvanne, T., Koistinen, K.M., Tarasov, K., Kauhanen, D., and Ekroos, K. (1811) High throughput quantitative molecular lipidomics. *Biochim. Biophys. Acta*, 925–934.

13 Haynes, C.A., Allegood, J.C., Park, H., and Sullards, M.C. (2009) Sphingolipidomics: methods for the comprehensive analysis of sphingolipids. *J. Chromatogr. B Analyt. Technol. Biomed. Life Sci.*, **877**, 2696–2708.

14 Svennerholm, L. (1964) The gangliosides. *J. Lipid Res.*, **5**, 145–155.

15 Sugiyama, E., Hara, A., Uemura, K., and Taketomi, T. (1997) Application of matrix-assisted laser desorption ionization time-of-flight mass spectrometry with delayed ion extraction to ganglioside analyses. *Glycobiology*, **7**, 719–724.

16 Suzuki, A., Hiraoka, N., Suzuki, M., Angata, K., Misra, A.K., McAuliffe, J., Hindsgaul, O., and Fukuda, M. (2001) Molecular cloning and expression of a novel human beta-Gal-3-*O*-sulfotransferase that acts preferentially on *N*-acetyllactosamine in *N*- and *O*-glycans. *J. Biol. Chem.*, **276**, 24388–24395.

17 O'Connor, P.B., Budnik, B.A., Ivleva, V.B., Kaur, P., Moyer, S.C., Pittman, J.L., and Costello, C.E. (2004) A high pressure matrix-assisted laser desorption ion source for Fourier transform mass spectrometry designed to accommodate large targets with diverse surfaces. *J. Am. Soc. Mass Spectrom.*, **15**, 128–132.

18 Houjou, T., Yamatani, K., Nakanishi, H., Imagawa, M., Shimizu, T., and Taguchi, R. (2004) Rapid and selective identification of molecular species in phosphatidylcholine and sphingomyelin by conditional neutral loss scanning and MS3. *Rapid. Commun. Mass Spectrom.*, **18**, 3123–3130.

19 Han, X. and Cheng, H. (2005) Characterization and direct quantitation of cerebroside molecular species from lipid extracts by shotgun lipidomics. *J. Lipid Res.*, **46**, 163–175.

20 Hsu, F.F. and Turk, J. (2001) Structural determination of glycosphingolipids as lithiated adducts by electrospray ionization mass spectrometry using low-energy collisional-activated dissociation on a triple stage quadrupole instrument. *J. Am. Soc. Mass Spectrom.*, **12**, 61–79.

21 Sullards, M.C., Wang, E., Peng, Q., and Merrill, A.H., Jr. (2003) Metabolomic profiling of sphingolipids in human glioma cell lines by liquid chromatography tandem mass spectrometry. *Cell Mol. Biol.*, **49**, 789–797.

22 Metelmann, W., Peter-Katalinic, J., and Muthing, J. (2001) Gangliosides from human granulocytes: a nano-ESI QTOF mass spectrometry fucosylation study of low abundance species in complex mixtures. *J. Am. Soc. Mass Spectrom.*, **12**, 964–973.

23 Metelmann, W., Vukelic, Z., and Peter-Katalinic, J. (2001) Nano-electrospray ionization time-of-flight mass spectrometry of gangliosides from human brain tissue. *J. Mass Spectrom.*, **36**, 21–29.

24 Shaner, R.L., Allegood, J.C., Park, H., Wang, E., Kelly, S., Haynes, C.A., Sullards, M.C., and Merrill, A.H., Jr. (2009) Quantitative analysis of sphingolipids for lipidomics using triple quadrupole and quadrupole linear ion trap mass spectrometers. *J. Lipid Res.*, **50**, 1692–1707.

25 Sullards, M.C., Liu, Y., Chen, Y., and Merrill, A.H., Jr. (2011) Analysis of mammalian sphingolipids by liquid chromatography tandem mass spectrometry (LC-MS/MS) and tissue imaging mass spectrometry (TIMS). *Biochim. Biophys. Acta*, **1811** (11), 838–853.

26 Folch, J., Lees, M., and Sloane Stanley, G. H. (1957) A simple method for the isolation and purification of total lipides from animal tissues. *J. Biol. Chem.*, **226**, 497–509.

27 Bligh, E.G. and Dyer, W.J. (1959) A rapid method of total lipid extraction and purification. *Can. J. Biochem. Physiol.*, **37**, 911–917.

28 van Echten-Deckert, G. (2000) Sphingolipid extraction and analysis by thin-layer chromatography. *Methods Enzymol.*, **312**, 64–79.

29 Sullards, M.C. and Merrill, A.H., Jr. (2001) Analysis of sphingosine 1-phosphate, ceramides, and other bioactive sphingolipids by high-performance liquid chromatography-tandem mass spectrometry. *Sci. STKE*, **2001**, pl1.

30 Lee, M.H., Lee, G.H., and Yoo, J.S. (2003) Analysis of ceramides in cosmetics by reversed-phase liquid chromatography/electrospray ionization mass spectrometry with collision-induced dissociation. *Rapid. Commun. Mass Spectrom.*, **17**, 64–75.
31 Merrill, A.H., Jr., Sullards, M.C., Allegood, J.C., Kelly, S., and Wang, E. (2005) Sphingolipidomics: high-throughput, structure-specific, and quantitative analysis of sphingolipids by liquid chromatography tandem mass spectrometry. *Methods*, **36**, 207–224.
32 Pettus, B.J., Kroesen, B.J., Szulc, Z.M., Bielawska, A., Bielawski, J., Hannun, Y.A., and Busman, M. (2004) Quantitative measurement of different ceramide species from crude cellular extracts by normal-phase high-performance liquid chromatography coupled to atmospheric pressure ionization mass spectrometry. *Rapid. Commun. Mass Spectrom.*, **18**, 577–583.
33 Kebarle, P. and Verkerk, U.H. (2009) Electrospray: from ions in solution to ions in the gas phase, what we know now. *Mass Spectrom. Rev.*, **28**, 898–917.
34 Merrill, A.H., Jr., Wang, E., Vales, T.R., Smith, E.R., Schroeder, J.J., Menaldino, D. S., Alexander, C., Crane, H.M., Xia, J., Liotta, D.C., Meredith, F.I., and Riley, R.T. (1996) Fumonisin toxicity and sphingolipid biosynthesis. *Adv. Exp. Med. Biol.*, **392**, 297–306.
35 Haynes, C.A., Allegood, J.C., Park, H., and Sullards, M.C. (2009) Sphingolipidomics: methods for the comprehensive analysis of sphingolipids. *J. Chromatogr. B Analyt. Technol. Biomed. Life Sci.*, **877**, 2696–2708.
36 Medler, T.R., Petrusca, D.N., Lee, P.J., Hubbard, W.C., Berdyshev, E.V., Skirball, J., Kamocki, K., Schuchman, E., Tuder, R. M., and Petrache, I. (2008) Apoptotic sphingolipid signaling by ceramides in lung endothelial cells. *Am. J. Respir. Cell Mol. Biol.*, **38**, 639–646.
37 Cowart, L.A. (2009) Sphingolipids: players in the pathology of metabolic disease. *Trends Endocrinol. Metab.*, **20**, 34–42.
38 Ogretmen, B. (2006) Sphingolipids in cancer: regulation of pathogenesis and therapy. *FEBS Lett.*, **580**, 5467–5476.
39 Schiffmann, S., Sandner, J., Birod, K., Wobst, I., Angioni, C., Ruckhaberle, E., Kaufmann, M., Ackermann, H., Lotsch, J., Schmidt, H., Geisslinger, G., and Grosch, S. (2009) Ceramide synthases and ceramide levels are increased in breast cancer tissue. *Carcinogenesis*, **30**, 745–752.
40 Hu, C., van der Heijden, R., Wang, M., van der Greef, J., Hankemeier, T., and Xu, G. (2009) Analytical strategies in lipidomics and applications in disease biomarker discovery. *J. Chromatogr. B Analyt. Technol. Biomed. Life Sci.*, **877**, 2836–2846.
41 Scherer, M., Schmitz, G., and Liebisch, G. (2009) High-throughput analysis of sphingosine 1-phosphate, sphinganine 1-phosphate, and lysophosphatidic acid in plasma samples by liquid chromatography–tandem mass spectrometry. *Clin. Chem.*, **55**, 1218–1222.
42 Quehenberger, O., Armando, A.M., Brown, A.H., Milne, S.B., Myers, D.S., Merrill, A.H., Bandyopadhyay, S., Jones, K.N., Kelly, S., Shaner, R.L., Sullards, C. M., Wang, E., Murphy, R.C., Barkley, R. M., Leiker, T.J., Raetz, C.R.H., Guan, Z.Q., Laird, G.M., Six, D.A., Russell, D.W., McDonald, J.G., Subramaniam, S., Fahy, E., and Dennis, E.A. (2010) Lipidomics reveals a remarkable diversity of lipids in human plasma. *J. Lipid Res.*, **51**, 3299–3305.
43 Sandra, K., Pereira, A.D., Vanhoenacker, G., David, F., and Sandra, P. (2010) Comprehensive blood plasma lipidomics by liquid chromatography/quadrupole time-of-flight mass spectrometry. *J. Chromatogr. A*, **1217**, 4087–4099.
44 Argraves, K.M., Sethi, A.A., Gazzolo, P.J., Wilkerson, B.A., Remaley, A.T., Tybjaerg-Hansen, A., Nordestgaard, B.G., Yeatts, S. D., Nicholas, K.S., Barth, J.L., and Argraves, W.S. (2011) S1P, dihydro-S1P and C24:1-ceramide levels in the HDL-containing fraction of serum inversely correlate with occurrence of ischemic heart disease. *Lipids Health Dis.*, **10**, 70.
45 Perman, J.C., Bostrom, P., Lindbom, M., Lidberg, U., Stahlman, M., Hagg, D., Lindskog, H., Scharin Tang, M., Omerovic, E., Mattsson Hulten, L., Jeppsson, A., Petursson, P., Herlitz, J.,

Olivecrona, G., Strickland, D.K., Ekroos, K., Olofsson, S.O., and Boren, J. (2011) The VLDL receptor promotes lipotoxicity and increases mortality in mice following an acute myocardial infarction. *J. Clin. Invest.*, **121**, 2625–2640.

46 Fox, T.E., Bewley, M.C., Unrath, K.A., Pedersen, M.M., Anderson, R.E., Jung, D. Y., Jefferson, L.S., Kim, J.K., Bronson, S. K., Flanagan, J.M., and Kester, M. (2011) Circulating sphingolipid biomarkers in models of type 1 diabetes. *J. Lipid Res.*, **52**, 509–517.

47 Haus, J.M., Kashyap, S.R., Kasumov, T., Zhang, R., Kelly, K.R., Defronzo, R.A., and Kirwan, J.P. (2009) Plasma ceramides are elevated in obese subjects with type 2 diabetes and correlate with the severity of insulin resistance. *Diabetes*, **58**, 337–343.

48 Liu, Y., Chen, Y., Momin, A., Shaner, R., Wang, E., Bowen, N.J., Matyunina, L.V., Walker, L.D., McDonald, J.F., Sullards, M.C., and Merrill, A.H., Jr. (2010) Elevation of sulfatides in ovarian cancer: an integrated transcriptomic and lipidomic analysis including tissue-imaging mass spectrometry. *Mol. Cancer*, **9**, 186.

49 Chen, Y., Allegood, J., Liu, Y., Wang, E., Cachon-Gonzalez, B., Cox, T.M., Merrill, A.H., Jr., and Sullards, M.C. (2008) Imaging MALDI mass spectrometry using an oscillating capillary nebulizer matrix coating system and its application to analysis of lipids in brain from a mouse model of Tay-Sachs/Sandhoff disease. *Anal. Chem.*, **80**, 2780–2788.

6
Structural Lipidomics

Todd W. Mitchell, Simon H.J. Brown, and Stephen J. Blanksby

6.1
Introduction

The term lipid encompasses a broad range of structurally diverse molecules with varied physical and chemical properties. Such diversity provides a significant analytical challenge. To overcome this, many of the traditional analytical techniques analyze lipids as a class (e.g., the analysis of phospholipid classes by thin-layer chromatography) or break down lipids into more simple molecules (e.g., gas chromatography (GC) of fatty acid methyl esters). While these and similar techniques have provided great insight into lipid biochemistry over the decades (and still do), molecular level information is lost and structural diversity underestimated.

The importance of molecular structure becomes apparent when one considers the vast array of cellular functions in which lipids are involved. Membrane trafficking is regulated through interaction of specific lipid species with protein complexes [1]. Two common functions of lipid–protein interactions are for enzyme targeting and messenger signaling. A protein domain can recognize a membrane lipid and be targeted to a specific location in the cell in order to localize protein function. Conversely, a lipid itself may act as a signaling molecule. For example, diacylglycerol (DAG) is a lipid second messenger that can modulate the function of protein kinase C (PKC) [2], while lysophosphatidic acid (LPA) is involved in the regulation of cell proliferation, migration, and survival to name but a few of its signaling functions [3].

Modern mass spectrometry techniques allow us to delve much deeper into lipid molecular structure affording a greater level of understanding of the role of lipids in cellular function and as such have become methods of choice in modern lipidomic analysis. Nevertheless, mass spectrometry is purely a measure of mass and charge and therefore suffers limitations in its ability to distinguish many structural isomers commonly found among lipids. In this chapter, we will discuss the capabilities and limitations of mass spectrometry in the structural characterization of lipids and some of the techniques currently being explored to overcome these limitations.

Lipidomics, First Edition. Edited by Kim Ekroos.

6.2
Lipid Structure

The traditional classification of a lipid, a biological substance that is soluble in organic solvents, is ambiguous and describes a plethora of biological molecules, including numerous proteins. More recently, the LIPID MAPS (LIPID Metabolites and Pathways Strategy) consortium has devised a new classification system that is based on the structural characteristics of a lipid rather than on its physical or chemical properties. Specifically, this new classification defines lipids as "hydrophobic or amphipathic small molecules that may originate entirely or in part by carbanion-based condensations of thioesters (fatty acids, polyketides, etc.) and/or by carbocation-based condensations of isoprene units (prenols, sterols, etc.)" [4]. Under this system, there are two main classes of lipids, the fatty acid-based lipids, for example, glycerolipids, glycerophospholipids, sphingolipids, and fatty acids themselves, and the prenols or sterols, for example, cholesterol, steroids, and isoprenoids. Some representative structures from these lipid classes are shown in Figure 6.1.

6.3
Structural Analysis of Lipids by Mass Spectrometry

The field of lipidomics has expanded rapidly in recent years, driven by the application of electrospray ionization- (ESI) and matrix-assisted laser desorption ionization- (MALDI) mass spectrometry to the analysis of lipids. These soft ionization techniques provide the ability to ionize molecular lipids with little to no fragmentation and can be combined with a tandem mass spectrometer (MS/MS) to selectively fragment these ions via collision-induced dissociation (CID). For example, Figure 6.2a shows the negative ion mass spectrum obtained from a human brain lipid extract. Based on the mass-to-charge ratio (m/z) alone, we can make some assumptions as to the possible identity of the ions present. Choosing the ion at m/z 788, we could make a preliminary assignment of a phosphatidylserine with two fatty acyl chains having a combined chain length of 36 carbons with 1 double bond PS(36:1). Indeed, utilizing accurate mass measurements on a contemporary high-resolution mass spectrometer, such as an orbitrap or a Fourier transform ion cyclotron resonance (FTICR) instrument, could confirm this assignment [5].

The CID spectrum obtained from this ion is shown in Figure 6.2b. The ion observed at m/z 701.5 arises from the neutral loss of the serine head group. The ions at m/z 417, 419, 435, and 437 are formed from the neutral losses of the two acyl chains as fatty acids or ketenes, respectively. The ions observed at m/z 281 and 283 are the 18:1 and 18:0 carboxylate ions and the ion at m/z 153 is a dehydrated glycerophosphate anion. Based on the ions present and their relative abundances, this lipid would generally be identified as 1-stearoyl-2-oleoyl-phosphatidylserine or PS(18:0/9*Z*-18:1) as shown in Figure 6.3a. There are, however, several assumptions in this interpretation that cannot be resolved based on the CID spectrum alone. They are as follows:

Fatty Acids

Fatty acids and conjugates
Eicosanoids
Fatty alcohols
Fatty esters
Fatty amides

Oleic Acid, 18:1 (18 carbons:1 double bond)

Glycerolipids

Monoacylglycerols
Diacylglycerols
Triacylglycerols

Triolein, TAG(18:1/18:1/18:1)

Glycerophospholipids

Glycerophosphocholines
Glycerophosphoathanolamines
Glycerophosphoserines
Glycerophosphoglycerols
Glycerophosphoinositols
Glycerophosphates

Phosphatidylserine, PS(16:0/18:1)

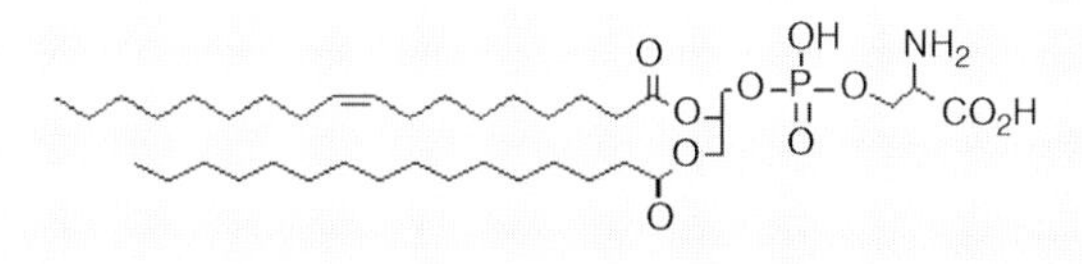

Shingolipids

Sphingoid bases
Ceramides
Phosphosphingolipids
Glycosphingolipids

Sphingomyelin, SM(18:1d/16:0)

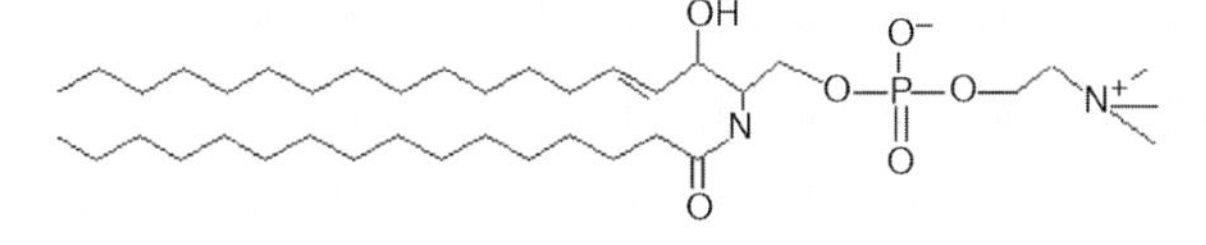

Sterols

Cholesterol and derivatives
Cholesteryl esters

Cholesteryl ester, CE(18:1)

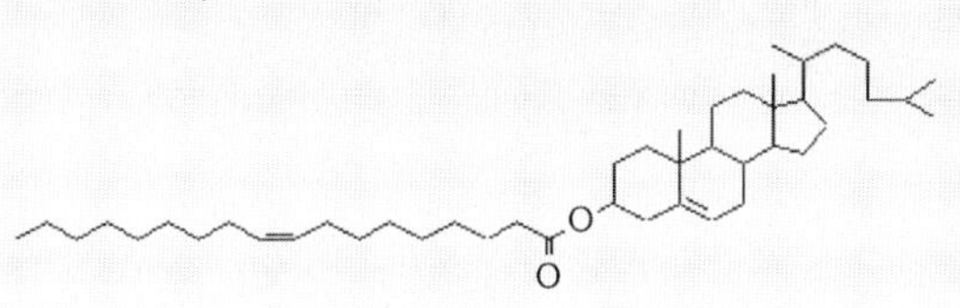

Figure 6.1 Representative structures of various lipid classes.

1) **Position of fatty acid esterification to the glycerol backbone (*sn* position).** Using the phosphatidylserine example in Figure 6.3 it can be seen that the structure in Figure 6.3a has 18:0 esterified at the *sn*-1 position and 18:1 esterified at the *sn*-2 position, that is, PS(18:0/18:1), while the structure in Figure 6.3b has the opposite, that is, PS(18:1/18:0). The *sn* terminology refers to stereospecific number where the *sn*-3 position in phospholipids is determined by the phosphodiester. Although this terminology is also utilized in triacylglycerol (TAG) terminology, assignment of *sn*-1 and *sn*-3 positions is arbitrary.

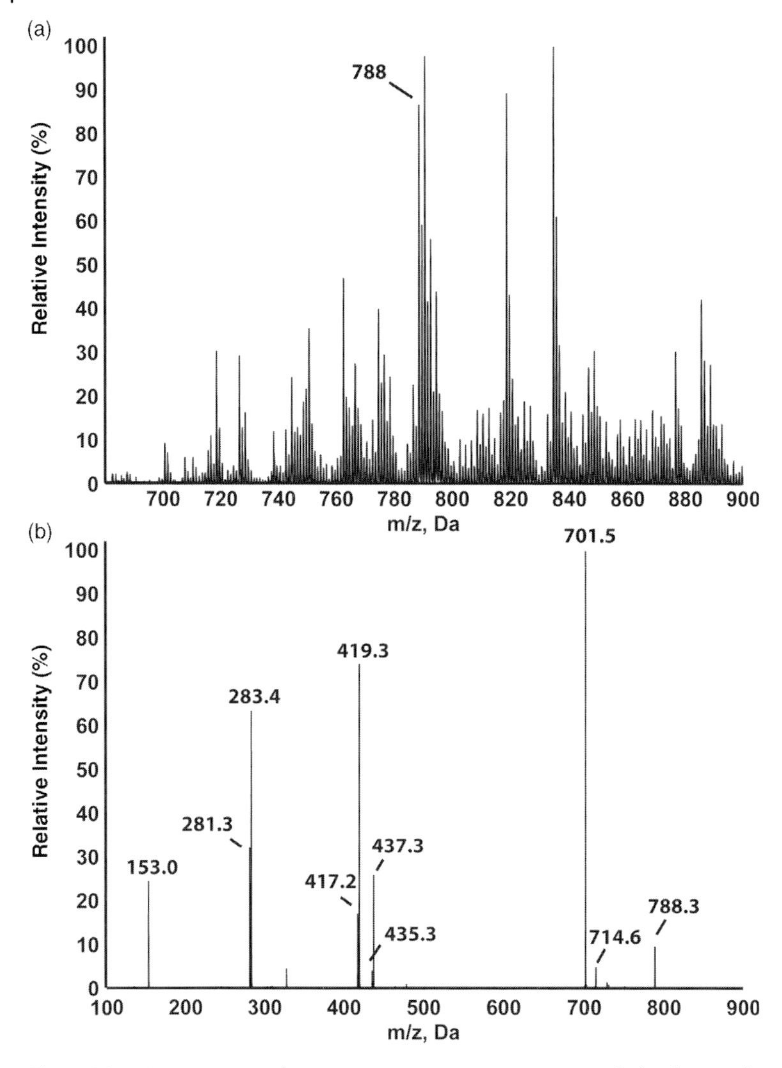

Figure 6.2 (a) Negative ion electrospray ionization mass spectrum of a lipid extract from human brain. (b) CID spectrum produced from the fragmentation of the ion at *m/z* 788.

2) **Position of the double bond within the fatty acyl chain.** Again, using Figure 6.3 as an example, the structures in Figure 6.3a and c are identical with the exception that the double bond position in the 18:1 acyl chain is in the *n*-9 and *n*-7 positions, respectively. The *n*- nomenclature refers to the position of the double bond in reference to the methyl end of the chain, that is, 18:1 *n*-7 identifies a fatty acid with 18 carbons and 1 double bond that is found at the 7th carbon from the methyl end. The opposing nomenclature is when the double bond position is identified from the carboxyl end, for example, 11Z–18:1 is the equivalent to 18:1 *n*-7 where the stereochemistry is not defined in the later.

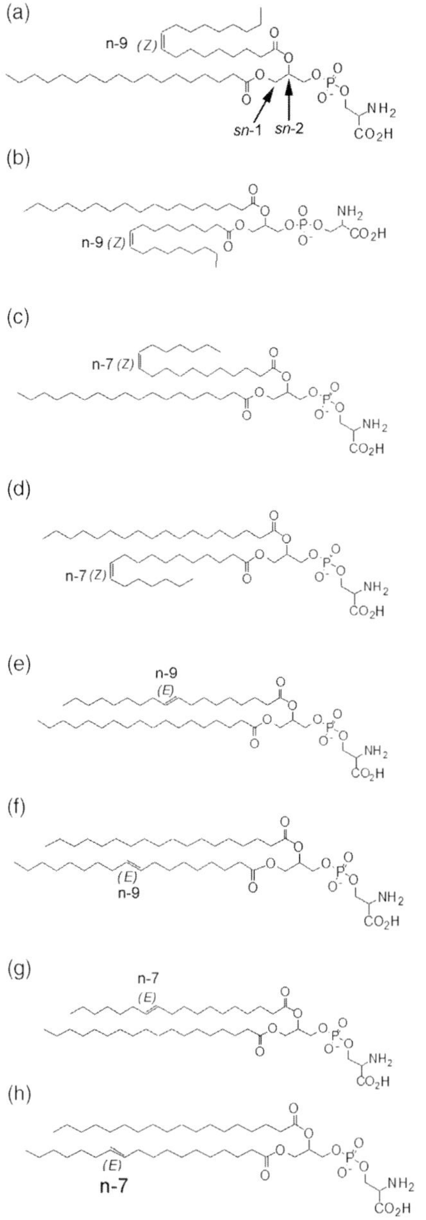

Figure 6.3 Isomeric lipids that could be assigned to the CID spectrum shown in Figure 6.2b. (a) PS(18:0/9*Z*-18:1). (b) PS(9*Z*-18:1/18:0). (c) PS(18:0/11*Z*-18:1). (d) PS(11*Z*-18:1/18:0). (e) PS (18:0/9*E*-18:1). (f) PS(9*E*-18:1/18:0). (g) PS(18:0/11*E*-18:1). (h) PS(11*E*-18:1/18:0).

3) **Stereochemistry of the double bond (*cis*-(Z) or *trans*-(E)).** An example of double bond stereoisomerism can be seen in Figure 6.3a and e where the only structural difference between the two is the *cis*-double bond in (a) and the *trans*-double bond in (e). In this case, the nomenclature for the different lipids would be PS(18:0/9*Z*-18:1) and PS(18:0/9E-18:1), respectively.

If these common isomeric motifs were considered, then the identification of the *m/z* 788 ion from the human brain would quickly expand to eight possible lipids (Figure 6.3b–h). In reality, it is likely that a mixture of several of these isomers is present. The problem of structural isomerism is further exacerbated when analyzing prokaryotic organisms where there is also the possibility of chain branching.

Although such information can be obtained using techniques that are currently available, it is an extremely laborious process and requires a significant amount of sample. An elegant example of this is the recent study by Shinzawa-Itoh *et al.* who were able to characterize the interaction of 13 integral lipids with bovine cytochrome c oxidase [6]. In this work, the X-ray crystal structures of the 13 lipids bound to bovine cytochrome c oxidase were complemented by full chemical identification of each lipid species. As summarized in Figure 6.4, six stages of lipid analysis were required to achieve this level of structural characterization:

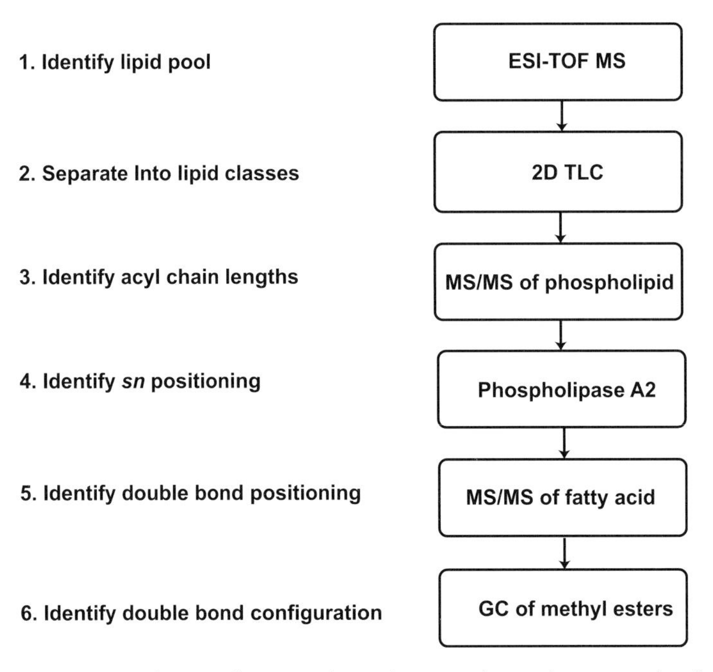

Figure 6.4 A schematic diagram outlining the comprehensive bottom-up identification of lipid molecular structure as performed by Shinzawa-Itoh *et al.* [5].

1) ESI-TOF MS analysis identified possible molecular phospholipid ions in both negative and positive ion modes.
2) Two-dimensional thin-layer chromatography (2D-TLC) was utilized to separate the polar lipids into lipid classes.
3) CID of phospholipids from the relevant fractions from 2D-TLC identified fatty acid chain length and degree of unsaturation in each lipid.
4) Treatment of isolated lipids with phospholipase A_2 released the *sn*-2 acyl chain, with subsequent MS analysis to assign *sn* position on the lipid backbone.
5) High-energy CID of fatty acids cleaved by phospholipase A_2 determined the double bond position(s).
6) Finally, gas chromatography analysis of fatty acid methyl esters determined the stereochemistry of double bonds in the fatty acyl chains.

This study is an outstanding example of tenacious and careful "bottom-up" lipid identification and demonstrates what is currently required to obtain (or to claim) complete structural characterization. This workflow is obviously time-consuming, requires abundant sample (due to limited sensitivity of the combined techniques), and is only possible when a small number of lipids are targeted. As a consequence, such an approach is incompatible with high-throughput lipidomic analyses.

While conventional CID is a powerful tool for the structural characterization of lipids, its limitations need to be more broadly recognized. Accordingly, there is an obvious need for complementary techniques that will allow full structural characterization without the need for large sample volumes and laborious experimentation. In the following sections, we will discuss some of the techniques currently being explored to resolve these limitations.

6.4 *sn* Position

It has long been accepted that the relative abundances of fragment ions produced from the CID of phospholipids can be used for structure elucidation, including *sn* position [7]. For example, the greater abundance of ions with a m/z of 419 compared to those of m/z 417 in Figure 6.2b is consistent with the location of the 18:0 and 18:1 fatty acids substituted on the *sn*-1 and *sn*-2 positions, respectively [8]. These assignments are based on studies of synthetic phospholipids, but such conclusions can be complicated by the difficulties in obtaining regiopure standards and the high likelihood of isomeric mixtures in naturally derived lipid extracts. Moreover, the ratio of the ion abundances is influenced by instrument configuration [9]. Ekroos *et al.* have undertaken the only systematic study on the application of CID for the relative quantification of phospholipid *sn* positional isomers in biological samples [10]. In this study, the authors observed a linear correlation between the relative abundance of the product ions arising from ketene elimination of acyl chains from $[M\text{-}CH_3]^-$ ions formed by ESI of phosphatidylcholines and the products of phospholipase A_2 hydrolysis. This calibration was then used to identify the presence of several regioisomeric PCs in human erythrocytes. This study demonstrates that, without careful calibration, the abundances of fragment ions in CID spectra provide a guide to the major *sn* positional isomer present, but do not exclude the presence of regioisomers in the extract. It also demonstrates that nonconventional regioisomers can be present up to at least 30% in nature and 20% in synthetic standards.

A similar approach has been recently applied to the quantification of TAG *sn* positional isomers in various oils using either $[M + NH_4]^+$ or $[M + Na]^+$ ions produced by ESI or $[M - H]^-$ ions produced by atmospheric pressure chemical ionization (APCI). In these experiments, calibration curves were produced for numerous TAGs using the ratio of ions produced from the neutral loss of *sn*-1/3 versus *sn*-2 during CID. The calibration curves where then used to quantify regioisomers in several edible oils, again uncovering an extensive level of *sn* positional isomerism [11–13]. It should be pointed out that three to nine TAG *sn* positional isomers were used to produce calibration curves in these studies and that their slopes were dependent on fatty acid composition. This suggests that calibration curves for every TAG present within a mixture would be required for their accurate quantification.

Alternatively, high-energy CID (>1 keV) is capable of inducing charge-remote fragmentation of sodiated triacylglycerols that produce fragment ions characteristic of either the *sn*-2 or the *sn*-1/3 acyl chain [14]. In these fragmentation pathways, it has been proposed that the *sn*-2 fatty acid is eliminated as either a sodiated vinyl ester (Scheme 6.1a) or as a neutral fragment. If lost as a neutral, secondary fragmentation of the remaining dehydrated diacylglycerol produces several fragments characteristic of either the *sn*-1 or the *sn*-3 fatty acid (Scheme 6.1b). In this case, regioisomer identification is based on the presence of diagnostic ions rather than on the ion abundance ratios. While original experiments were performed on ions produced by fast atom bombardment in a sector instrument, similar results have also been obtained in recent years using MALDI produced ions in a dual time-of-flight mass spectrometer (TOF/TOF) [15, 16]. A key limitation of TOF/TOF instruments is that most have a minimum ion selection window of approximately 4 Th. The inability to mass select ions at unit resolution is very important for lipid analysis as many species are separated by only 2 Th. The recent development of a MALDI-TOF/TOF that is able to achieve such resolution during ion selection shows considerable promise for lipid structural analysis. Kubo *et al.* were able to identify TAG *sn* positional isomers in olive oil and margarine using this contemporary MALDI-TOF/TOF instrument [16]; however, its ability to quantify or provide relative ratios of *sn* positional isomers is yet to be tested.

Another approach that has been described recently is to combine CID with ion molecule reactions to selective cleave *sn* positions. This approach, combining CID with ozone-induced dissociation (OzID) (described later) was recently used to structurally characterize TAGs extracted from human very low-density lipoprotein [17]. This process is shown for TAG(50:1) in Figure 6.5 and described in Scheme 6.2. In Figure 6.5b, TAG(50:1) containing two 16:0 and one 18:1 fatty acids (m/z 855 in Figure 6.5a) has been mass selected in an ion trap mass spectrometer and subjected to CID. This results in the loss of the 16:0 fatty acid, producing ions

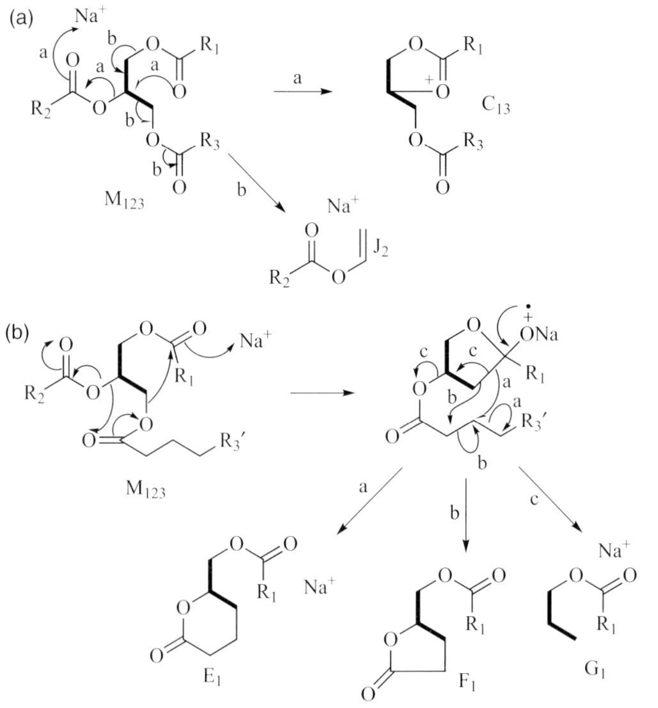

Scheme 6.1 Formation of fragment ions from high-energy CID of sodiated triacylglycerols. The pathways shown produce structurally characteristic ions for (a) *sn*-2 or (b) *sn*-1/3 fatty acids. Reproduced with permission from Ref. [14].

at *m*/*z* 599 (Scheme 6.2a). As shown in Figure 6.5c, subsequent isolation and reaction of the *m*/*z* 599 ions with ozone produces two pairs of ions, *m*/*z* 405 and 421 from the loss of the remaining 16:0 chain, and *m*/*z* 379 and 395 from the loss of the 18:1 chain. When the 18:1 fatty acid is at the *sn*-2 position, the *m*/*z* 379/395 ion pair will be produced exclusively; when it is at the *sn*-1/3 position, both pairs of ions are produced (Scheme 6.2b) (note only the structure of the *m*/*z* 405 and 379 ions have been shown for simplicity). As with high-energy CID, the ions produced are exclusively for *sn* position. Nevertheless, calibration curves for each TAG are still required if quantification is to be performed.

6.5
Double Bond Position

Determination of double bond position in simple lipids has long been the domain of gas chromatography and other specialized chromatographic techniques, such as silver ion chromatography [18]. It should be pointed out that these techniques use

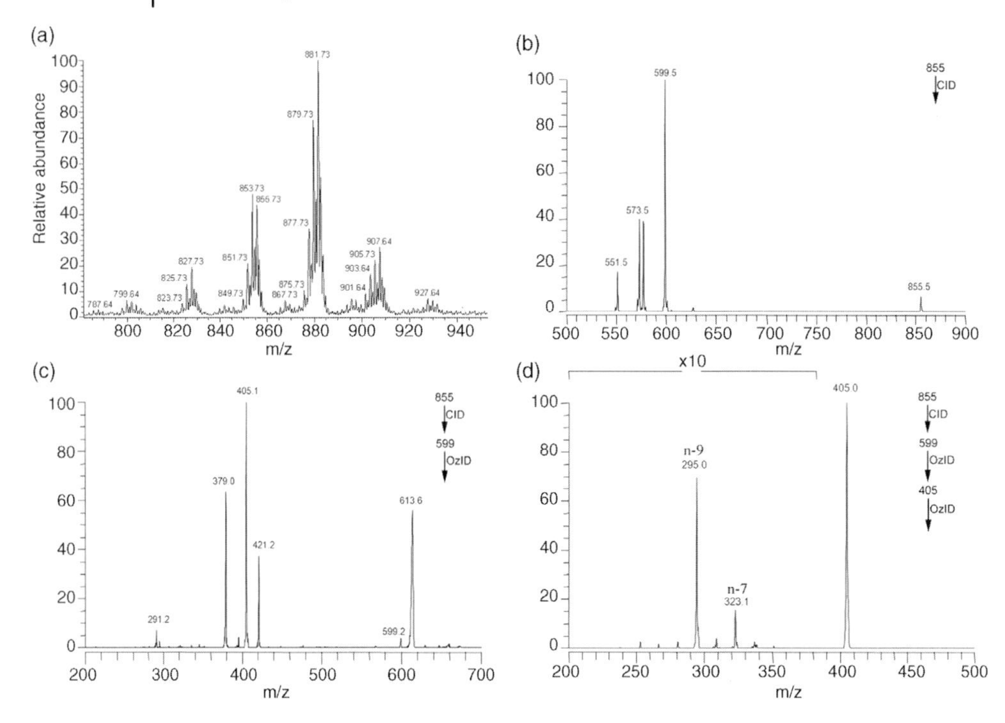

Figure 6.5 (a) An ESI-MS spectrum of triacylglycerols isolated from human very low-density lipoprotein (VLDL). (b) The CID spectrum of sodiated TAG(50:1) seen at *m/z* 855.7 in the MS spectrum. (c) The CID/OzID spectrum of the $[M + Na - 16{:}0]^+$ ion (855 → 599). (d) The CID/OzID/OzID spectrum of $[M + Na - 16{:}0\text{–}194]^+$ ion (855 → 599 → 405). The proposed structures of some of the key ions are shown in Scheme 6.2.

comparison of retention times or migration distance with that of an authentic standard and therefore *de novo* identification of double bond position is not possible. It is far more challenging to determine double bond position in more complex lipids, particularly those within a complex biological extract. In this section, we will focus solely on those techniques compatible with "top-down" mass spectrometry-based lipidomic analysis. A more comprehensive overview of the various techniques used for double bond determination in lipids can be found in Mitchell *et al.* [19].

The simplest way to obtain information regarding the position of double bonds within acyl chains is to induce fragmentation within the chain. This can be achieved in two ways; impart enough energy into the chain so that all carbon–carbon bonds fragment (untargeted) or specifically cleave the chain at the double bond (targeted fragmentation).

6.5.1
Untargeted Fragmentation

Several mass spectrometry-based methods can be used to obtain untargeted backbone fragmentation. The first is the use of high-energy CID. This technique was

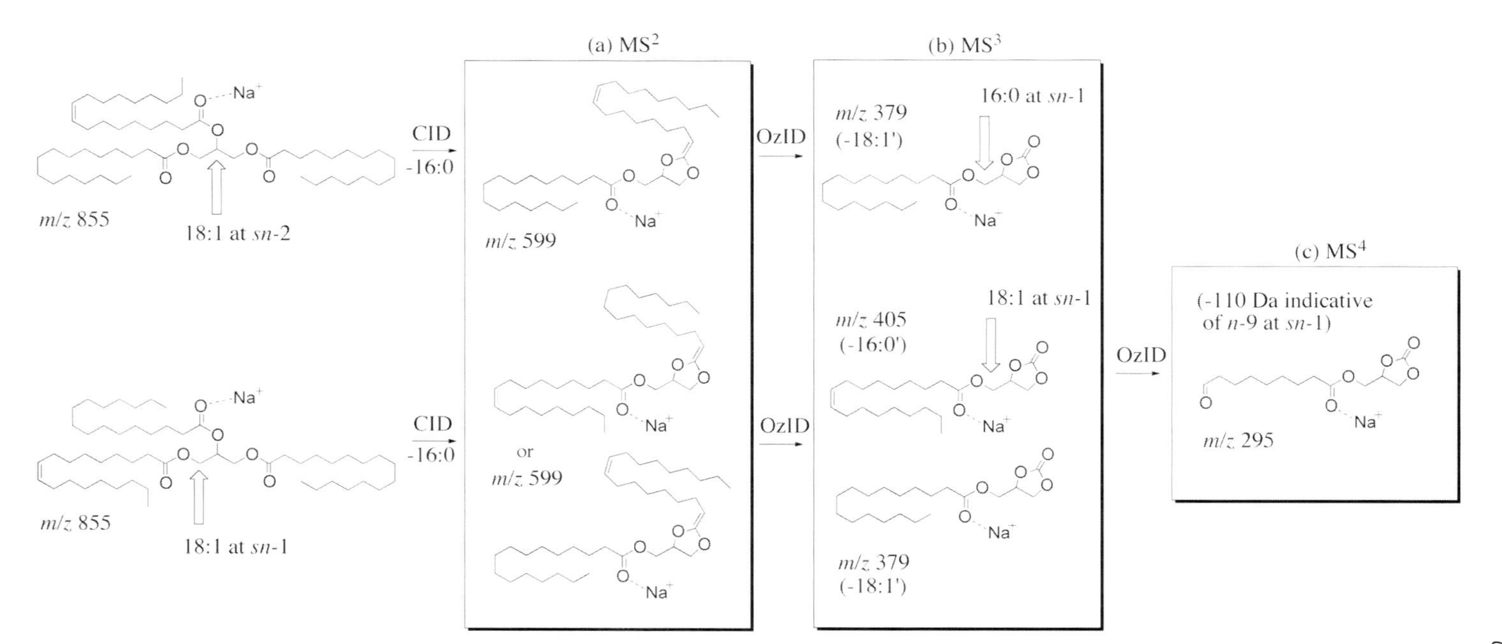

Scheme 6.2 A reaction scheme showing (a) the formation of product ions at *m/z* 599 from collision-induced dissociation (CID) of sodium adduct ions of the *sn* positional isomers TAG(16:0/18:1/16:0) (top) and TAG(18:1/16:0/16:0) (bottom); (b) the formation of diagnostic, *sn* position-dependent product ions following subsequent isolation of the isomeric *m/z* 599 ions and reaction with ozone via a combination of CID/OzID experiments; and (c) further isolation and reaction of the *m/z* 405 ion – unique to the TAG(18:1/16:0/16:0) positional isomer – with ozone (CID/OzID/OzID) yields products identifying the double bond position exclusively on the *sn*-1 fatty acyl moiety.

pioneered by Gross and coworkers to determine double bond position in fatty acid carboxylate ions [20]; however, it has also been applied to the analysis of phospholipids [21, 22] and triacylglycerols [14]. Although high-energy CID of phospholipid anions can produce fragment ions indicative of double bond position, these spectra are complex and the exact origin of the product ions, that is, whether they arise from the *sn*-1 or *sn*-2 radyl, or both, is difficult to ascertain [21]. A similar dilemma is encountered when attempting to characterize triacylglycerols. The position of unsaturation in very simple molecules may be determined using high-energy CID [14, 23]; however, the complexity of the spectra ensures that interpretation, particularly from highly unsaturated triacylglycerols and/or isomeric triacylglycerols, is extremely difficult. It should also be noted that these collision energy regimes are beyond the reach of most contemporary mass spectrometers used in lipid analysis (i.e., triple quadrupoles, ion traps, and quadrupole time-of-flight instruments generally operate in the low-energy <100 eV regime). In a promising recent development, a new generation of commercial MALDI-TOF/TOF mass spectrometers operate at collision energies above 1 keV that are capable of determining double bond position of fatty acids from $[M - H + Li_2]^+$ ions [24]. In these experiments, there was a clear reduction in the abundance of ions produced from chain fragmentation at the double bond site as demonstrated for linoleic acid, α-linolenic acid, and docosahexaenoic acid in Figure 6.6b–d, respectively.

Unfortunately, double bond position in both triacylglycerols [15] and phospholipids [25] could only be determined in favorable cases. As mentioned above, this limitation arises from the wide ion selection window of most TOF/TOF instruments. Recent work indicates that this can be overcome with the next-generation TOF/TOF instruments [26]; however, key diagnostic ions are often of low abundance and the ability of high-energy CID to identify double bond positional isomers from a complex biological mixture remains to be demonstrated.

A second approach to untargeted fragmentation is to perform multistage mass spectrometry in the low-energy regime. Hsu and Turk have recently used this technique for the determination of double bond positions in ESI-produced lithiated ions of triacylglycerols and phospholipids in a linear ion trap mass spectrometer [27, 28]. CID of $[M + Li]^+$ TAG ions results in the loss of one fatty acid (Scheme 6.2a), a second stage of CID (MS^3) produces a low-abundance series of ions from allylic, vinylic, or β-cleavage of the fatty acid chains indicating double bond position. This technique is also applicable to most phospholipid classes when analyzed as $[M + Li]^+$, $[M - H + Li_2]^+$, and even $[M - H_2 + Li_3]^+$ ions. When monolithiated ions of phospholipids are analyzed using two or three rounds of subsequent ion selection and fragmentation (MS^3 and MS^4, respectively), a dehydrated diacyl glycerol ion is produced. This ion can be further fragmented in an MS^4 or MS^5 experiment to produce the typical allylic and vinylic series of fatty acid fragment ions, allowing identification of double bonds within the remaining acyl chains (Scheme 6.3) [27]. Nevertheless, the determination of double bond position could become very difficult if two unsaturated fatty acids, particularly polyunsaturated fatty acids, are present on the dehydrated diacylglycerol ion. The identification of double bond position in di- or trilithiated glycerophospholipids resolves

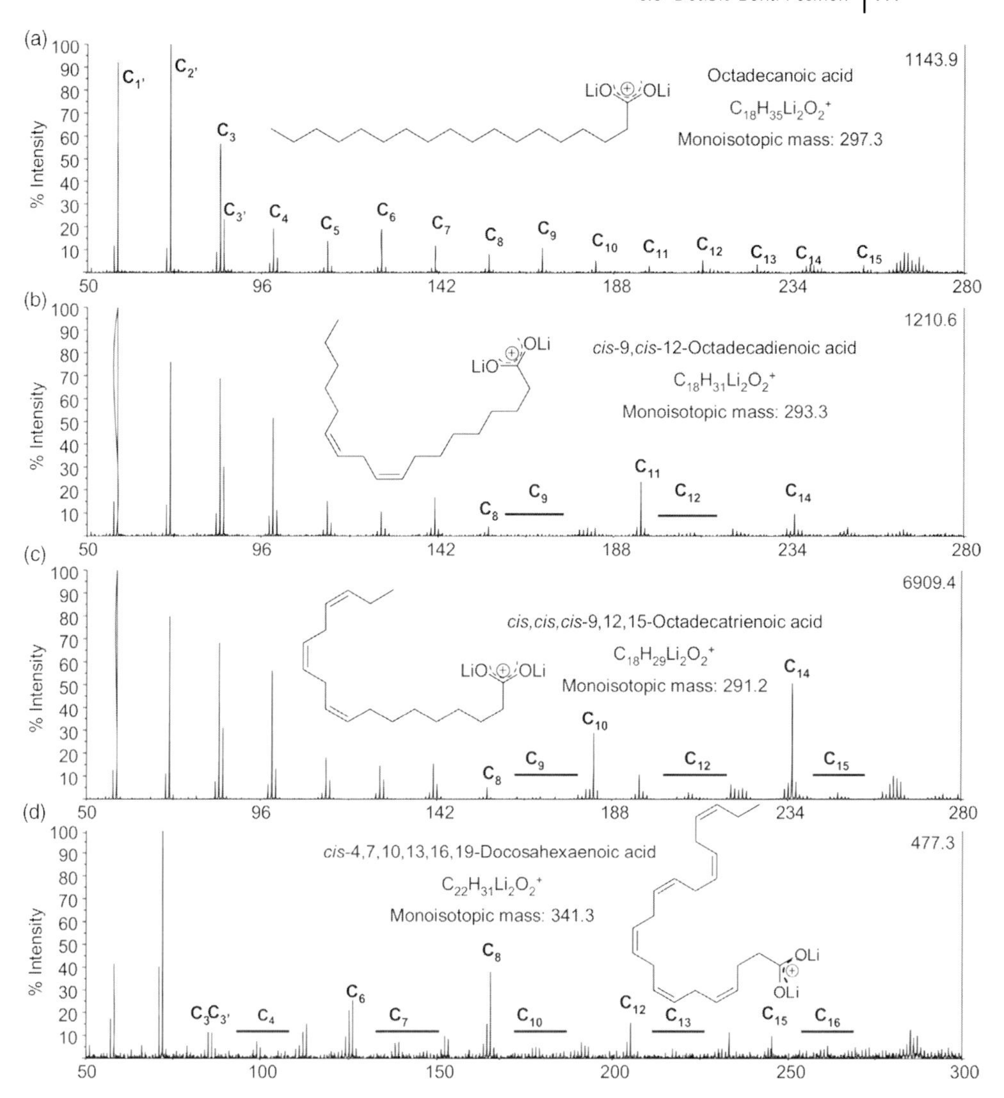

Figure 6.6 6.6High-energy CID spectra of $[M - H + 2Li]^+$ ions of (a) stearic acid (18:0), (b) *cis,cis*-9,12-octadecadienoic acid (9*Z*,12*Z*-18:2), (c) *cis,cis,cis*-9,12,15-octadecatrienoic acid (9*Z*,12*Z*,15*Z*-18:3), and (d) all *cis*-4,7,10,13,16,19-docosahexaenoic acid (4*Z*,7*Z*,10*Z*,13*Z*,16*Z*,19*Z*-22:6). The peaks labeled C_3–C_{15} are charge-remote fragment ions starting at the carboxy terminus (ions labeled C1′–C3′ are radical cleavage products). Adapted from Ref. [24] with permission.

this problem as MS^3 or MS^4 experiments performed on these ions produce dilithiated cations of each fatty acid or dehydrated lysophospholipid. MS^4 or MS^5 of these fragment ions produces a CID spectrum characteristic of double bond position(s) in the remaining fatty acid (Scheme 6.3).

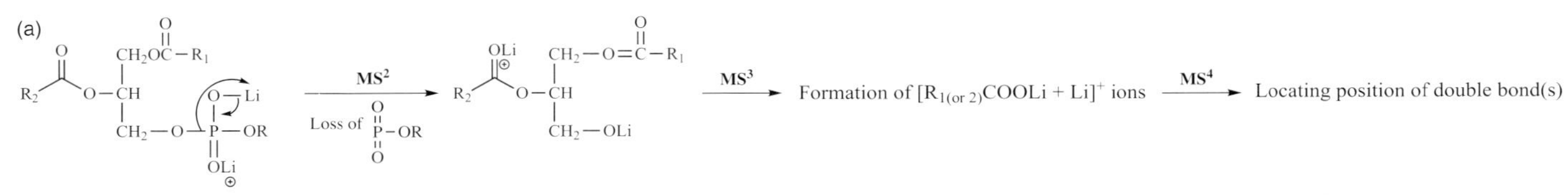

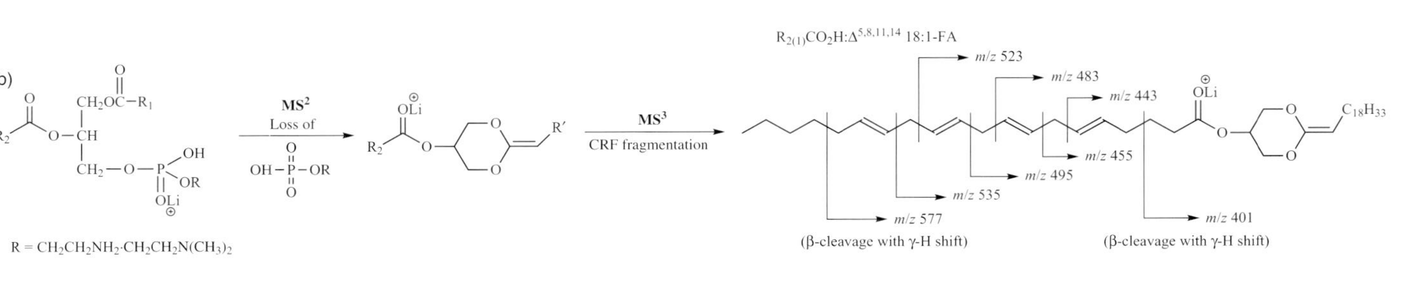

Scheme 6.3 Proposed pathways in the genesis of the major fragment ions observed in the multistage CID of lithiated adducts of glycerophospholipids. (a) Dilithiated adduct ions: PE, PI, and PG (including plasmalogens). (b) Monolithiated adduct ions: applicable to all the GPL classes that form $(M + Li)^*$ ions. (c) Dilithiated and *trilithiated adduct ions: PS, PA, and lyso PPL (*same fragment ions were produced by MS^2 on trilithiated ions). Reproduced with permission from Ref. [27].

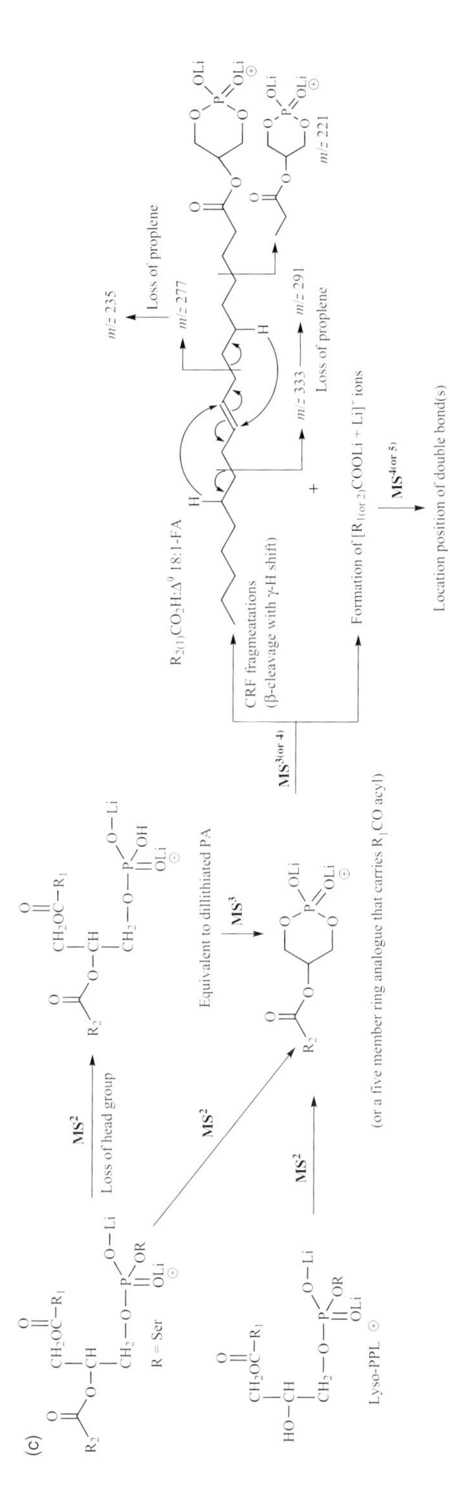

Scheme 6.3 *(Continued)*

In summary, high-energy CID is capable of determining the position of double bonds in mass-selected, regiopure fatty acids, but the excessive fragmentation makes interpretation difficult. The use of multistage MS for the determination of double bond position in intact lipids has an advantage over high-energy CID in that it is compatibile with ESI instrumentation (the method of choice for modern lipidomic applications) and allows the determination of double bond position in complex lipids and its relationship to *sn* position. Nevertheless, the loss of ions during each round of isolation and fragmentation means that only high-abundance ions can be effectively characterized. Notwithstanding the fact, the low-energy CID is dominated by C—O bond cleavage, not by the desired C—C bond cleavage. While these techniques are relatively simple to perform and can be implemented on commercial instrumentation, the fact that their fragmentation is nonspecific means that the spectra used to determine double bond position are complex with the specific ions of interest generally of very low abundance. This means that interpretation of spectra from double bond positional isomers would be extremely difficult to perform when a complex mixture is present. Recently, CID has been combined with traveling-wave ion mobility mass spectrometry to identify double bond position in plasma phosphatidylcholine after HPLC separation [29]. Although combining these techniques does provide some improvement in sensitivity compared to multistage CID, it is still untargeted fragmentation, meaning that interpretation of spectra is likely to be difficult when isomers of polyunsaturated fatty acids are analyzed. Furthermore, it needs to be clarified that to this point isomeric lipids cannot be discriminated by mobility alone.

An alternative to CID, ultraviolet photodissociation has also been explored for the structural characterization of biomolecules [30, 31]. A recent example of its application to lipid analysis is that of Reilly and coworkers who utilized an excimer laser (157 nm) to fragment $[M + H]^+$, $[M + Na]^+$, and $[M\text{-}H]^-$ ions of isomeric leukotriene lipids in an ion trap mass spectrometer [32]. The product ions observed for these compounds upon photoexcitation were a combination of those obtained in low- and high-energy CID experiments, allowing the differentiation of the isomeric forms of the lipid. Unfortunately, lipids have relatively poor absorption cross sections at visible and ultraviolet wavelengths and although there may be niche applications for lipids with particular structural motifs, for example, conjugated double bonds, widespread use of direct photodissociation in lipidomics is unlikely.

The major limitation of untargeted fragmentation is that the available charge is distributed over numerous product ion channels. This means that diagnostic ions are often of low abundance and therefore it is unlikely to possess the required sensitivity to detect double bond positional isomers of low abundance that are often found in biological samples. Moreover, with the increase in the number of isomers present, the number of ions also increases, further reducing sensitivity and increasing spectral complexity. Indeed, the utility of untargeted methods is yet to be demonstrated for isometric mixtures of the type found in nature. To overcome these limitations, several researchers have been developing techniques capable of cleavage at the specific site of the double bond/s, that is, targeted fragmentation.

6.5.2
Targeted Fragmentation

Targeted approaches generally require some form of chemical reaction with the double bond resulting in its direct cleavage or in an increased susceptibility to fragmentation under energetic conditions. These reactions can be performed in the laboratory prior to MS analysis or during the MS process.

A common approach for the identification of double bond position in fatty acids by mass spectrometry has been off-line derivatization followed by CID of the derivatized ions. Some promising recent applications to molecular phospholipids include the use of osmium tetroxide or ozone as derivatization reagents. Treating phospholipids and free fatty acids with osmium tetroxide produces phospholipids with acyl chains containing hydroxyl groups on both sides of the initial site(s) of unsaturation [33, 34]. Subsequent ESI-MS/MS, whereby the dihydroxylated lipid ions are isolated and subjected to CID, produces characteristic fragment ions identifying the position of the initial double bond [35]. Exposing a thin film of phospholipids to ozone results in an almost quantitative conversion of olefinic bonds to ozonides. Subsequent analysis of these ozonides by ESI-MS/MS, utilizing CID in either positive or negative ion mode, leads to dissociation of the ozonide moiety producing fragment ions uniquely identifying the double bond position [36]. More recent data demonstrate that ESI-MS of products derived from cross-metathesis of methyl acrylate with olefins (catalyzed with second-generation Grubbs catalyst) can be used to identify double bond position [37]. Although ESI-MS or MS/MS of derivatized lipids can be used for locating double bond position, it has the undesirable requirement of additional sample preparation prior to analysis. This has led to several groups exploring online methods capable of uniquely identifying double bond position.

Van Pelt and Brenna [43] have demonstrated an online chemical ionization method for the elucidation of double bond position [38–44]. In this method, the presence of acetonitrile during ionization results in the modification of double bonds producing an $[M + 54]^+$ ion (Scheme 6.4) [45]. Collision-induced dissociation of these modified ions yields products that are indicative of the position of the double bond(s) within the molecule (e.g., m/z 252 in Scheme 6.4). Moreover, these ions can be predicted *de novo* [40], a major advantage in any natural product investigation. The use of APCI allows the determination of double bond position in more complex lipids such as triacylglycerols [44] and wax esters [46] using acetonitrile. An APCI source also improves compatibility with modern lipidomic methods as it allows liquid-phase sample introduction of intact lipids, that is, no prior lipid saponification is required. Moreover, it allows coupling to high-performance liquid chromatography (HPLC) to simplify the sample matrix prior to analysis [46].

Another approach is the reaction of lipids with ozone in the source of the mass spectrometer [47–49]. Several methods for the production of ozone have been described. The first involves using oxygen as the nebulizing gas during electrospray ionization and increasing the voltage applied to the spray capillary until a corona discharge is induced, a technique termed ozone electrospray ionization (OzESI) [47]. The advantage of this method is that no specialized equipment is required. Nevertheless, this technique is limited to the analysis of anionic lipids as production of ozone from a positive ion capillary discharge is an order of magnitude lower than that produced from a negative discharge [50]. In a similar approach, Cooks and coworkers have utilized the ozone produced during low-temperature plasma (LTP) ionization to determine the double bond position in fatty acids and fatty acid methyl esters, as shown in Figure 6.7 [49]. This technique has

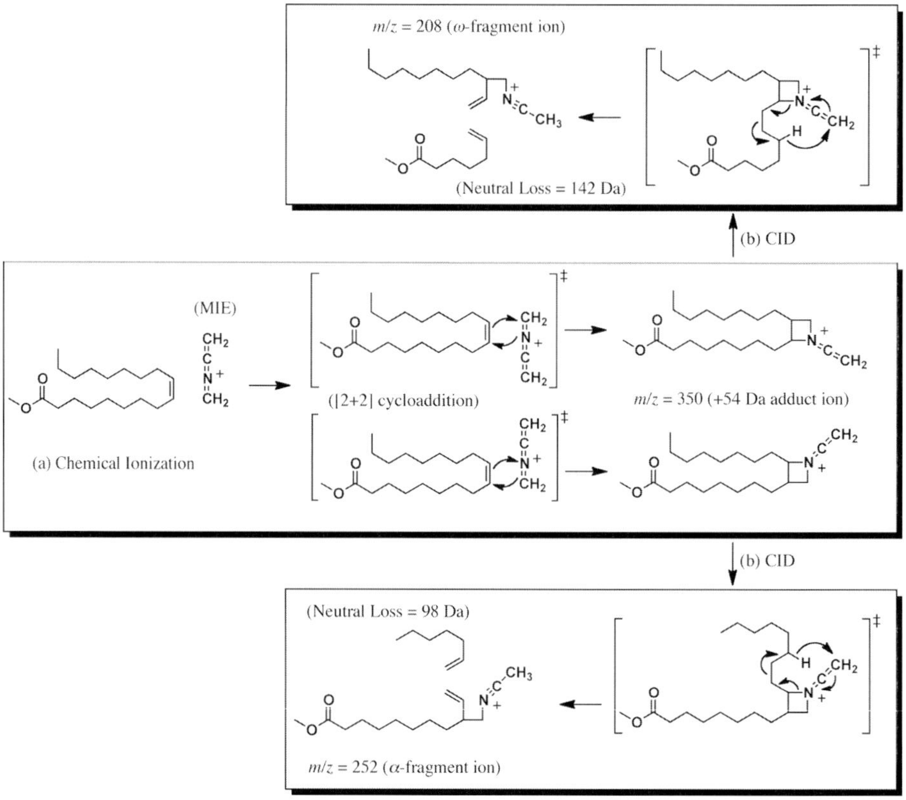

Scheme 6.4 The proposed mechanism to rationalize (a) the in-source formation of an $[M + 54]^+$ adduct ion via [2 + 2] cycloaddition of FAME of 18:1 (*n*-9) with the (1-methyleneimino)-1-ethenylium (MIE) reagent ion produced from chemical ionization in the presence of acetonitrile, and (b) the dissociation of each of the isomeric $[M + 54]^+$ adduct ions upon collision-induced dissociation (CID) to yield prominent α- and ω-product ions that are diagnostic of the position of the double bond in the FAME.

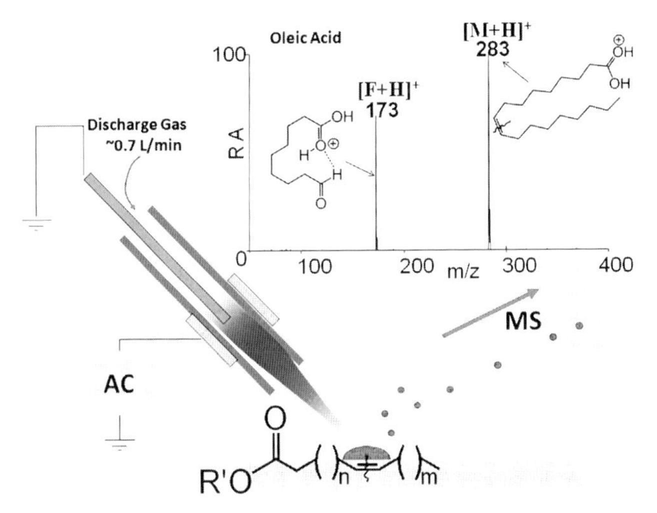

Figure 6.7 Ozonolysis of fatty acids during low-temperature plasma ionization. Reproduced from Ref. [49] with permission.

the distinct advantage over OzESI (when ozone is produced via corona discharge) in that positive ions can also be analyzed.

A third technique is to supply ozone to the ESI source directly from an ozone generator [48]. This also overcomes the problem of analyzing lipids in positive ion mode. In addition, this method produces a higher ozone concentration allowing a more efficient cleavage of double bonds. The positive ion OzESI-MS spectra of the regioisomeric *n*-9 and *n*-12 glycerophospholipids, PC(9*Z*-18:1/9*Z*-18:1) and PC(6*Z*-18:1/6*Z*-18:1), are shown in Figure 6.8a and b, respectively. These isomers produce distinctive OzESI spectra with chemically induced fragment ions showing neutral losses of 62 Da (*m/z* 724) and 110 Da (*m/z* 676) identifying the *n*-9 double bonds, while the 104 Da (*m/z* 682) and 152 Da (*m/z* 634) losses are characteristic of an *n*-12 double bond. The proposed mechanisms involved in OzESI are shown in Scheme 6.5 using the $[M + H]^+$ ion of PC(16:0/9*Z*-18:1) as an example.

More recently, the reaction of ambient ozone with lipids dried on a surface has been exploited for the identification of double bond position. Ellis *et al.* [51] spotted lipids onto a PFTE-coated microscope slide and were able to detect lipid ozonides and their fragment ions, thus identifying double bond position directly from the slide using desorption electrospray ionization (DESI)-MS. Moreover, this ambient ozonolysis was performed using TLC plates allowing the separation of complex lipid mixtures prior to ozonolysis and DESI analysis.

Unlike the untargeted methods described above, fragmentation via ozonolysis is not excessive with two product ions present resulting from the cleavage of each double bond. This simplifies the spectra significantly and also allows the systematic prediction of fragments even without conducting the experiment on an authentic standard.

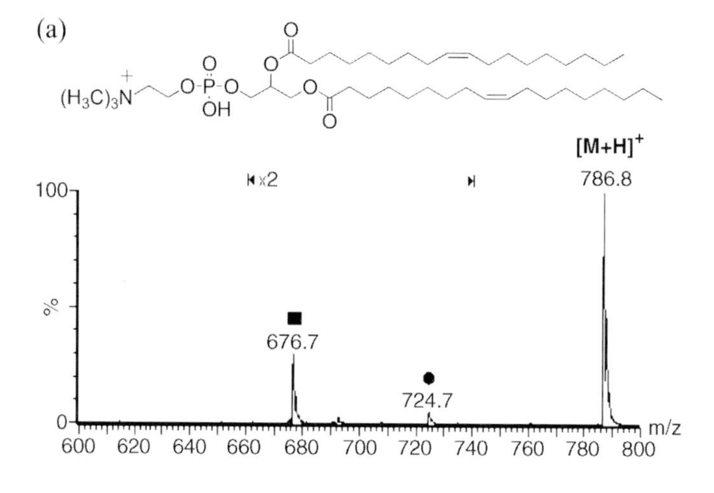

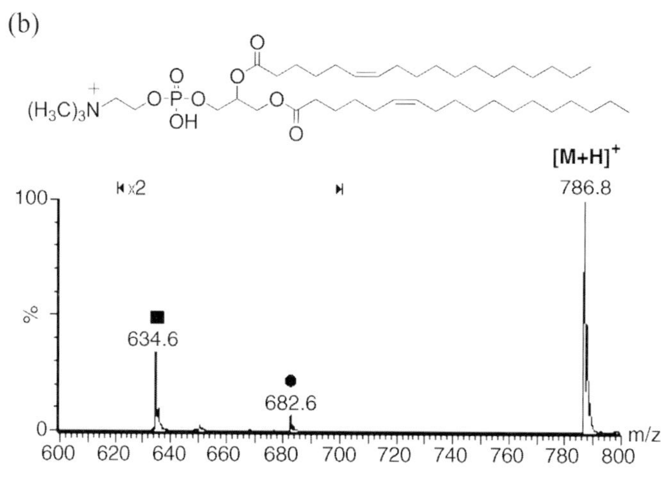

Figure 6.8 The positive ion OzESI spectra for (a) PC(9*Z*-18:1/9*Z*-18:1) and (b) PC(6*Z*-18:1/6*Z*-18:1). The aldehyde and α-methoxyhydroperoxide ozonolysis product ions are labeled with the ■ and • symbols, respectively. Reproduced with permission from Ref. [48].

In-source ozonolysis (or ambient ozonolysis) technologies are simple to use and highly effective for individual lipids or simple lipid mixtures; however, interpretation of source ozonolysis spectra of complex mixtures is difficult. This situation arises because ozone-induced fragments are often isobaric with other lipid ions and the assignment of fragments to their respective precursor ions becomes ambiguous as the complexity of the mixture increases. To overcome this limitation, a new method has been developed whereby mass-selected lipid ions are exposed to ozone within an ion trap mass spectrometer [52, 53]. In this approach, the instrument buffer gas (helium or nitrogen) is seeded with a small amount of an ozone/oxygen mixture (~12% O_3 in O_2) produced by a commercial high-concentration

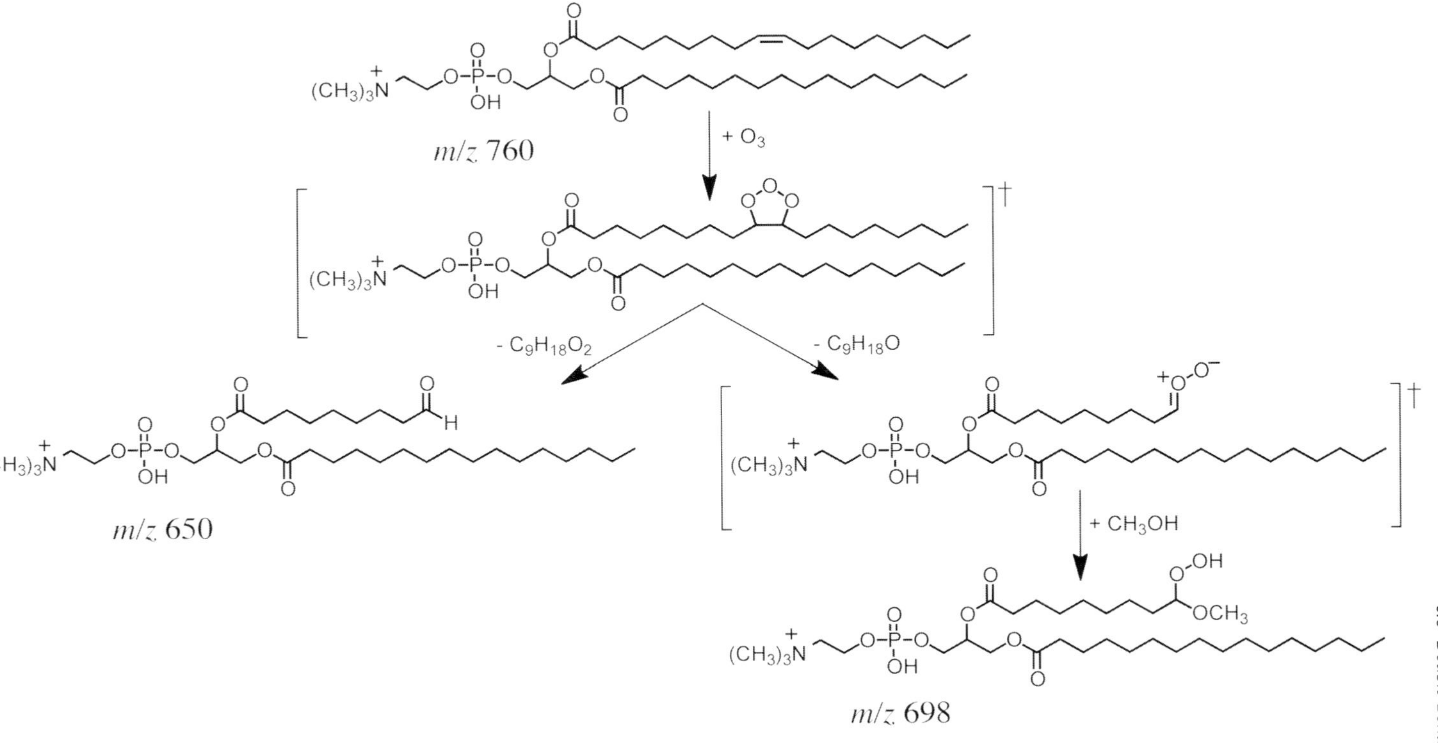

Scheme 6.5 Proposed reactions leading to formation of aldehyde (m/z 650) and α-methoxyhydroperoxide (m/z 698) ions during ozone electrospray ionization mass spectrometry (OzESI-MS) of the phosphatidylcholine, PC(16:0/18:1(n-9)), when methanol is used as the electrospray solvent. The presence of these ions in the OzESI mass spectrum is diagnostic of the presence of the double bond at the n-9 position in the intact phospholipid.

ozone generator. The distinct advantage of this approach is that each lipid of interest can be isolated by mass selection within the trap before ozonolysis, allowing the analysis of complex mixtures. As with the OzESI-MS experiments, two fragment ions are produced from the cleavage of each double bond. In the absence of methanol however, conversion of the proposed Criegee intermediate to an α-methoxyhydroperoxide cannot take place (Scheme 6.5). Instead, it has been proposed that the Criegee intermediate rearranges to form a more stable carboxylic acid or vinyl hydroperoxide (Scheme 6.6). Knowledge of this mechanism allows *de novo* identification of double bond position based on the neutral losses listed in Table 6.1.

Another advantage of this technique is the ability to perform multiple stages of MS, utilizing different combinations of CID and OzID. For example, performing CID/OzID allows the targeting of each fatty acid bound to a parent phospholipid. This is achieved by first performing CID to remove one of the fatty acids and then performing OzID on the remaining lipid ion containing only one fatty acid [54]. Further stages of MS, that is, CID/OzID/OzID may also be employed for the identification of each fatty acid esterified to a triacylglycerol. An example of this approach is the characterization of an abundant triacylglycerol found in human very low-density lipoprotein discussed earlier (Figure 6.5, Scheme 6.2). Once *sn* position has been determined using CID/OzID, a second stage of OzID (Scheme 6.2c) can then be performed to identify the double bond position in the final remaining fatty acid. In the example in Figure 6.5d, both 18:1 *n*-9 and *n*-7 double bond isomers were observed at the *sn*-1/3 position of TAG(50:1).

The limitation of the OzID technique is the requirement of specialized equipment and instrument modification. Nevertheless, this technique has been demonstrated to be capable of identifying lipid double bond positions in complex biological mixtures, providing easy-to-interpret data, and identifying double bond isomers even if complex polyunsaturated fatty acids are present. Thus, it displays particular promise for application in complex online lipidomic analyses, where, when used in conjunction with established CID and MS^3 protocols, it may provide near-complete structure elucidation of complex lipids directly from electrospray analysis of lipid extracts. For further discussion on the use of ozone in lipid structural analysis, see Ref. [55].

In what could be described as a semitargeted approach, Yang *et al.* have demonstrated the ability to identify and quantify double bond isomers using the fragment ions produced from CO_2 or H_2O loss during CID of unsaturated fatty acids [56]. The authors found that the abundance of these ions was influenced by the number of double bonds, position of the first double bond (relative to the carbonyl end), and chain length. As this identification is based solely on ion abundance, a standard curve of fragment ion intensity versus isomeric concentration is required for all isomers of interest. Similar fragment ions were also produced from the secondary fragmentation of phospholipids, allowing its application to complex molecular lipids [56]. Nevertheless, this would further expand the number of calibration curves required, and at present the availability of isomeric lipid standards required to construct such standard curves is limited. It should also be considered when analyzing phospholipids that the abundance of the carboxylate ion and hence the ion

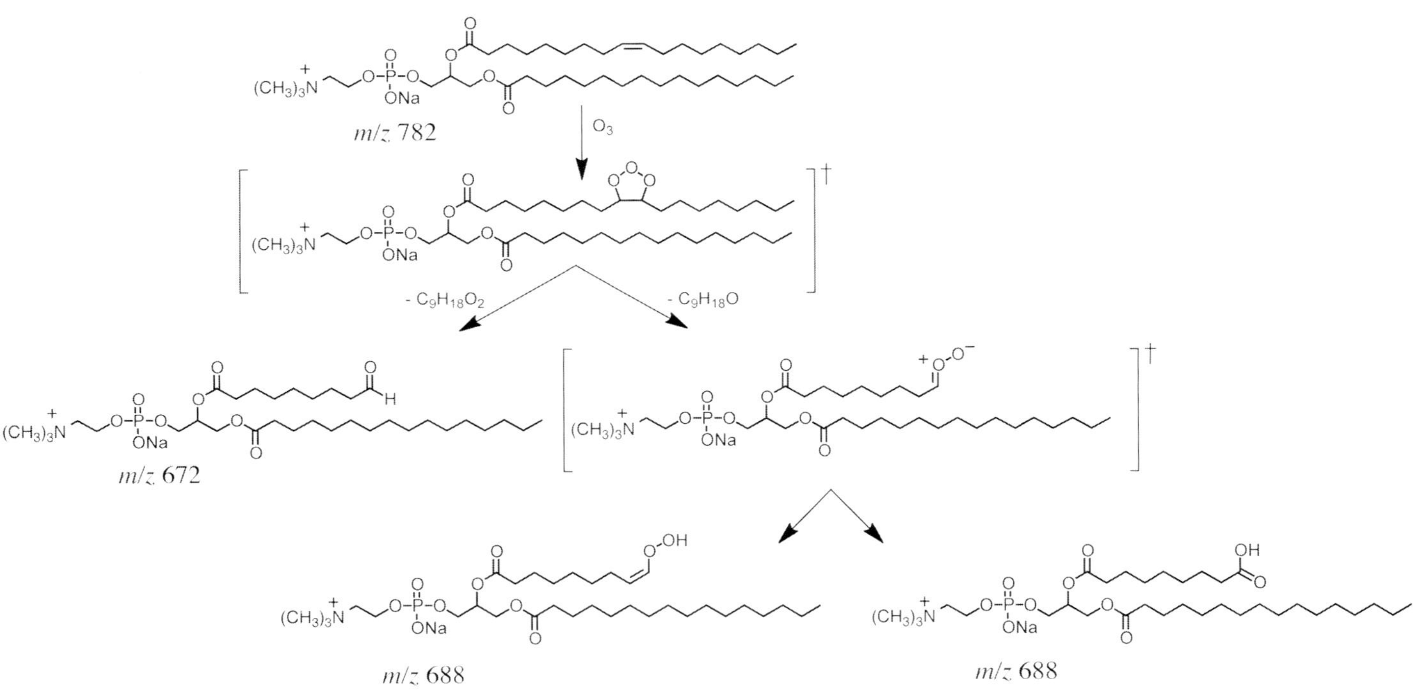

Scheme 6.6 Proposed reactions leading to formation of aldehyde (*m/z* 650) and "Criegee" ions (*m/z* 688) during ozone-induced dissociation (OzID) of mass-selected phosphatidylcholine cations, $[PC(16:0/18:1(n\text{-}9)) + H]^+$, in an ion trap mass spectrometer. The presence of these ions in the OzID mass spectrum is diagnostic of the presence of the double bond at the *n*-9 position in the intact phospholipid.

Table 6.1 Predicted neutral losses (or gains) for OzID product ions showing the dependence on position and degree of unsaturation.

Monounsaturated			Polyunsaturated			
DB position	Neutral loss		Class	DB position	Neutral loss	
n-	Aldehyde	Criegee		*n*-	Aldehyde	Criegee
1	−2	−18	*n*-3	3	26	10
2	12	−4		6	66	50
3	26	10		9	106	90
4	40	24		12	146	130
5	54	38		15	186	170
6	68	52		18	226	210
7	82	66	*n*-6	6	68	52
8	96	80		9	108	92
9	110	94		12	148	132
10	124	108		15	188	172
11	138	122		18	228	212
12	152	136	*n*-9	9	110	94
13	166	150		12	150	134
14	180	164		15	190	174
15	194	178				

produced from its loss of CO_2 will be influenced by its attachment site on the glycerol backbone, that is, *sn*-1 or *sn*-2 (Figure 6.2b). This would further complicate interpretation of the spectrum if *sn* positional isomers of the unsaturated lipids are present. Quantification would be particularly difficult if one double bond isomer was more prevalent at the *sn*-1 position while the other was more prevalent at the *sn*-2 position, as previously observed [54].

The targeted approaches for identifying double bond position in molecular lipids described above have the distinct advantage over untargeted methods in that far fewer fragment ions are produced. This means that the spectra are far simpler to interpret and sensitivity is improved by the concentration of product ion abundance in diagnostic ions. Nevertheless, considerable work is still required to prove the ability of these techniques to quantify double bond positional isomers in complex samples.

6.6 Double Bond Stereochemistry

Gas chromatography has long been used in the separation of *cis*- and *trans*-double bonds in fatty acid methyl esters. Determination of double bond stereochemistry in more complex molecular lipids is, however, a much greater

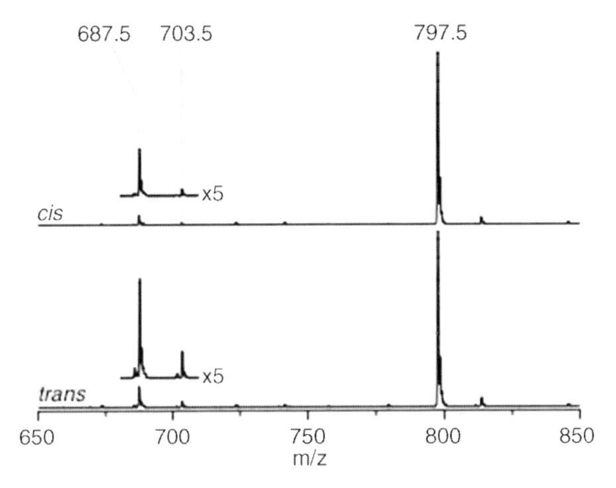

Figure 6.9 Two OzID spectra illustrating the different reactivity of the $[M + Na]^+$ adduct ions of the phosphatidylcholine stereoisomers PG(9*Z*-18:1/9*Z*-18:1) (top trace) and PG(9*E*-18:1/9*E*18:1) (bottom trace). Reproduced with permission from Ref. [53].

challenge, particularly using techniques compatible with modern lipidomic analyses. Chiral LC-MS has been used for the characterization of eicosanoid [57] and resolvin [58] enantiomers. However, the determination of double bond stereochemistry is likely to be more challenging and the utility of such an approach is yet to be demonstrated.

It has recently been suggested that the differing rate of ozone reaction between *cis*- and *trans*-double bond isomers may be able to resolve them. As shown in Figure 6.9, OzID of sodiated PG(9*E*-18:1/9*E*-18:1) ions produced ozonolysis product ions (*m/z* 687.5 and 703.5) twice as abundant as those produced from sodiated PG(9*Z*-18:1/9*Z*-18:1) [53]. Although this shows some promise, the ions produced are identical and it is only ion abundance that differs. Therefore, like many of the techniques discussed above, appropriate standard curves for mixtures of each set of isomers would need to be constructed if double bond stereoisomers were to be identified and quantified in complex mixtures.

Determining double bond stereochemistry using a "top-down" mass spectrometry approach is rather more difficult than determining double bond position and this is a likely reason for the lack of development in this area. Nevertheless, *trans*-fatty acids are known to impact human health [59, 60] and therefore the ability to rapidly detect them in biological samples would be an important breakthrough for lipidomics.

6.7 Conclusions

Modern MS analysis of lipids provides an unprecedented level of analytical power. In short, it provides excellent sensitivity, the ability to observe hundreds of individual lipid molecules directly from crude mixtures without prior fractionation, in very short analysis times, and detailed molecular structure of individual lipids (compared to traditional chromatographic techniques). Nevertheless, a concerted effort is still required in several areas if we wish to fully characterize the lipidome and its role in health and disease.

We are still unable to determine the complete molecular structure (i.e., *sn* position, double bond position, and double bond stereoisomerism) of many naturally occurring complex lipids without large amounts of sample and exhaustive fractionation. As a consequence, the possibility of greater molecular complexity is typically ignored.

As can be seen from the discussion above, although considerable effort is now being made to address these issues, they are yet to be fully resolved. For example, multistage CID/OzID is the only technique that has been demonstrated to identify and fully characterize double bond positional isomers in complex lipids from a crude biological extract, but its ability to quantify these isomers has not been proven. Also, high-energy CID on a MALDI-TOF/TOF is simple to perform and produces ions diagnostic of both *sn* position and double bond position; however, the "smearing" of charge across a large number of product ions, many of which are either not diagnostic or provide redundant information, means that this technique is unlikely to posses the sensitivity to characterize isomers in a true biological sample.

With the increasing understanding of the importance of lipid molecular structure, for example, lipid signaling and protein–lipid interactions, our requirement to rapidly characterize and quantify lipids at the molecular level also increases. Therefore, further development of the techniques mentioned above and/or new and innovative MS methods for the comprehensive structural analyses that are compatible with modern lipidomic techniques will become increasingly important.

References

1 Lemmon, M.A. (2008) Membrane recognition by phospholipid-binding domains. *Nat. Rev. Mol. Cell Biol.*, **9**, 99–111.

2 Newton, A.C. (2004) Diacylglycerol's affair with protein kinase C turns 25. *Trends Pharmacol. Sci.*, **25**, 175–177.

3 Moolenaar, W.H., van Meeteren, L.A., and Giepmans, B.N.G. (2004) The ins and outs of lysophosphatidic acid signaling. *Bioessays*, **26**, 870–881.

4 Fahy, E., Subramaniam, S., Brown, H.A., Glass, C.K., Merrill, A.H., Jr., Murphy, R. C., Raetz, C.R.H., Russell, D.W., Seyama, Y., Shaw, W., Shimizu, T., Spener, F., van Meer, G., VanNieuwenhze, M.S., White, S.H., Witztum, J.L., and Dennis, E.A. (2005) A comprehensive classification system for lipids. *J. Lipid Res.*, **46**, 839–861.

5 Schwudke, D., Schuhmann, K., Herzog, R., Bornstein, S.R., and Shevchenko, A. (2011) Shotgun lipidomics on high resolution mass spectrometers. *Cold Spring Harb. Perspect. Biol.*, **3**. doi: 10.1101/cshperspect.a004614.

6 Shinzawa-Itoh, K., Aoyama, H., Muramoto, K., Terada, H., Kurauchi, T., Tadehara, Y., Yamasaki, A., Sugimura, T., Kurono, S., Tsujimoto, K., Mizushima, T.,

Yamashita, E., Tsukihara, T., and Yoshikawa, S. (2007) Structures and physiological roles of 13 integral lipids of bovine heart cytochrome c oxidase. *EMBO J.*, **26**, 1713–1725.

7 Huang, Z.-H., Gage, D., and Sweeley, C. (1992) Characterization of diacylglycerylphosphocholine molecular species by FAB-CAD-MS/MS: a general method not sensitive to the nature of the fatty acyl groups. *J. Am. Soc. Mass Spectrom.*, **3**, 71–78.

8 Hsu, F.-F. and Turk, J. (2005) Studies on phosphatidylserine by tandem quadrupole and multiple stage quadrupole ion-trap mass spectrometry with electrospray ionization: structural characterization and the fragmentation processes. *J. Am. Soc. Mass Spectrom.*, **16**, 1510–1522.

9 Hou, W., Zhou, H., Khalil, M.B., Seebun, D., Bennett, S.A.L., and Figeys, D. (2011) Lyso-form fragment ions facilitate the determination of stereospecificity of diacyl glycerophospholipids. *Rapid. Commun. Mass Spectrom.*, **25**, 205–217.

10 Ekroos, K., Ejsing, C.S., Bahr, U., Karas, M., Simons, K., and Shevchenko, A. (2003) Charting molecular composition of phosphatidylcholines by fatty acid scanning and ion trap MS3 fragmentation. *J. Lipid Res.*, **44**, 2181–2192.

11 Leskinen, H., Suomela, J.P., and Kallio, H. (2007) Quantification of triacylglycerol regioisomers in oils and fat using different mass spectrometric and liquid chromatographic methods. *Rapid. Commun. Mass Spectrom.*, **21**, 2361–2373.

12 Herrera, L.C., Potvin, M.A., and Melanson, J.E. (2010) Quantitative analysis of positional isomers of triacylglycerols via electrospray ionization tandem mass spectrometry of sodiated adducts. *Rapid. Commun. Mass Spectrom.*, **24**, 2745–2752.

13 Leskinen, H.M., Suomela, J.-P., and Kallio, H.P. (2010) Quantification of triacylglycerol regioisomers by ultra-high-performance liquid chromatography and ammonia negative ion atmospheric pressure chemical ionization tandem mass spectrometry. *Rapid. Commun. Mass Spectrom.*, **24**, 1–5.

14 Cheng, C., Gross, M., and Pittenauer, E. (1998) Complete structural elucidation of triacylglycerols by tandem sector mass spectrometry. *Anal. Chem.*, **70**, 4417–4426.

15 Pittenauer, E. and Allmaier, G. (2009) The renaissance of high-energy CID for structural elucidation of complex lipids: MALDI-TOF/RTOF-MS of alkali cationized triacylglycerols. *J. Am. Soc. Mass Spectrom.*, **20**, 1037–1047.

16 Kubo, A., Itoh, Y., Ubukata, M., Hashimoto, M., Tamura, J., Onodera, J., and DiPasquale, R.A. (2011) Structural analysis of complex lipids using MALDI-TOF-TOF tandem MS with high precursor-ion selectivity. 59th ASMS Conference on Mass Spectrometry and Allied Topics, Denver.

17 Ståhlman, M., Pham, H., Adiels, M., Mitchell, T., Blanksby, S., Fagerberg, B., Ekroos, K., and Borén, J. (2012) Clinical dyslipidaemia is associated with changes in the lipid composition and inflammatory properties of apolipoprotein-B-containing lipoproteins from women with type 2 diabetes. *Diabetologia*, **55**, 1156–1166.

18 Nikolova-Damyanova, B. and Momchilova, S. (2002) Silver ion HPLC for the analysis of positionally isomeric fatty acids. *J. Liq. Chromatogr. Relat. Technol.*, **25**, 1947–1965.

19 Mitchell, T.W., Pham, H., Thomas, M.C., and Blanksby, S.J. (2009) Identification of double bond position in lipids: from GC to OzID. *J. Chromatogr. B*, **877**, 2722–2735.

20 Tomer, K.B., Crow, F.W., and Gross, M.L. (1983) Location of double-bond position in unsaturated fatty acids by negative ion MS/MS. *J. Am. Chem. Soc.*, **105**, 5487–5488.

21 Jensen, N.J., Tomer, K.B., and Gross, M.L. (1986) Fast-atom-bombardment and tandem mass-spectrometry of phosphatidylserine and phosphatidylcholine. *Lipids*, **21**, 580–588.

22 Jensen, N.J., Tomer, K.B., and Gross, M.L. (1987) Fab MS/MS for phosphatidylinositol, -glycerol. phosphatidylethanolamine and other complex phospholipids. *Lipids*, **22**, 480–489.

23 Cheng, C., Pittenauer, E., and Gross, M.L. (1998) Charge-remote fragmentations are energy-dependent processes. *J. Am. Soc. Mass Spectrom.*, **9**, 840–844.
24 Trimpin, S., Clemmer, D.E., and McEwen, C.N. (2007) Charge-remote fragmentation of lithiated fatty acids on a TOF–TOF instrument using matrix-ionization. *J. Am. Soc. Mass Spectrom.*, **18**, 1967–1972.
25 Simões, C., Simões, V., Reis, A., Domingues, P., and Domingues, M.R.M. (2008) Determination of the fatty acyl profiles of phosphatidylethanolamines by tandem mass spectrometry of sodium adducts. *Rapid Commun. Mass Spectrom.*, **22**, 3238–3244.
26 Satoh, T., Sato, T., Kubo, A., and Tamura, J. (2011) Tandem time-of-flight mass spectrometer with high precursor ion selectivity employing spiral ion trajectory and improved offset parabolic reflectron. *J. Am. Soc. Mass Spectrom.*, **22**, 797–803.
27 Hsu, F.-F. and Turk, J. (2008) Structural characterization of unsaturated glycerophospholipids by multiple-stage linear ion-trap mass spectrometry with electrospray ionization. *J. Am. Soc. Mass Spectrom.*, **19**, 1681–1691.
28 Hsu, F.-F. and Turk, J. (2010) Electrospray ionization multiple-stage linear ion-trap mass spectrometry for structural elucidation of triacylglycerols: assignment of fatty acyl groups on the glycerol backbone and location of double bonds. *J. Am. Soc. Mass Spectrom.*, **21**, 657–669.
29 Castro-Perez, J., Roddy, T., Nibbering, N., Shah, V., McLaren, D., Previs, S., Attygalle, A., Herath, K., Chen, Z., Wang, S.-P., Mitnaul, L., Hubbard, B., Vreeken, R., Johns, D., and Hankemeier, T. (2011) Localization of fatty acyl and double bond positions in phosphatidylcholines using a dual stage CID fragmentation coupled with ion mobility mass spectrometry. *J. Am. Soc. Mass Spectrom.*, **22**, 1552–1567.
30 Ly, T. and Julian, R.R. (2009) Ultraviolet photodissociation: developments towards applications for mass-spectrometry-based proteomics. *Angew. Chem., Int. Ed.*, **48**, 7130–7137.
31 Reilly, J.P. (2009) Ultraviolet photofragmentation of biomolecular ions. *Mass Spectrom. Rev.*, **28**, 425–447.
32 Devakumar, A., O'Dell, D., Walker, J., and Reilly, J. (2008) Structural analysis of leukotriene C_4 isomers using collisional activation and 157nm photodissociation. *J. Am. Soc. Mass Spectrom.*, **19**, 14–26.
33 Moe, M.K., Anderssen, T., Strøm, M.B., and Jensen, E. (2004) Vicinal hydroxylation of unsaturated fatty acids for structural characterization of intact neutral phospholipids by negative electrospray ionization tandem quadrupole mass spectrometry. *Rapid. Commun. Mass Spectrom.*, **18**, 2121–2130.
34 Moe, M.K., Anderssen, T., Strøm, M.B., and Jensen, E. (2005) Total structure characterization of unsaturated acidic phospholipids provided by vicinal di-hydroxylation of fatty acid double bonds and negative electrospray ionization mass spectrometry. *J. Am. Soc. Mass Spectrom.*, **16**, 46–59.
35 Moe, M.K., Strøm, M.B., Jensen, E., and Claeys, M. (2004) Negative electrospray ionization low-energy tandem mass spectrometry of hydroxylated fatty acids: a mechanistic study. *Rapid. Commun. Mass Spectrom.*, **18**, 1731–1740.
36 Harrison, K.A. and Murphy, R.C. (1996) Direct mass spectrometric analysis of ozonides: application to unsaturated glycerophosphocholine lipids. *Anal. Chem.*, **68**, 3224–3230.
37 Kwon, Y., Lee, S., Oh, D.-C., and Kim, S. (2011) Simple determination of double-bond positions in long-chain olefins by cross-metathesis. *Angew. Chem., Int. Ed.*, **50**, 8275–8278.
38 Gomez-Cortes, P., Tyburczy, C., Brenna, J. T., Juarez, M., and de la Fuente, M.A. (2009) Characterization of *cis*-9 *trans*-11 *trans*-15 C18:3 in milk fat by GC and covalent adduct chemical ionization tandem MS. *J. Lipid Res.*, **50**, 2412–2420.
39 Lawrence, P. and Brenna, J.T. (2006) Acetonitrile covalent adduct chemical ionization mass spectrometry for double bond localization in non-methylene-interrupted polyene fatty acid methyl esters. *Anal. Chem.*, **78**, 1312–1317.
40 Michaud, A.L., Diau, G.-Y., Abril, R., and Brenna, J.T. (2002) Double bond localization in minor homoallylic fatty acid methyl esters using acetonitrile

chemical ionization tandem mass spectrometry. *Anal. Biochem.*, **307**, 348–360.

41 Michaud, A.L., Lawrence, P., Adlof, R., and Brenna, J.T. (2005) On the formation of conjugated linoleic acid diagnostic ions with acetonitrile chemical ionization tandem mass spectrometry. *Rapid. Commun. Mass Spectrom.*, **19**, 363–368.

42 Michaud, A.L., Yurawecz, M.P., Delmonte, P., Corl, B.A., Bauman, D.E., and Brenna, J.T. (2003) Identification and characterization of conjugated fatty acid methyl esters of mixed double bond geometry by acetonitrile chemical ionization tandem mass spectrometry. *Anal. Chem.*, **75**, 4925–4930.

43 Van Pelt, C.K. and Brenna, J.T. (1999) Acetonitrile chemical ionization tandem mass spectrometry to locate double bonds in polyunsaturated fatty acid methyl esters. *Anal. Chem.*, **71**, 1981–1989.

44 Xu, Y. and Brenna, J.T. (2007) Atmospheric pressure covalent adduct chemical ionization tandem mass spectrometry for double bond localization in monoene- and diene-containing triacylglycerols. *Anal. Chem.*, **79**, 2525–2536.

45 Oldham, N.J. and Svatoscaron, A. (1999) Determination of the double bond position in functionalized monoenes by chemical ionization ion-trap mass spectrometry using acetonitrile as a reagent gas. *Rapid Commun. Mass Spectrom.*, **13**, 331–336.

46 Vrkoslav, V., Háková, M., Pecková, K., Urbanová, K., and Cvacka, J. (2011) Localization of double bonds in wax esters by high-performance liquid chromatography/atmospheric pressure chemical ionization mass spectrometry utilizing the fragmentation of acetonitrile-related adducts. *Anal. Chem.*, **83**, 2978–2986.

47 Thomas, M.C., Mitchell, T.W., and Blanksby, S.J. (2006) Ozonolysis of phospholipid double bonds during electrospray ionization: a new tool for structure determination. *J. Am. Chem. Soc.*, **128**, 58–59.

48 Thomas, M.C., Mitchell, T.W., Harman, D.G., Deeley, J.M., Murphy, R.C., and Blanksby, S.J. (2007) Elucidation of double bond position in unsaturated lipids by ozone electrospray ionization mass spectrometry. *Anal. Chem.*, **79**, 5013–5022.

49 Zhang, J.I., Tao, W.A., and Cooks, R.G. (2011) Facile determination of double bond position in unsaturated fatty acids and esters by low temperature plasma ionization mass spectrometry. *Anal. Chem.*, **83**, 4738–4744.

50 Chen, J. and Davidson, J.H. (2003) Ozone production in the negative DC corona: the dependence of discharge polarity. *Plasma Chem. Plasma Process.*, **23**, 501–518.

51 Ellis, S.R., Hughes, J.R., Mitchell, T.W., Panhuis, M.I.H., and Blanksby, S.J. (2012) Using ambient ozone for assignment of double bond position in unsaturated lipids. *Analyst*, **137**, 1100–1110.

52 Thomas, M.C., Mitchell, T.W., Harman, D.G., Deeley, J.M., Nealon, J.R., and Blanksby, S.J. (2008) Ozone-induced dissociation: elucidation of double bond position within mass-selected lipid ions. *Anal. Chem.*, **80**, 303–311.

53 Poad, B.L.J., Pham, H.T., Thomas, M.C., Nealon, J.R., Campbell, J.L., Mitchell, T. W., and Blanksby, S.J. (2010) Ozone-induced dissociation on a modified tandem linear ion-trap: observations of different reactivity for isomeric lipids. *J. Am. Soc. Mass Spectrom.*, **21**, 1989–1999.

54 Deeley, J.M., Thomas, M.C., Truscott, R.J. W., Mitchell, T.W., and Blanksby, S.J. (2009) Identification of abundant alkyl ether glycerophospholipids in the human lens by tandem mass spectrometry techniques. *Anal. Chem.*, **81**, 1920–1930.

55 Brown, S.H.J., Mitchell, T.W., and Blanksby, S.J. (2011) Analysis of unsaturated lipids by ozone-induced dissociation. *Biochim. Biophys. Acta Mol. Cell Biol. Lipids*, **1811**, 807–817.

56 Yang, K., Zhao, Z., Gross, R.W., and Han, X. (2011) Identification and quantitation of unsaturated fatty acid isomers by electrospray ionization tandem mass spectrometry: a shotgun lipidomics approach. *Anal. Chem.*, **83**, 4243–4250.

57 Seon, H.L., Williams, M.V., and Blair, I.A. (2005) Targeted chiral lipidomics analysis. *Prostaglandins Other Lipid Mediat.*, **77**, 141–157.

58 Oh, S.F., Vickery, T.W., and Serhan, C.N. (2011) Chiral lipidomics of E-series resolvins: aspirin and the biosynthesis of novel mediators. *Biochim. Biophys. Acta Mol. Cell Biol. Lipids*, **1811**, 737–747.

59 Bray, G.A., Lovejoy, J.C., Smith, S.R., DeLany, J.P., Lefevre, M., Hwang, D., Ryan, D.H., and York, D.A. (2002) The influence of different fats and fatty acids on obesity: insulin resistance and inflammation. *J. Nutr.*, **132**, 2488–2491.

60 Odegaard, A.O. and Pereira, M.A. (2006) *Trans* fatty acids, insulin resistance, and type 2 diabetes. *Nutr. Rev.*, **64**, 364–372.

7
Imaging Lipids in Tissues by Matrix-Assisted Laser Desorption/Ionization Mass Spectrometry

Robert M. Barkley, Joseph A. Hankin, Karin A. Zemski Berry, and Robert C. Murphy

7.1
Introduction

Matrix-assisted laser desorption/ionization (MALDI) can be used to reveal the distribution of complex lipids in tissues with the technique of imaging mass spectrometry (IMS). Figure 7.1 shows an example of negative and positive ion images obtained from sections of mouse brain for a sulfatide (ST), a glycerophosphoserine (PS), and two different glycerophosphocholine (PC) lipids. Since lipids are present at high local concentrations in cellular membranes and lipid storage vesicles (lipid droplets), these molecules are often the most abundant molecules desorbed as ions from a tissue surface by the MALDI process. One of the unique advantages of the analysis of lipids is their relative low molecular weight, typically below 1000 Da, making them well suited for this mode of ionization/desorption. Only a few naturally abundant lipids, such as the cardiolipins or the complex glycosphingolipids, occur at molecular weights above 1000 Da. There are also a number of lipids that are of such low, local abundance that often they are not observed in the MALDI IMS spectrum.

The overall strategy of MALDI IMS is to obtain information relevant to the local distribution of lipids as they occur in tissues as a result of the complex events of lipid biosynthesis, transport, and metabolism. A substantially different workflow is employed to obtain such information when compared to the typical steps engaged to obtain more traditional mass spectral data, such as by liquid chromatography/mass spectrometry (LC/MS) and LC/MS/MS strategies. The major steps for MALDI IMS involve techniques to (i) collect and prepare the sample, (ii) apply matrix, (iii) acquire data, and (iv) process data with specialized software to enable image visualization. A remaining challenge that has not been successfully met is to convert the observed ion abundances for molecular species into quantitative measures of local lipid concentrations.

Lipidomics, First Edition. Edited by Kim Ekroos.

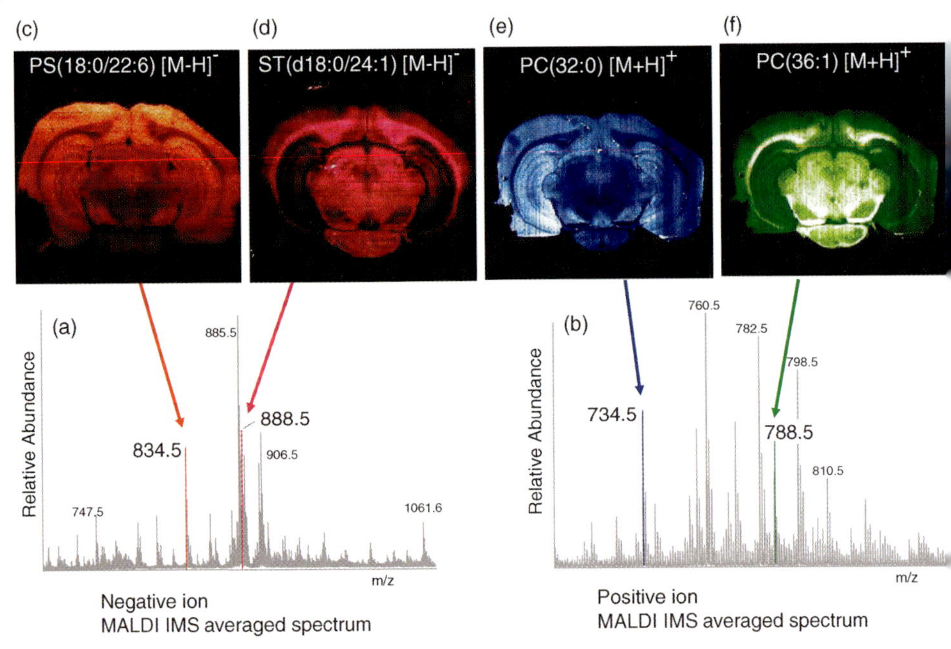

Figure 7.1 MALDI mass spectra averaged across the entire tissue surface of sagittal sections of mouse brain, acquired as (a) negative ions with 5-aminoquinoline matrix, and (b) positive ions with DHB matrix. Representative MALDI mass spectra images are displayed for the corresponding ions: (c) (orange) PS(18:0/22:6), $[M-H]^-$ m/z 834.5; (d) (red) ST(d18:0/24:1), $[M-H]^-$ m/z 888.5; (e) (blue) PC(32:1), $[M+H]^+$ m/z 734.5; and (f) (green) PC(36:1), $[M+H]^+$ m/z 788.5.

7.2 Sample Preparation

The preparation of tissues for MALDI IMS is the most important step to ensure a high-quality image. Care must be taken to prepare the sample specifically for MALDI IMS because fixation of the tissue can obliterate the signal from lipid molecular species [1]. Typically, thin sections of frozen tissue are obtained using a cryostat. Homogeneous tissues, like brain and liver, can simply be flash frozen directly after dissection from the animal and can be mounted directly to the cryostat using a small amount of optimal cutting temperature compound (OCT). However, porous tissues such as lung and aorta need to be embedded in a supporting medium before freezing because these tissues do not withstand the knife cutting pressure and collapse during sectioning. In standard histological sample preparation, OCT is used to embed tissues; however, commercial preparations of OCT contain a benzalkonium salt as a preservative. Such preservatives make OCT less than optimal for MALDI imaging due to the highly abundant m/z 332 for the benzalkonium ion and for adducts of benzalkonium ion with endogenous lipids, which unnecessarily complicate the MALDI spectrum [2]. In order to circumvent this problem, a modified OCT (mOCT) was formulated of 10% polyvinyl alcohol, 8% polypropylene glycol (average MW = 2000 g/mol), and 0.1% NaN_3. We found that eliminating the benzalkonium salt leads to a suitably modified material that is useful to embed porous tissue, and yet it is compatible with MALDI IMS [1].

MALDI IMS is carried out on tissue slices $<20\,\mu m$ in thickness and these can be placed onto a metal MALDI plate, indium tin oxide (ITO)-coated glass microscope slide, or a glass cover slip. Tissues can be used immediately or stored frozen. Enhanced signals have been observed when samples are placed in a desiccator for 15 min prior to coating with matrix in order to remove any moisture from the sample. In addition, prior to desiccation and application of matrix, water or a solution of ammonium acetate can be used to rinse salt from the tissue [3]. The images and spectra in Figure 7.2 indicate the dramatic differences that can occur when tissue is rinsed before application of matrix.

7.3 Matrix

One aspect of MALDI IMS method development that has received considerable attention is the choice of matrix material and the method by which it is applied. The chemical composition and method for application of the matrix onto a tissue slice have been explored in numerous studies with the desired endpoint being improved image quality.

7.3.1 Techniques for Matrix Application

Most matrices can be physically applied by several methods, with the choice depending on the analyte of interest, specific matrix, cost, and convenience. The usual methods for matrix application to tissue can be classified as wet methods, which use solutions of matrix in solvents, or dry methods, which are solvent free. Wet methods include droplets [4], electrospray [5], airbrush [6–9], inkjet [9], and microspotting [10, 11]. The major drawbacks of wet matrix application methods are that they may lead to delocalization of lipids, a decrease in spatial resolution, and addition of Na^+ or K^+ from exogenous sources, such as impure matrix and solvents stored in borosilicate glass. However, one distinct advantage of the wet methods is that it is easy to modify the matrix with desired adducting species such as Li^+ [12] and other reagents [13].

Solvent-free, direct application of dry matrix onto tissue by sublimation [14] or by dry coating [15] of 2,5-dihydroxybenzoic acid (DHB) prevents lipid delocalization. These methods yield an even, reproducible coating of matrix, increasing the MALDI MS signal and producing high-quality images of phospholipids. Sublimation also purifies the matrix as it is applied and eliminates the possibility of

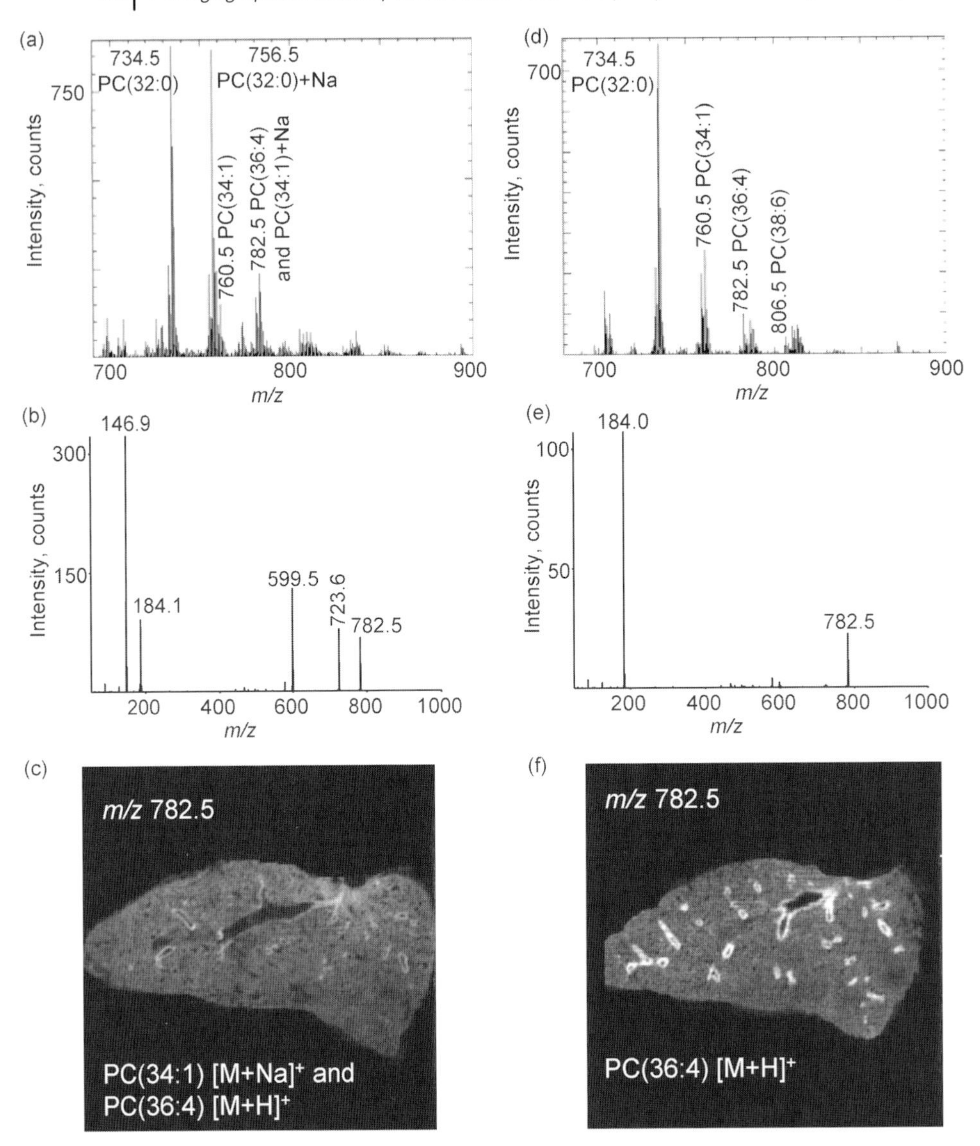

Figure 7.2 Effect of rinsing with ammonium acetate on MALDI IMS data. (a) The positive ion MALDI IMS spectrum from an unrinsed section of mouse lung contains $[M + H]^+$, $[M + Na]^+$, and $[M + K]^+$ from lipid molecular species found in the lung. (b) The CID spectrum of m/z 782.5 from unrinsed lung tissue yields product ions that indicate the signal arises from a mixture of $[M + H]^+$ for PC (36:4) and $[M + Na]^+$ for PC(34:1). (c) Positive ion MALDI image of m/z 782.5 in a slice of unrinsed lung tissue indicates that this species is present in the alveolar region and is a slightly stronger signal at the edges of some airways. (d) The positive ion MALDI MS spectrum of a mouse lung rinsed with 150 mM ammonium acetate before matrix application contains just $[M + H]^+$ from pulmonary lipid molecular species since the sodium and potassium salts were rinsed away. (e) CID spectrum of m/z 782.5 from a mouse lung rinsed with 150 mM ammonium acetate contains product ions that suggest this ion is composed of only $[M + H]^+$ from PC(36:4). (f) Positive ion MALDI image of m/z 782.5 from a lung slice rinsed with 150 mM ammonium acetate indicates the presence of PC(36:4), primarily at the airway edges.

transferring impurities to the tissue. Sublimation, however, is not a panacea since matrix sublimation during data acquisition is a problem when the matrix disappears midanalysis in a high-vacuum source region of some instruments [16]. In this case, either the imaging experiment needs to be performed quickly or a less-volatile matrix needs to be used. Matrix-free MALDI IMS can be carried out using a nanostructure surface sample plate (NALDI). Lipid analytes can be transferred by direct contact of a tissue slice, then washed, and subsequently imaged without application of any matrix [17].

7.3.2
Matrix Compounds

A variety of matrices have been used for MALDI IMS of lipids from tissue slices and the structures and properties of some of the common matrices used are shown in Figure 7.3. Factors influencing the choice of a matrix compound include selective ionization of a specific lipid class, uniformity of coating, crystal size, and the ability to persist in the MALDI source without sublimating away during the MALDI IMS experiment. DHB has been the most commonly used matrix for positive ion MALDI IMS analysis of phospholipids in tissues, with PC and sphingomyelin (SM) molecules yielding intense signals as mixtures of $[M+H]^+$, $[M+Na]^+$, and $[M+K]^+$ adduct ions. Collisional activation of these ions typically reveals information only about the polar head group [18, 19].

Individual molecular species from other classes of phospholipids, including glycerophosphates, glycerophosphoethanolamines, glycerophosphoglycerols, glycerophosphoinositols, and PSs, are easier to detect as negative ions, with the added advantage that acyl chain information can be derived from collisional activation of negative precursor ions [18]. Several MALDI matrices, such as 2,6-dihydroxyacetophenone (DHAP) [20], 9-aminoacridine [21], and

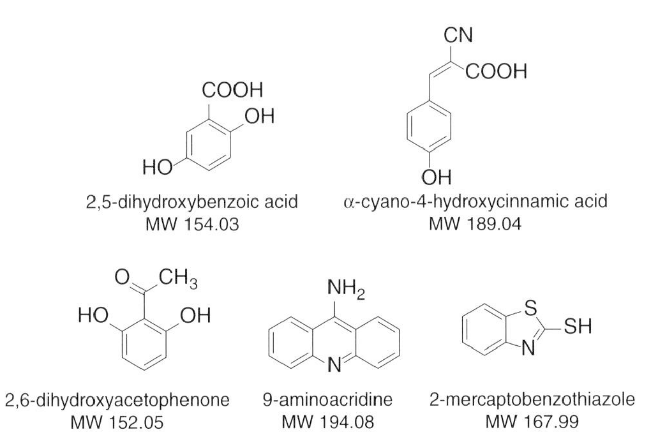

Figure 7.3 MALDI matrices commonly used for MALDI IMS of lipids.

2-mercaptobenzothiazole [22], all of which are less acidic than DHB, have been applied to enhance negative ion yield of these phospholipids.

In addition, ionic liquid matrices, prepared from an acidic matrix and a base, have been found useful for MALDI IMS because they generate good ion yields from tissues. Salts of DHB or α-cyano-4-hydroxycinnamic acid (CHCA) with *n*-butyl amine [23] and a combination of DHAP with ammonium sulfate [13] are matrix mixtures that have been successfully used in MALDI IMS. The inclusion of alkali metals with matrices has also been a means to control volatility, as well as the mass action-driven formation of cation adducts in the MALDI plume. Lithium trifluoroacetate was found to be an excellent matrix additive to CHCA [12] leading to phospholipids observed as $[M+Li]^+$ species. Since the collision-induced dissociation (CID) behavior of many lipids as the lithiated adduct has been studied in great detail [24], the potential for structural characterization is significantly enhanced with this simple modification of the matrix.

7.4 Instrumentation

Once the sample is prepared onto a suitable surface and matrix has been applied, the sample is introduced into the source of the imaging mass spectrometer. There are several configurations of lasers, different orientations of laser to sample, and different strategies for moving the sample plate sequentially across the laser beam developed by instrument manufacturers. The formation of ions by the MALDI process is similar regardless of instrument configuration, although some MALDI interfaces operate at high vacuum and others at slightly higher pressures in order to reduce the kinetic energy and stabilize the ions produced. Several different types of mass analyzers have been used for MALDI IMS experiments.

7.4.1 Lasers and Rastering

The next step in the MALDI IMS process is to place the MALDI plate holder containing the tissue slice with matrix into the appropriate imaging mass spectrometer. Such instruments have been designed to enable computer-controlled MALDI sample plate movement so that the *x*,*y*-position of the tissue is easily controlled in relatively small incremental steps (e.g., 50 μm). New instruments have been designed to step at even finer increments in a very reproducible manner. A video camera is usually built into the instrument ion source region to directly view the tissue under vacuum and to set up the positional limits used to control the rastering experiment. The next step involves rastering the laser beam across the tissue in a manner to generate pixels of mass spectra at specific locations. These positionally indexed mass spectra become the image pixels. While "rastering" of the laser is often stated as the means by which the tissue is sampled, in MALDI IMS it is the sample stage and tissue that is "rastered" and the focused laser beam is stationary.

There are two general approaches for moving the MALDI sample stage. One is a stepwise rastering process where an *x*,*y*-position is established and the laser beam is fired for a certain number of times (e.g., 5–10 times per pixel for a typical nitrogen laser) to ablate the matrix. During this time, the ions generated from several laser pulses are accumulated into discrete packets, separated according to m/z value, detected, and ion abundances stored. The mass spectra are then summed to generate a single mass spectrum at that *x*,*y*-index position before moving the laser to the next *x*,*y*-position (e.g., $x + 50\,\mu m$, y).

A refinement of this process to improve the lateral resolution of the image is to use a plate step increment that is smaller than the actual MALDI spot size. Many of the early MALDI IMS instruments had a relatively large laser spot size, for example, an elliptical shape that is approximately 200 μm × 350 μm. This is a much larger laser spot size than desirable for detailed imaging, if the step increment employed was equal to the laser spot size. However, by stepping the tissue plate only a fraction of the laser spot size and completely ablating the matrix at each *x*,*y*-index position, the tissue plate can be incremented by some fraction of the spot size, exposing only a small amount of matrix-coated tissue to the next laser pulse. Using this oversampling technique [25] to improve lateral resolution, rastering can continue, for example, left to right, until the right limit has been reached. The sample stage is then adjusted by 50 μm below the first raster line and the incremental steps now proceed right to left (in the opposite direction) until the left limit is reached. This produces a serpentine set of pixel data where the image is then reconstructed from the *x*,*y*-index positional data going left to right in alternate lines followed by right to left.

An alternative rastering approach that improves the speed of the overall imaging process involves continuous movement (as opposed to stepping) of the MALDI plate holder while firing the laser at a reasonably fast repetition rate (50–250 times per pixel for a high repetition rate laser). In such cases, a pixel can be defined either by the number of laser shots or by the number of resulting spectra that are stored and captured during a unit of time while the plate is moved. Since the laser pulse and mass spectral data acquisition are considerably faster than the plate movement, an accurate location of the "average" *x*,*y*-index position is obtained. Starting from the left side and going to the right side, this continuous laser process can lead to the acquisition of 50 μm pixels at a much more rapid pace, but upon reaching the right limit the laser firing is stopped and the plate holder is repositioned back to the left limit now 50 μm lower than the previous scan line. This process is then repeated until the entire tissue is covered, thus leading to an image that is generated from raster lines always made in the same direction (e.g., from left to right). Regardless of the rastering technique, optimum lateral resolution with the oversampling approach is obtained when the matrix is fully ablated for each pixel.

Nitrogen and solid-state lasers have been used in MALDI imaging instrumentation. The large number of laser shots required for a single-imaging experiment (40 000 pixels across a 1 cm^2 tissue slice acquired at 50 μm plate movement) places a heavy demand on the performance and endurance of the laser. The UV light absorbance profile of the matrix should ideally match with the wavelength of the laser light (N_2 337 nm; solid state 355 nm) for efficiency of ion generation, although other practical considerations weigh into matrix choice, and laser power can be arbitrarily increased to compensate for inefficient matching of energy absorbance.

Successful MALDI IMS requires mass spectrometer software that is capable of rapidly recording the mass spectral data, including mass-to-charge ratio and intensity of each ion formed by MALDI and mass separation (e.g., time-of-flight (TOF) or other mass separation strategy), summation of the spectra during a unit of time that defines a pixel, and notation of the location of the pixel.

7.4.2
Ion Formation

The MALDI matrix is the essential component required for ionization and desorption of lipid molecules from tissues by this laser-dependent process, with its primary purpose being to couple the energy of the laser photon to the lipid for efficient ionization and desorption from the surface. While no precise description of the exact mechanism of this complex desorption process currently exists, a theoretical desorption mechanism that invokes phase explosion and ejection of large clusters from the crystal lattice has emerged [26]. These ideas and related mechanisms by which ions are emitted during MALDI have been developed largely considering hydrophilic molecules (e.g., peptides) and reasonable mixing of the matrix molecule with the analyte. These conditions are different from those that actually exist when considering formation of phospholipid ions from the surface of a tissue, but some extensions are reasonable. Most theory requires the analyte molecule to be within the crystal lattice of the applied matrix and the laser shots then induce a rapid adiabatic rise in crystal temperature, which causes the phase explosion [26]. If these assumptions are correct, then the lipid ions observed in MALDI IMS must arise from the movement of lipid molecules out of the bilayer environment and into the matrix crystal. Although this would seem reasonable when matrix is applied in a solvent by, for example, dried droplet application [27], it is a less obvious process when matrix is sublimed onto the tissue. During the gas-phase deposition of matrix by sublimation, it is possible that the concentration of matrix can disturb lipid bilayers sufficiently to “extract” lipids into the growing matrix crystal. It has been suggested that the hydrophobicity of the matrix determines the efficiency of extraction of these hydrophobic molecules [28]. The ion abundance for a particular type of lipid observed in the MALDI IMS experiment might be a result of this initial discriminatory event in that the most efficiently extracted lipids are the source of the most abundant ions. However, the lipid present in the expanding MALDI plume during phase explosion must form an ion in order to be observed in the mass spectrometer. For preionized lipids, such as phospholipids or sphingolipids, this would be a process of charge separation since these lipids are ionized in the tissue and the counterion would be separated by the voltage bias of the ion source. For a neutral lipid such as cholesterol, the molecule must be ionized by either attachment of a charging ion or formal loss of a radical species such as hydroxyl radical. While radical loss is not particularly favorable for cholesterol, the protonation of the carbinol oxygen atom $[M+H]^+$, followed by loss of H_2O, would explain the cholesteryl carbocation observed at m/z 369.3.

The exact processes that take place when the matrix is applied to a tissue surface and result in lipids appearing in the MALDI plume are poorly understood at best, but clearly are important [29]. Even less understood is the influence that the alkali metal ions Na^+ and K^+, which are present in tissues at millimolar concentrations, have on the MALDI process, let alone how they appear in the crystal lattice. The combination of a highly complex mixture of organic and inorganic molecules, a complicated physical arrangement of lipids within cells, and the chemical diversity of all molecules at the surface of a tissue slice, makes this a difficult system to model precisely, leaving empirical studies to dominate the field of investigation.

7.4.3
Mass Analyzers and Ion Detection

Early work in MALDI IMS was carried out on commercial MALDI-TOF instruments that were modified to permit repetitive rastering of the MALDI target. The time-of-flight mass analyzer was very useful for rapid detection of a broad range of biomolecules, including proteins, peptides, and lipids, largely because all ions that formed following the laser shot could be collected in the time-of-flight experiment as a result of fast timing electronic circuits that were readily available [5, 30]. In addition, modern TOF instruments are capable of high-resolution performance, enabling these mass spectrometers to attain mass accuracies of 5–10 ppm. The development of hybrid Q-TOF instrumentation [31] led to advancement in MALDI IMS technology with the addition of CID capability to assist in structural identification of ions. This instrument permitted unit mass resolution of a precursor ion derived from the tissue sample, which, as discussed above, is one of the most important parameters for a detailed analysis of lipids by CID experiments. This requirement is due in part to the existence of closely related lipid molecular species that differ by one double bond in a fatty acyl chain and thus differ from the molecule of interest by 2 Da. If CID of a mixture of molecular species takes place, it is impossible to uniquely characterize the lipid even if high-resolution measurement of product ions is sufficient for elemental composition analysis.

Other instrument configurations have been developed for MALDI IMS to help decipher the myriad mixture of ions that are observed in the analysis of biological samples. A vacuum MALDI source interfaced to an ion trap has been used to identify phospholipids in rat spinal cord and brain using the MS^n capabilities of the ion trap [32, 33]. A MALDI quadrupole ion trap (QIT)-TOF instrument has been used to carry out MALDI IMS of lipids on mouse brain sections [34]. This instrument provided structural elucidation capabilities by CID as well as mass accuracy of fragments separated by the TOF mass analyzer.

The ability to establish the molecular identity of a lipid species and to resolve isobaric mixtures through use of instrumentation with high mass resolution has

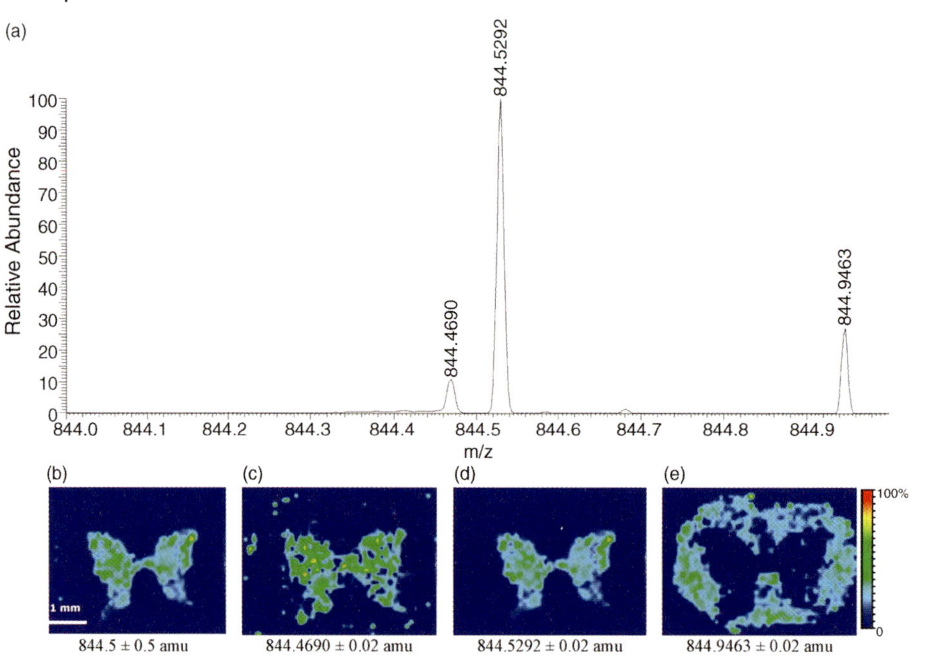

Figure 7.4 Advantage of high resolution in MALDI IMS. (a) High-resolution MALDI mass spectrum from a rat spinal cord, which shows that three ions are detected at nominal *m/z* 844. (b) MALDI image of *m/z* values centered at *m/z* 844.5 ± 0.5 using a linear ion trap with unit mass resolution. By plotting a 1 Da range of *m/z* values, an image dominated by the base peak at *m/z* 844.5292 is obtained. MALDI images of (c) *m/z* 844.4690, [PS (36:4) + Na + K − H]$^+$, (d) *m/z* 844.5292, [PC (38:6) + K]$^+$, and (e) *m/z* 844.9463, unidentified. The high-resolution spectrum indicates that both phospholipid molecular species appear in the gray matter of the spinal cord; however, the ion at *m/z* 844.4690 has a greater distribution throughout the gray matter. In addition, the unidentified ion is localized in the white matter. This figure clearly illustrates the advantages of high resolution in MALDI IMS. Reproduced with permission from Ref. [37].

led to MALDI IMS applications that utilize FT-ICR to accurately mass analyze the ions generated from tissue slices. Although first applied to the analysis of peptides on rat brain [35], the high mass accuracy, to 1 ppm, of the FT-ICR was used to diminish ambiguity about the identity of a lipid species from a MALDI image of mouse brain [36]. The images shown in Figure 7.4 are derived from high-resolution mass spectra obtained from an orbitrap [37].

Another emerging instrumental innovation has been the use of ion mobility to separate ions desorbed from a tissue sample as a function of velocity through an electrical drift field. This ancillary technique permits a level of separation, although generally less powerful than online chromatography, based on differences in the apparent surface area of the ion [38]. Most separations have been demonstrated for highly divergent ions emerging from a tissue, such as separation of protein ions from lipid ions, with the lipids having the highest velocity in the drift experiment. This additional component of ion separation prior to mass analysis has been configured between a MALDI source and TOF mass analyzer and tested by imaging lipids in rat brain slices. The ion mobility component easily separated isobaric peptides and phospholipids present in the mixture of ions ablated off the tissue and generated images of brain phospholipids with faithful representation of tissue morphology [39–41]. Given the limitations of data file size and drift tube resolution, this configuration of complex components has shown potential to solve the difficult problem of structural identification of lipids within a mixture of isobaric species.

One very important aspect of MALDI IMS of lipids is the fact that an additional experiment can be carried out with the tissues that have been previously imaged and then reinvestigated after reapplication of matrix. This step involves tandem mass spectrometry where collisional activation of specific ions observed at a specific region can then be performed in order to obtain structural information as to the type of lipid and unique molecular species identification of that lipid. This feature is somewhat unique for lipids since a robust literature is now available to describe the decomposition ion chemistry of lipids generated either by MALDI IMS or by electrospray ionization [18]. Some unique complications do exist in the MALDI IMS experiment since minimal sample preparation is employed, leaving the lipids embedded in a local tissue environment that contains cellular and extracellular fluids with high concentrations of sodium and potassium alkali metal ions. This leads to the appearance of alkali attachment ions with phospholipids [19]. Nonetheless, it is possible to reproduce these adducts with standard molecules in order to understand collisional activation of alkali adducts of the lipid ion. More sophisticated instruments have high-resolution capability that enables generation of separate ion images from molecular species that have very different elemental compositions, but the same nominal m/z value [42]. Collisional activation of MALDI ions and measurement of exact mass of the desorbed ion are critical pieces of information to generate a more complete understanding of the ions observed in the MALDI IMS process.

7.5 Data Processing

The final step for obtaining an image by MALDI IMS is data processing, including conversion of the mass spectral data into a four-dimensional database with x,y-coordinates that correspond to the location (x,y-index position) and m/z and intensity values in the third and fourth dimensions to define each pixel. Some investigators have relied on a software program called BioMap [43], which is written in IDL [44], a data visualization and analysis programming language. BioMap requires IMS data to be in a standard format called ANALYZE [45], which uses three different data files that are rapidly processed into images. The raw data files collected during the data acquisition steps are unique to each instrument manufacturer, but a separate manufacturer-supplied program is used to convert

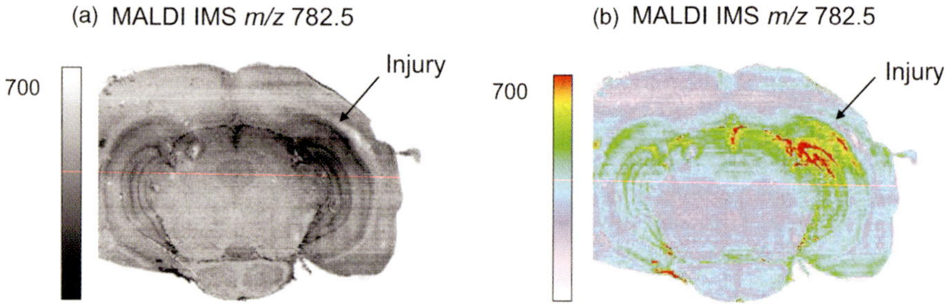

(a) MALDI IMS m/z 782.5

(b) MALDI IMS m/z 782.5

Figure 7.5 MALDI mass spectrometric images of coronal sections of rat brain representing the distribution of lipid PC(34:1), $[M + Na]^+$ at m/z 782.5. The image was generated in a study of traumatic brain injury [46], and the images were processed in Biomap in both (a) gray scale and (b) color to demonstrate the benefits of viewing images using false colors. In this example, the accumulation of this ion in the injured region is more apparent in the colored display.

the raw files into proper ANALYZE data files. Although BioMap is freely available, it has also been optimized for use with particular instruments in order to more directly connect the acquisition with the processing of data. BioMap processes ANALYZE image data files (.img) at the mass resolution specified when the raw data file was converted. This file contains arrays of sequential x,y coordinates of each pixel and within each pixel the sequential abundance of the ions. The BioMap header file (.hdr) contains information about the data structure in the image file and other sample parameters. The third file is a mass-to-charge calibration file (.t2m) for mass-to-charge ratio conversion of the sequential list of mass peaks in each spectrum. Once these three data files are available, the viewing software program allows one to quickly interact with this four-dimensional database to view the data as a tissue image either by integrating all ion abundances or by extracting specific ion abundances relative to the x,y-index position. The intensity of observed ions is typically presented in a map that can be visually interpreted. This could be black and white with ion abundance units in a gray scale or a scaled color bar where color wavelengths are defined to indicate specific ion abundances, as shown in Figure 7.5. This step is very fast and allows the viewer to quickly observe images and move across the image with a cursor to individual pixels and observe the local mass spectrum. It is also possible to extract individual ions and observe the entire image from that particular ion perspective. Oftentimes it is helpful to superimpose the images of different ions in order to interpret the MALDI IMS data. ImageJ software [47] has been used to easily overlay red, green, and blue images of different phospholipid molecular species in order to determine the colocalization of phospholipids in a tissue slice, as shown in Figure 7.6.

A recent approach to visualize the four-dimensional database in terms of specific and important images is the use of information processing and a multivariate statistical analysis approach to reveal unique m/z values that can define specific regions. This facilitates the discovery of unique m/z values in a more unbiased manner [48] and is an area of intense development from the standpoint of not

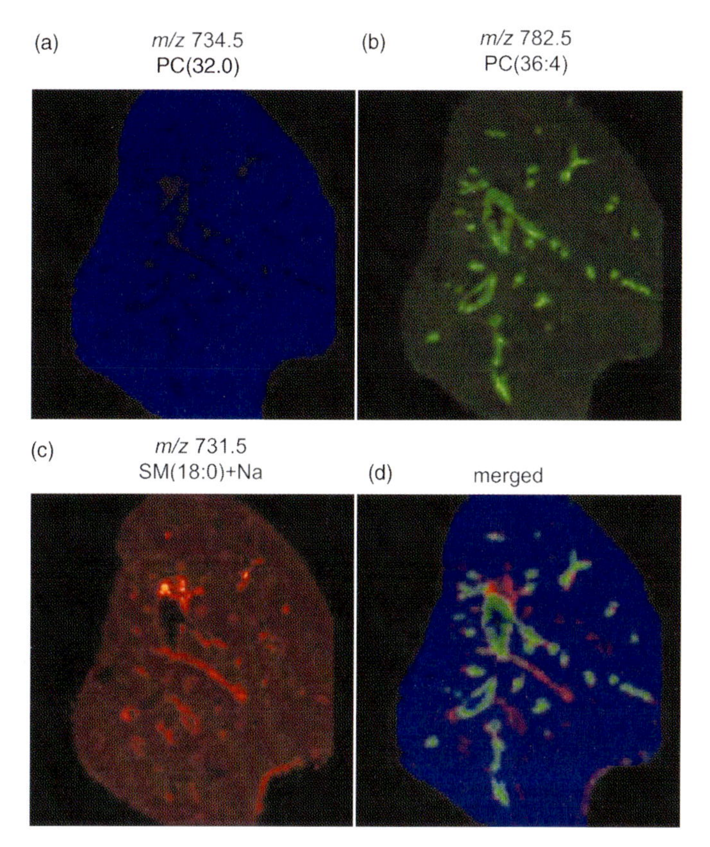

Figure 7.6 Merging of MALDI images. Positive ion MALDI images of (a) m/z 734.5 (PC(32:0)), (b) m/z 782.5 (PC(36:4)), and (c) m/z 731.5 (SM(d18:1/18:0)) from a lung slice rinsed with 150 mM ammonium acetate. The individual MALDI images were merged to determine colocalization of the ions (d). This merged figure shows that PC(32:0) is present uniformly throughout the alveolar region, while PC(36:4) is present at the airway edges and SM(d18:1/18:0) is present at the edges of the blood vessels and that these ions only have minimal overlap in the alveolar region.

only the instrument manufacturers but also the investigators employing MALDI IMS and other mass spectrometry-based imaging techniques such as desorption electrospray mass spectrometry (DESI) [49] and secondary ion mass spectrometry (SIMS) [50].

7.6 Conclusions

An extraordinary number of impressive MALDI mass spectral images of lipid molecular species from tissue have been obtained in just a few years because rapid advances in tissue preparation, techniques for matrix application, instrumentation, and data analysis have coincided. Although the current state of the art is sufficient for detailed studies of lipid distribution, lipid metabolism, and characterization of disease states and biological processes, such reports have only recently emerged. As technical advances continue to be made, the number, scope, and quality of such studies by MALDI IMS will grow rapidly.

Acknowledgments

This work was supported in part by grants HL034303 and GM069338 from the National Institutes of Health.

References

1 Berry, K.A.Z., Li, B., Reynolds, S.D., Barkley, R.M., Gijón, M.A., Hankin, J.A., Henson, P.M., and Murphy, R.C. (2011) MALDI imaging MS of phospholipids in the mouse lung. *J. Lipid Res.*, **52** (8), 1551–1560.

2 Schwartz, S.A., Reyzer, M.L., and Caprioli, R.M. (2003) Direct tissue analysis using matrix-assisted laser desorption/ionization mass spectrometry: practical aspects of sample preparation. *J. Mass Spectrom.*, **38** (7), 699–708.

3 Wang, Y.J., Liu, C.B., and Wu, H.W. (2011) A simple desalting method for direct MALDI mass spectrometry profiling of tissue lipids. *J. Lipid Res.*, **52** (4), 840–849.

4 Luxembourg, S.L., McDonnell, L.A., Duursma, M.C., Guo, X., and Heeren, R.M. (2003) Effect of local matrix crystal variations in matrix-assisted ionization techniques for mass spectrometry. *Anal. Chem.*, **75** (10), 2333–2341.

5 Caprioli, R.M., Farmer, T.B., and Gile, J. (1997) Molecular imaging of biological samples: localization of peptides and proteins using MALDI-TOF MS. *Anal. Chem.*, **69** (23), 4751–4760.

6 Miliotis, T., Kjellstrom, S., Nilsson, J., Laurell, T., Edholm, L.E., and Marko-Varga, G. (2002) Ready-made matrix-assisted laser desorption/ionization target plates coated with thin matrix layer for automated sample deposition in high-density array format. *Rapid. Commun. Mass Spectrom.*, **16** (2), 117–126.

7 Wang, H.Y., Jackson, S.N., Post, J., and Woods, A.S. (2008) A minimalist approach to MALDI imaging of glycerophospholipids and sphingolipids in rat brain sections. *Int. J. Mass Spectrom.*, **278** (2–3), 143–149.

8 Shimma, S., Sugiura, Y., Hayasaka, T., Hoshikawa, Y., Noda, T., and Setou, M. (2007) MALDI-based imaging mass spectrometry revealed abnormal distribution of phospholipids in colon cancer liver metastasis. *J. Chromatogr. B*, **855** (1), 98–103.

9 Baluya, D.L., Garrett, T.J., and Yost, R.A. (2007) Automated MALDI matrix deposition method with inkjet printing for imaging mass spectrometry. *Anal. Chem.*, **79** (17), 6862–6867.

10 Aerni, H.R., Cornett, D.S., and Caprioli, R.M. (2006) Automated acoustic matrix deposition for MALDI sample preparation. *Anal. Chem.*, **78** (3), 827–834.

11 Franck, J., Arafah, K., Barnes, A., Wisztorski, M., Salzet, M., and Fournier, I. (2009) Improving tissue preparation for matrix-assisted laser desorption ionization mass spectrometry imaging. Part 1: using microspotting. *Anal. Chem.*, **81** (19), 8193–8202.

12 Cerruti, C.D., Touboul, D., Guerineau, V., Petit, V.W., Laprevote, O., and Brunelle, A. (2011) MALDI imaging mass

spectrometry of lipids by adding lithium salts to the matrix solution. *Anal. Bioanal. Chem.*, **401** (1), 75–87.

13 Colsch, B. and Woods, A.S. (2010) Localization and imaging of sialylated glycosphingolipids in brain tissue sections by MALDI mass spectrometry. *Glycobiology*, **20** (6), 661–667.

14 Hankin, J.A., Barkley, R.M., and Murphy, R.C. (2007) Sublimation as a method of matrix application for mass spectrometric imaging. *J. Am. Soc. Mass Spectrom.*, **18** (9), 1646–1652.

15 Puolitaival, S.M., Burnum, K.E., Cornett, D.S., and Caprioli, R.M. (2008) Solvent-free matrix dry-coating for MALDI imaging of phospholipids. *J. Am. Soc. Mass Spectrom.*, **19** (6), 882–886.

16 Fernandez, J.A., Ochoa, B., Fresnedo, O., Giralt, M.T., and Rodriguez-Puertas, R. (2011) Matrix-assisted laser desorption ionization imaging mass spectrometry in lipidomics. *Anal. Bioanal. Chem.*, **401** (1), 29–51.

17 Vidova, V., Novak, P., Strohalm, M., Pol, J., Havlicek, V., and Volny, M. (2010) Laser desorption–ionization of lipid transfers: tissue mass spectrometry imaging without MALDI matrix. *Anal. Chem.*, **82** (12), 4994–4997.

18 Pulfer, M. and Murphy, R.C. (2003) Electrospray mass spectrometry of phospholipids. *Mass Spectrom. Rev.*, **22** (5), 332–364.

19 Jackson, S.N., Wang, H.-Y.J., and Woods, A.S. (2005) *In situ* structural characterization of phosphatidylcholines in brain tissue using MALDI-MS/MS. *J. Am. Soc. Mass Spectrom.*, **16** (12), 2052–2056.

20 Burnum, K.E., Cornett, D.S., Puolitaival, S.M., Milne, S.B., Myers, D.S., Tranguch, S., Brown, H.A., Dey, S.K., and Caprioli, R.M. (2009) Spatial and temporal alterations of phospholipids determined by mass spectrometry during mouse embryo implantation. *J. Lipid Res.*, **50** (11), 2290–2298.

21 Marsching, C., Eckhardt, M., Gröne, H.J., Sandhoff, R., and Hopf, C. (2011) Imaging of complex sulfatides SM3 and SB1a in mouse kidney using MALDI-TOF/TOF mass spectrometry. *Anal. Bioanal. Chem.*, **401** (1), 53–64.

22 Astigarraga, E., Barreda-Gomez, G., Lombardero, L., Fresnedo, O., Castano, F., Giralt, M.T., Ochoa, B., Rodriguez-Puertas, R., and Fernandez, J.A. (2008) Profiling and imaging of lipids on brain and liver tissue by matrix-assisted laser desorption/ionization mass spectrometry using 2-mercaptobenzothiazole as a matrix. *Anal. Chem.*, **80** (23), 9105–9114.

23 Shrivas, K., Hayasaka, T., Goto-Inoue, N., Sugiura, Y., Zaima, N., and Setou, M. (2010) Ionic matrix for enhanced MALDI imaging mass spectrometry for identification of phospholipids in mouse liver and cerebellum tissue sections. *Anal. Chem.*, **82** (21), 8800–8806.

24 Hsu, F.F. and Turk, J. (2008) Structural characterization of unsaturated glycerophospholipids by multiple-stage linear ion-trap mass spectrometry with electrospray ionization. *J. Am. Soc. Mass Spectrom.*, **19** (11), 1681–1691.

25 Jurchen, J.C., Rubakhin, S.S., and Sweedler, J.V. (2005) MALDI-MS imaging of features smaller than the size of the laser beam. *J. Am. Soc. Mass Spectrom.*, **16** (10), 1654–1659.

26 Garrison, B.J. and Postawa, Z. (2008) Computational view of surface based organic mass spectrometry. *Mass Spectrom. Rev.*, **27** (4), 289–315.

27 Juhasz, P., Costello, C.E., and Biemann, K. (1993) Matrix-assisted laser desorption ionization mass spectrometry with 2-(4-hydroxyphenylazo)benzoic acid matrix. *J. Am. Soc. Mass Spectrom.*, **4** (5), 399–409.

28 Bouschen, W., Schulz, O., Eikel, D., and Spengler, B. (2010) Matrix vapor deposition/recrystallization and dedicated spray preparation for high-resolution scanning microprobe matrix-assisted laser desorption/ionization imaging mass spectrometry (SMALDI-MS) of tissue and single cells. *Rapid. Commun. Mass Spectrom.*, **24** (3), 355–364.

29 Dreisewerd, K., Schurenberg, M., Karas, M., and Hillenkamp, F. (1995) Influence of the laser intensity and spot size on the desorption of molecules and ions in matrix-assisted laser-desorption/ionization with a uniform

beam profile. *Int. J. Mass Spectrom. Ion Process.*, **141** (2), 127–148.

30 Stoeckli, M., Farmer, T.B., and Caprioli, R.M. (1999) Automated mass spectrometry imaging with a matrix-assisted laser desorption ionization time-of-flight instrument. *J. Am. Soc. Mass Spectrom.*, **10** (1), 67–71.

31 Loboda, A.V., Ackloo, S., and Chernushevich, I.V. (2003) A high-performance matrix-assisted laser desorption/ionization orthogonal time-of-flight mass spectrometer with collisional cooling. *Rapid. Commun. Mass Spectrom.*, **17** (22), 2508–2516.

32 Garrett, T.J. and Yost, R.A. (2006) Analysis of intact tissue by intermediate-pressure MALDI on a linear ion trap mass spectrometer. *Anal. Chem.*, **78** (7), 2465–2469.

33 Garrett, T.J., Prieto Conaway, M.C., Kovtoun, V., Bui, H., Izgarian, N., Stafford, G., and Yost, R.A. (2007) Imaging of small molecules in tissue sections with a new intermediate-pressure MALDI linear ion trap mass spectrometer. *Int. J. Mass Spectrom.*, **260** (2–3), 166–176.

34 Shimma, S., Sugiura, Y., Hayasaka, T., Zaima, N., Matsumoto, M., and Setou, M. (2008) Mass imaging and identification of biomolecules with MALDI-QIT-TOF-based system. *Anal. Chem.*, **80** (3), 878–885.

35 Taban, I.M., Altelaar, A.F., van der Burgt, Y.E., McDonnell, L.A., Heeren, R.M., Fuchser, J., and Baykut, G. (2007) Imaging of peptides in the rat brain using MALDI-FTICR mass spectrometry. *J. Am. Soc. Mass Spectrom.*, **18** (1), 145–151.

36 Pol, J., Vidova, V., Kruppa, G., Kobliha, V., Novak, P., Lemr, K., Kotiaho, T., Kostiainen, R., Havlicek, V., and Volny, M. (2009) Automated ambient desorption–ionization platform for surface imaging integrated with a commercial Fourier transform ion cyclotron resonance mass spectrometer. *Anal. Chem.*, **81** (20), 8479–8487.

37 Landgraf, R.R., Prieto Conaway, M.C., Garrett, T.J., Stacpoole, P.W., and Yost, R. A. (2009) Imaging of lipids in spinal cord using intermediate pressure matrix-assisted laser desorption-linear ion trap/orbitrap MS. *Anal. Chem.*, **81** (20), 8488–8495.

38 Bohrer, B.C., Merenbloom, S.I., Koeniger, S.L., Hilderbrand, A.E., and Clemmer, D.E. (2008) Biomolecule analysis by ion mobility spectrometry. *Annu. Rev. Anal. Chem.*, **1**, 293–327.

39 Jackson, S.N., Ugarov, M., Egan, T., Post, J.D., Langlais, D., Albert, S.J., and Woods, A.S. (2007) MALDI-ion mobility-TOFMS imaging of lipids in rat brain tissue. *J. Mass Spectrom.*, **42** (8), 1093–1098.

40 McLean, J.A., Ridenour, W.B., and Caprioli, R.M. (2007) Profiling and imaging of tissues by imaging ion mobility-mass spectrometry. *J. Mass Spectrom.*, **42** (8), 1099–1105.

41 Stauber, J., MacAleese, L., Franck, J., Claude, E., Snel, M., Kaletas, B.K., Wiel, I.M., Wisztorski, M., Fournier, I., and Heeren, R.M. (2010) On-tissue protein identification and imaging by MALDI-ion mobility mass spectrometry. *J. Am. Soc. Mass Spectrom.*, **21** (3), 338–347.

42 Smith, D.F., Aizikov, K., Duursma, M.C., Giskes, F., Spaanderman, D.J., McDonnell, L.A., O'Connor, P.B., and Heeren, R.M. (2011) An external matrix-assisted laser desorption ionization source for flexible FT-ICR mass spectrometry imaging with internal calibration on adjacent samples. *J. Am. Soc. Mass Spectrom.*, **22** (1), 130–137.

43 Novartis, Basel, Switzerland (2011) www.maldi-msi.org (September 14).

44 ITT Visual Information Solutions (2011) www.ittvis.com (September 14).

45 Mayo Clinic (2011) www.mayo.edu/bir/Software/Analyze/Analyze.html (September 14).

46 Hankin, J.A., Farias, S.E., Barkley, R.M., Heidenreich, K., Frey, L.C., Hamazaki, K., Kim, H.Y., and Murphy, R.C. (2011) MALDI mass spectrometric imaging of lipids in rat brain injury models. *J. Am. Soc. Mass Spectrom.*, **22** (6), 1014–1021.

47 National Institutes of Health (2011) http://imagej.nih.gov/ij (September 15).

48 Trim, P.J., Atkinson, S.J., Princivalle, A.P., Marshall, P.S., West, A., and Clench, M.R. (2008) Matrix-assisted laser desorption/ionisation mass spectrometry imaging of lipids in rat brain tissue with

integrated unsupervised and supervised multivariant statistical analysis. *Rapid. Commun. Mass Spectrom.*, **22** (10), 1503–1509.

49 Cooks, R.G., Ouyang, Z., Takats, Z., and Wiseman, J.M. (2006) Detection technologies: ambient mass spectrometry. *Science*, **311** (5767), 1566–1570.

50 Amstalden van Hove, E.R., Smith, D.F., and Heeren, R.M. (2010) A concise review of mass spectrometry imaging. *J. Chromatogr. A*, **1217** (25), 3946–3954.

8
Lipid Informatics: From a Mass Spectrum to Interactomics

Christer S. Ejsing, Peter Husen, and Kirill Tarasov

8.1
Introduction

This chapter provides an overview of lipidomic data processing and strategies for integrating high-content lipidomic data sets with other resource data. First, we discuss the system for shorthand lipid species annotation that is at the core of lipid informatics. We then summarize the basic properties of lipid mass spectrometric data formats that are our primary resource for compiling lipid information. Subsequently, we outline strategies for data normalization and discuss tools and approaches for visualization and data mining of lipidomic data sets. Finally, we highlight approaches for integrating lipidomic data sets with other high-content resource data.

Mass spectrometry (MS)-based lipidomic workflows comprise a series of experimental routines that include the following: (i) Construction of a working hypothesis that is to be tested using lipidomics. (ii) Sample preparation that typically involves solvent-based lipid extraction of cell homogenates, tissue homogenates, subcellular organelles or biofluids, or tissue sectioning if spatial lipid distribution is being investigated. (iii) Ionization of lipid analytes using appropriate ionization techniques such as (nano)electrospray ionization or matrix-assisted laser desorption/ionization. Lipid analytes can be ionized by direct infusion (i.e., shotgun lipidomics) or, alternatively, be separated in time using liquid chromatography (LC) prior to ionization. (iv) Mass analysis using various approaches that include survey analysis by high-resolution Fourier transform mass spectrometry (FT MS) and tandem mass spectrometry (MS/MS, MS^n) by collision-induced dissociation using, for example, quadrupole or ion trap-based instruments. (v) Processing of recorded multidimensional spectral data sets by lipid-centric software tools for identification and quantification of detected lipid species followed by calculation of "lipidomic features" such as absolute abundance of lipid species, lipid class composition, and lipid double bond and fatty acyl chain length indexes. (vi) Interpretation of acquired lipidomic data by visualization, comparative multivariate data analysis, and statistics. A general overview of the lipidomic workflow is presented in Figure 8.1.

Lipidomics, First Edition. Edited by Kim Ekroos.

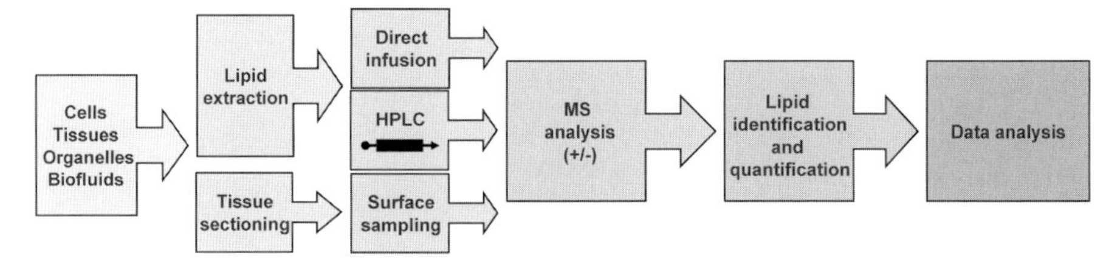

Figure 8.1 Schematic outline of lipidomic workflows.

8.2 Lipid Nomenclature

Lipidomes comprise a multitude of different lipid classes where each class may be composed of a large number of distinct molecular lipid species. According to different estimates, a eukaryotic lipidome might contain 9000–100 000 individual molecular lipid species in total (Figure 8.2a) [1–3]. This enormous compositional complexity necessitates an accurate, concise, and comprehensive system for shorthand lipid species annotation that links the analytical capabilities of mass spectrometric methodologies and the processing, databasing, and computational analysis of high-content lipidome data sets. In-depth structural analysis by mass spectrometry is required to annotate a lipid molecule with a "molecular composition" such as PC 16:0/18:1 that specifically denotes the molecule 1-palmitoyl-2-oleoyl-*sn*-glycero-3-phosphocholine (Figure 8.2b). The quantitative characterization of fatty acid positioning in molecular phosphatidylcholine (PC) species can be performed using multistage activation in negative ion mode on an ion trap mass spectrometer [4]. In comparison, tandem mass analysis on quadrupole-based instruments can only be used for semiquantitative positioning of fatty acid moieties in molecular PC species. When using this mode of analysis, a glycerophospholipid molecule can also be annotated with a "molecular composition," but using a "–" instead of "/" to denote the presence of fatty acid moieties (e.g., PC 16:0–18:1, which denotes PC molecules having both a 16:0 and a 18:1 moiety) [4, 5]. Lipid annotation is sometimes based on survey scans recorded by high-mass resolution analysis (e.g., FT MS). Using this mode of analysis, lipid species should be annotated with a "sum composition" such as PC 34:1 that denotes the total number of carbon atoms and double bonds in the fatty acid moieties of the monitored lipid molecule. The above-mentioned principles of lipid species annotation, which are based on the specificity of the lipidomic methodology, apply to all lipid categories. For example, a monitored triacylglycerol (TG) molecule can be annotated with the sum composition TG 52:3, which can correspond to isomeric and isobaric species with molecular compositions such as TG 16:0/18:1/18:2, TG 18:2/16:0/18:1, and TG 16:1/18:1/18:1. Notably, using tandem mass analysis it is difficult to accurately determine the exact positions of fatty acid moieties in TG species [6]. Thus,

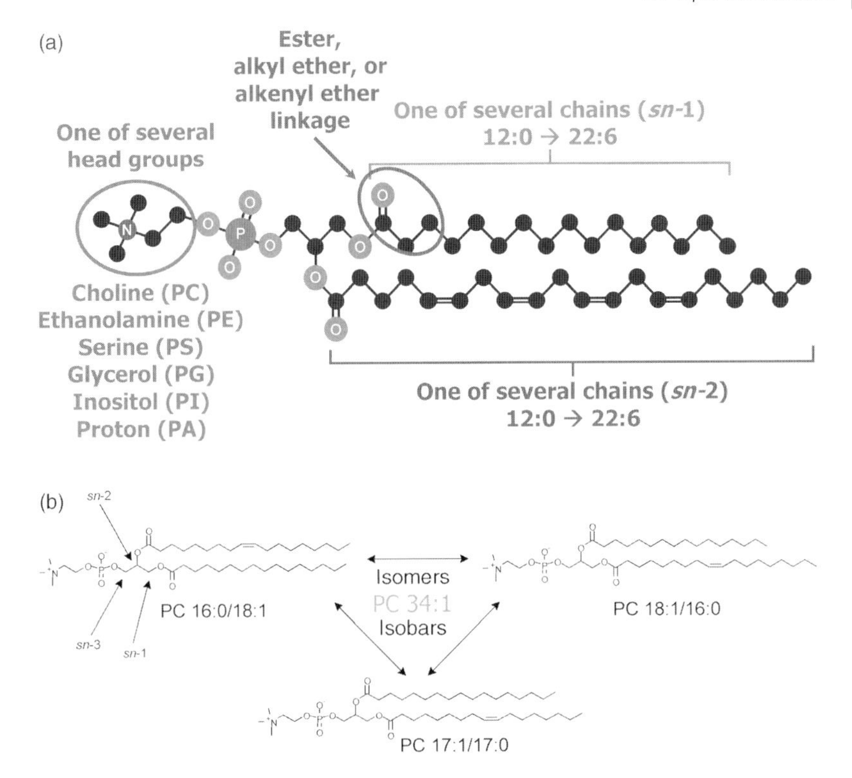

Figure 8.2 Lipid species complexity and annotation. (a) The structural diversity of molecular glycerophospholipid species derives from the multitude of different fatty acid moieties that can be attached to the *sn*-1 and *sn*-2 positions of the glycerophosphate backbone, the chemical linkages of these fatty acid moieties, and the different structures of the polar head groups linked to the glycerophosphate backbone. Combining all the indicated lipid features at random results in an array of several thousands of individual glycerophospholipid species. Note, however, that the molecular composition of glycerophospholipid species in biological systems is determined by the substrate specificities of lipid enzymes. (b) Molecular glycerophospholipid species can be isomeric and isobaric depending on the positioning and composition of the attached fatty acid moieties. Lipids can be annotated by either "molecular composition" that specifies the exact position and chemistry of the attached fatty acid moieties (e.g., PC 16:0/18:1) or by "sum composition" that outlines the total number of carbon atoms and double bonds in the lipid molecules (e.g., PC 34:1). The exact lipid species annotation depends on the analytical capabilities of the applied lipidomic methodology.

identified TG species can be annotated using the molecular composition TG 16:0–18:1–18:2, which denotes only the presence of 16:0, 18:1, and 18:2 moiety.

The exact annotation of all structural attributes in a lipid species necessitates the additional positioning and configuration of double bonds, chemical motifs (e.g., hydroxyl group), and in principle also the stereochemical relationships of the lipid molecule. Notably, no contemporary lipidomic platform can yet afford this level of analytical precision. Interestingly, a recently developed method using ozone-induced dissociation to localize double bonds may be one approach for establishing lipidomic methodology with absolute analytical precision for localization and quantification of all lipid structural attributes [7].

As a first step toward establishing a comprehensive lipid classification system and potentially a system for comprehensive lipid species annotation, the LIPID MAPS Consortium (www.lipidmaps.org) has recently developed the "comprehensive classification system for lipids" featuring an online lipid database [8, 9]. This classification system is generally based on the guidelines for lipid systematic names as defined by the International Union of Pure and Applied Chemists and the International Union of Biochemistry and Molecular Biology (IUPAC-IUBMB) Commission on Biochemical Nomenclature (www.chem.qmul.ac.uk/iupac). The lipid classification system comprises eight lipid categories with distinct chemical features: fatty acyls, glycerolipids, glycerophospholipids, sphingolipids, sterol lipids, prenol lipids, saccharolipids, and polyketides. Each lipid category has its own extensive subclassification hierarchy. For example, the glycerophospholipid category includes the lipid classes: PC, phosphatidylethanolamine (PE), phosphatidylserine (PS), phosphatidylinositol (PI), and phosphatidic acid (PA). In the online lipid database, each subclassification level is composed of molecular lipid species having specific features, including an identifier of 12–14 characters, the chemical structure, lipid name, systematic name, chemical formula, and mass. Although the lipid database features an extensive array of molecular lipid species common to mammalian lipidomes, the database still requires a more comprehensive entry of molecular lipid species from other phyla [9]. Notably, several features have recently been included to link the online lipid database with the analytical specificity of our contemporary lipidomic techniques. For example, abbreviations used for molecular glycerophospholipid species apply to the more universally used two-letter PC/PE/PS/PI/PA format. Since it is difficult to experimentally decipher the exact position of fatty acid moieties for most glycerolipid and glycerophospholipid species, the lipid database has been designed to explicitly highlight the structural isomers using an "[iso]" suffix together with the number of possible isomers (e.g., TG 16:0/17:0/17:1 [iso6] where the structure in the online lipid database corresponds to the fatty acid composition shown in the annotation; alternatively, the mix of molecules could collectively be annotated as TG 16:0–17:0–17:1). In cases where only the glycerophospholipid or glycerolipid species' sum composition is known but the attached fatty acid regiochemistry and stereochemistry are unknown, abbreviations such as TG 52:1 and diacylglycerol (DG) 34:2 are supported. Moreover, the LIPID MAPS annotation of glycerophospholipid and glycerolipid species with 1-alkyl and 1-alkenyl linkages by molecular composition is performed using "O-" and "P-" prefix, respectively (e.g., PE O-16:1/18:1 and PE P-16:1/18:1). We note here that the LIPID MAPS nomenclature has conflicting notations for molecular composition and sum composition of ether-linked species. Monitoring ether lipid species by high-resolution FT MS or LC MS analysis without fragmentation requires that ether lipid species are annotated by sum composition using the prefix "O-". In this mode of analysis, the prefix "O-" includes both 1-alkyl- and 1-alkenyl-linked species. For example, PE O-34:3 *m/z* 700.52757 detected by FT MS analysis can comprise a mixture of 1-alkyl-linked species PE O-16:2/18:1 and 1-alkenyl-linked species PE P-16:1/18:1 as annotated according to the LIPID MAPS nomenclature. Alternatively, the species composition for the 1-alkyl-linked species could be annotated as PE O-16:2a/18:1 (where "a" denotes the alkyl residue) and the 1-alkenyl-linked species could be annotated as PE O-16:2p/18:1 (where "p" denotes the alkenyl/plamenyl residue). Note that this alternative style of species annotation accurately accounts for all double bonds in a numerical format and thereby facilitates the comparison of lipid species annotated by sum composition (based on information from survey scans) and molecular composition (based on information from tandem mass spectra). Similar nomenclature conflicts exist for other lipid classes and modes of analysis. For example, profiling of sphingomyelin (SM) species by precursor ion scan analysis for the fragment ion *m/z* 184.07 in fact only allows sum composition annotation of detected species. However, SM species are typically annotated wrongly using species composition such as SM 16:0 (*m/z* 703.57485) based on the assumption that all sphingomyelin species comprise a C18 sphingosine backbone. This assumption, however, does not always hold true [10]. Thus, the shorthand nomenclature used for annotating lipid species should adequately match the analytical capabilities of the applied lipidomic methodology. Lipid databases should therefore contain a subnomenclature hierarchy to account for molecular compositions and sum compositions especially if the database is used for querying lipidomic data. As the field of lipidomics expands, we might consider developing a grand unified lipid nomenclature system that spans all lipidomic platforms and lipidomic resources.

8.3 Basic Properties of Lipid Mass Spectrometric Data

The mass spectrometric techniques applied in contemporary lipidomics are typically executed on triple quadrupole, ion trap, and various hybrid instrumentations, including quadrupole time-of-flight (TOF), ion trap-Fourier transform ion cyclotron resonance (FTICR), and ion trap–orbitrap machines. Each type of mass spectrometer provides a distinct set of mass analysis methods that typically include survey scans (MS^1) and tandem mass analysis, including single and multiple reaction monitoring, multiplexed MS/MS analysis, data-dependent acquisition, higher order fragmentation analysis (MS^n), and precursor ion and neutral loss scanning analysis for specific lipid fragment ions. The mass analysis can be executed in positive and negative ion modes that promote both distinct ionization efficiencies and fragmentation mechanisms of lipid analytes that can affect the downstream data processing routines.

8.3.1
Mass Spectrum

Common to all mass spectrometric techniques is the recording of mass spectra. A mass spectrum is composed of a series of m/z bin values with corresponding intensities that draw a histogram (Figure 8.3a). Mass spectral data can be stored either in profile mode or in centroid mode (Figure 8.3b). Profile mode data contain m/z bins with the corresponding intensity values that enable the investigator to

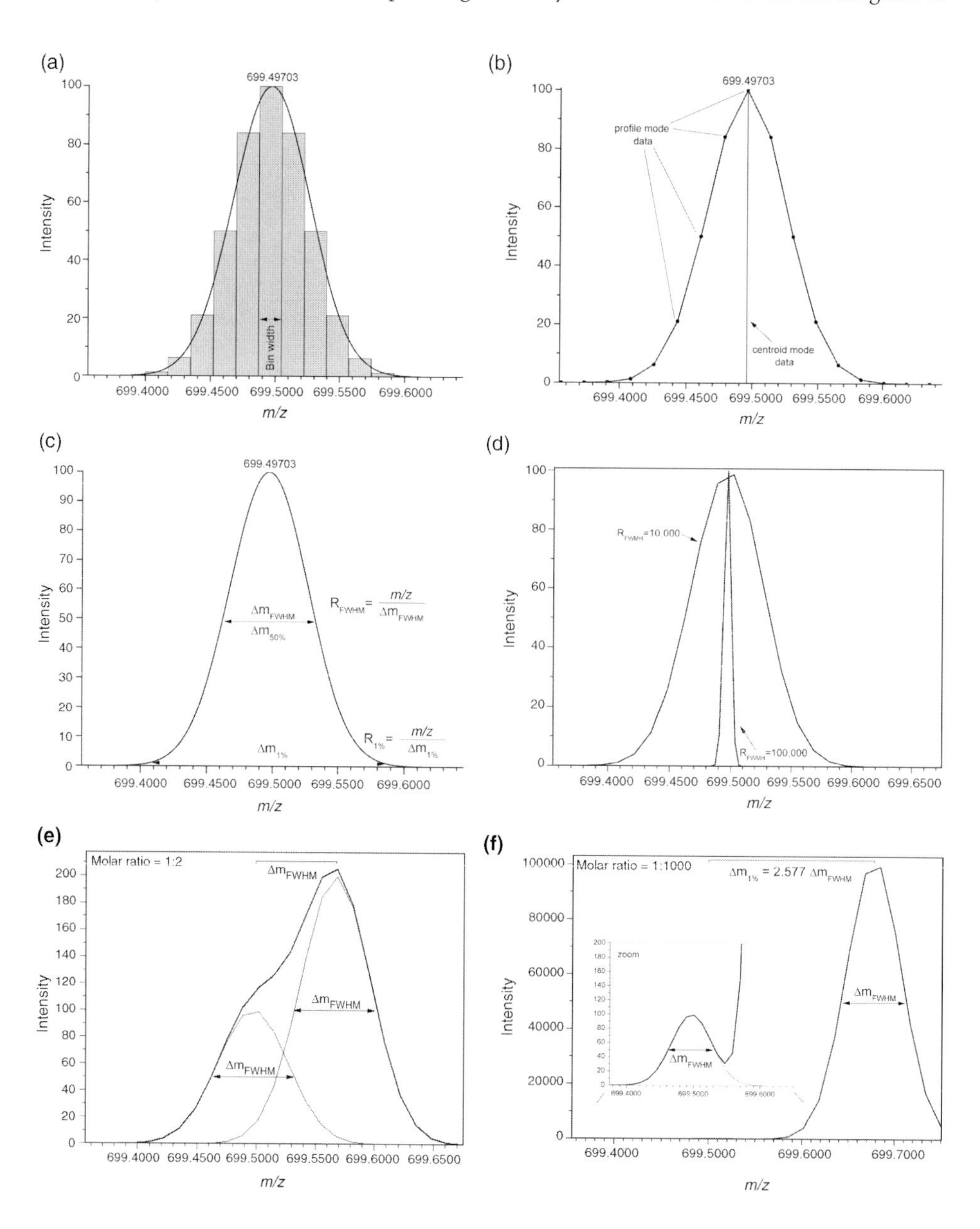

visualize the spectral profile. Centroid mode data contain only centroid m/z values and centroid peak intensities that are typically determined by proprietary algorithms embedded in the instrument software package. Importantly, saving mass spectra in the centroid mode reduces the spectral information content and prevents the investigator from fully assessing spectral data quality as when using profile mode data.

Mass resolution (R), also termed resolving power, is key to the specificity of lipid species identification. The mass resolution is typically specified using the parameter "full width at half maximum" (FWHM) (Figure 8.3c) (http://en.wikipedia.org/wiki/Full_width_at_half_maximum). The FWHM-based mass resolution (R_{FWHM}) is the ratio between the centroid m/z value and the peak width defined as the FWHM (Δm_{FWHM}) (Figure 8.3c). The R_{FWHM} of ion traps and quadrupoles is typically on the order of 1000, whereas TOF analyzers are on the order of 7000–40 000. Ultrahigh mass resolution analysis as achieved using the orbitrap and FTICR can range from 100 000 to above 1 000 000, respectively. Figure 8.3d illustrates the impact of differences in resolving power. Notably, the resolving power of quadrupoles, ion traps, and time-of-flight analyzers is largely independent of m/z. In contrast, ultrahigh mass resolution analyzers suffer from a decaying resolving power as a function of m/z due to the nonlinear dependence of the resonant frequency on m/z. Importantly, key to separating two ions in a mass spectrum is both the instrumental mass resolution and the molecular stoichiometry of the two neighboring ions (Figure 8.3d and e). It is important to note that the commonly applied resolving power R_{FWHM} is not sufficient for separating two molecular ions that differ by only the mass difference Δm_{FWHM} if the molar ratio is more than 1 : 2. If required, baseline separation of two neighboring ions with a molar ratio of 1 : 1000 (as can be observed in lipidomic experiments) can be achieved using a resolving power defined as "full width at 1/100th of maximum" ($\Delta m_{1\%} = \Delta m_{FWHM} \cdot 2.577$) (Figure 8.3e).

Figure 8.3 Illustration of mass resolution and related parameters. (a) A mass spectrum is composed of a series of m/z bins (x-axis) that contain recorded counts (intensity, y-axis). These bins can be displayed as a histogram where each neighboring bin is separated by the bin width. The profile of the histogram follows a Gaussian distribution. (b) A mass spectrum can be recorded and saved as either profile mode data or as centroid mode data. Profile mode data contain all data points (bins and intensities) that allow visualization of the spectral profile. Centroid mode data contain only centroid m/z values and corresponding peak intensities. (c) Definition of mass resolution based on the "full width at half maximum" (R_{FWHM}) and, alternatively, on peak width at 1% of maximum ($R_{1\%}$). (d) Simulated mass spectra with $R_{FWHM} = 10\,000$ and 100 000 as typically obtained using TOF instruments and orbitrap machines, respectively. (e) Simulated mass spectrum (in bold) of two monoisotopic ions that differ by Δm_{FWHM} (determined by the instrument-specific mass resolution). The molar ratio is specified as 1 : 2 in order to emphasize that the instrument-specific resolving power at FWHM is not always sufficient for separating two molecules. (f) Simulated mass spectrum (in bold) of two monoisotopic ions that differ by $\Delta m_{1\%}$. The molar ratio is specified as 1 : 1000 in order to emphasize that the two ions can be separated if using $R_{1\%}$. All plots were made using OriginPro 8.5.

8.3.2
Mass Accuracy and Reproducibility

Mass accuracy and mass reproducibility are also important parameters for the specificity of lipid species identification. Mass accuracy is commonly specified as the ratio of the m/z measurement error ($\Delta m/z$) to the true m/z. The mass accuracy is typically reported in parts per million (ppm, 10^{-6}). Mass reproducibility is the standard error of the m/z measurement. The mass accuracy and the mass reproducibility are closely linked to both the resolving power and the mass spectrometer hardware as both ultrahigh mass resolution machines and some TOF-based instruments can deliver sub-ppm mass accuracy and mass reproducibility. Effectively, a relative mass reproducibility of ± 1 ppm implies that the measured m/z error of a lipid ion with m/z 699.4970 can be within ± 0.0007. Consequently, the measured m/z value of the lipid ion will be in the range of m/z 699.4970 ± 0.0007. Thus, the identification of lipid ions and the concomitant automated export of peak intensities are typically performed using an "m/z tolerance window" that is set according to the mass resolution, mass accuracy, and mass reproducibility of the recorded spectral data. We note that accurate calibration of the mass analyzer is required for achieving sub-ppm mass accuracy and specific lipid identification. We also note that extensive lipidomic experiments with data acquisition over several days might be subject to significant drifts of mass analyzer calibration resulting in a *calibration m/z offset* that will affect the mass accuracy but not the mass reproducibility within each mass spectrometric experiment. The adverse effects of calibration m/z offset on lipid identification can be minimized by online lock-mass calibration using known internal lipid standards or spectral contaminants [11, 12]. Alternatively, sample-specific calibration mass offset can be accounted for after mass spectrometric data acquisition provided the lock-mass ions are specifically detected. Several lipid-centric software routines allow the implementation of offline determined m/z offsets for accurate identification of lipid species and specific export of peak intensities [5, 13].

8.3.3
Isotopes, Deisotoping, and Isotope Correction

Like all (bio)molecules, lipids also feature a distinct distribution of naturally occurring isotopes that has to be taken into account to avoid false-positive identification and biased quantification of lipid species. Isotopes are variants of atoms of a given chemical element, which have different numbers of neutrons (e.g., ^{12}C, ^{13}C, ^{1}H, ^{2}H, ^{6}Li, ^{7}Li). The difference in the number of neutrons yields a nominal mass difference (Δm) of 1 amu per additional neutron. Importantly, the exact mass of a neutron depends on the atomic binding energy of a given chemical element, which is related to the so-called mass defect. For example, the mass of an additional neutron in an H, C, and N atom is 1.00627, 1.00335, and 0.99704 amu, respectively (note that the mass of an electron is 0.00055 amu).

The isotope distribution of a given lipid species is governed by its chemical composition and the associated adduct ion that facilitates ionization (e.g., $^{35}Cl^-$ and $^{37}Cl^-$). Lipids are primarily composed of carbon, hydrogen, nitrogen, oxygen, phosphorus, and sometimes sulfur (e.g., sulfatides). The main contributors to isotopic variance are ^{13}C (1.07% of all C atoms), ^{18}O (0.205% of all O atoms), and ^{34}S (4.29% of all S atoms) (Figure 8.4). Although lipid species comprise a relatively higher number of hydrogen atoms, their impact on the isotope distribution is minor due to the relatively low abundance of ^2H (deuterium; 0.012% of all H atoms). Likewise, the isotope contribution from nitrogen is also relatively minor since lipid molecules typically contain only one or two nitrogen atoms (^{15}N comprise 0.368% of all N atoms). Phosphorus does not contribute to the isotopic variation. The isotope distribution of a molecule can be described mathematically as the product of binomial distributions with parameters that account for the number of all atomic constituents and their respective isotopic abundances (i.e., probabilities p). Importantly, most mass spectrometric software tools feature built-in algorithms (based on Eq. (8.1)) that allow users to simulate the isotopic distribution of molecules. The probability of a (lipid) molecule with a given isotope composition is given by the following expression:

$$
\begin{aligned}
& p(n_{12_C}, n_{13_C}; n_{14_N}, n_{15_N}; n_{16_O}, n_{17_O}, n_{18_O}; \ldots) \\
& = \left(p_{12_C}^{n_{12_C}} p_{13_C}^{n_{13_C}}\right) \frac{(n_{12_C} + n_{13_C})!}{n_{12_C}! n_{13_C}!} \cdot \left(p_{14_N}^{n_{14_N}} p_{15_N}^{n_{15_N}}\right) \frac{(n_{14_N} + n_{15_N})!}{n_{14_N}! n_{15_N}!} \\
& \cdot \left(p_{16_O}^{n_{16_O}} p_{17_O}^{n_{17_O}} p_{18_O}^{n_{18_O}}\right) \frac{(n_{16_O} + n_{17_O} + n_{18_O})!}{n_{16_O}! n_{17_O}! n_{18_O}!} \cdots
\end{aligned}
\tag{8.1}
$$

where n_i and p_i are the number and probability, respectively, of the indicated elements (e.g., as stated above, the probability of a ^{13}C atom (p_{13_C}) is 1.07%).

The isotope distribution of deprotonated and singly charged phosphatidylinositol shows a common trend for the majority of lipid molecules (Figure 8.4). The monoisotopic PI ion is the most abundant and corresponds to 57.89% of all PI molecules. The first isotope cluster of PI is offset by approximately 1 amu and corresponds to 30.26% of all PI molecules. The first PI isotope cluster is primarily composed of molecules with ^{13}C atoms (Figure 8.4c). The second isotope cluster is offset by approximately 2 amu compared to the monoisotopic PI ion. The second isotope cluster comprises 9.30% of all PI molecules, and 16.64% and 78.75% of these molecules contain an ^{18}O atom and two ^{13}C atoms, respectively (Figure 8.4d). The third and the fourth isotope clusters correspond to 2.10% and 0.37% of all PI molecules, respectively. Each isotope cluster is primarily composed of a combination of ^{13}C and ^{18}O atoms (Figure 8.4e and f). PI molecules with ^2H and ^{17}O atoms constitute a relatively low fraction of all PI molecules. Importantly, the isotope pattern in Figure 8.4 shows that isotope effects from ^{13}C atoms, ^{18}O atoms, and, where relevant, ^{34}S atoms should be considered when processing lipidomic data.

Isotopes can bias the identification and quantification of lipid species. The complicating effects of isotopes on lipidomic data are primarily governed by the spectral

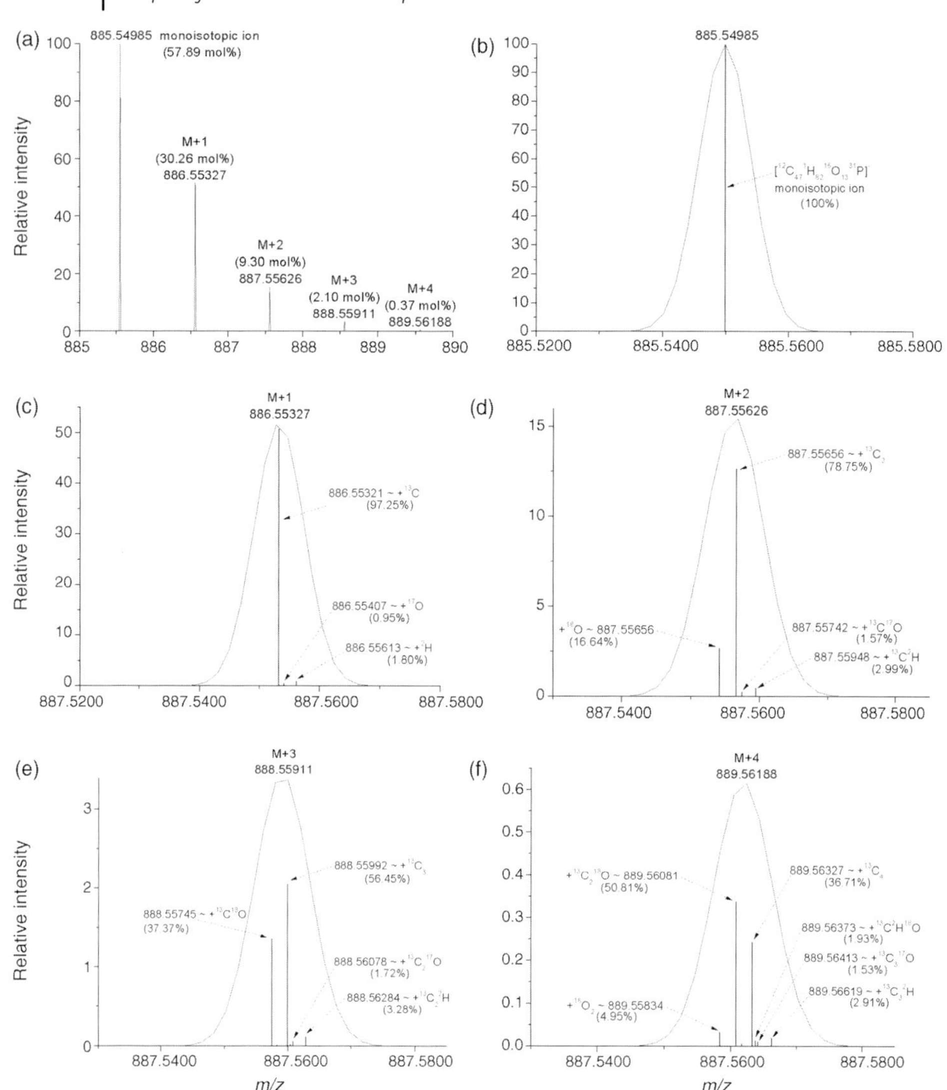

Figure 8.4 Calculated isotope distribution and isotope pattern for negatively charged 1-arachidonoyl-2-stearoyl-*sn*-glycero-3-(1′-myo-inositol) (PI 20:4/18:0, $C_{47}H_{82}O_{13}P_1$). (a) Simulated FT MS spectrum of PI 20:4/18:0 using $R_{FWHM} = 91\,000$ as typically obtained using a LTQ Orbitrap XL. The simulated spectrum is reproduced throughout all panels in the figure. The spectrum emphasizes that monoisotopic PI molecules comprise 57.89 mol% of the total PI molecules. The molar abundances of PI molecules within the first–fourth isotope clusters are given in parentheses. (b) Zoom of the monoisotopic PI 20:4/18:0 at *m/z* 885.54985. A single line is shown for monoisotopic PI 20:4/18:0, which comprises 100% of the isotopic cluster. (c) Zoom of the first isotopic cluster of PI 20:4/18:0. Note that 97.25% of the total PI molecules in the first isotope cluster comprise a ^{13}C atom. PI molecules with a ^{2}H atom and a ^{17}O atom comprise 1.80% and 0.95%, respectively. (d) Zoom of the second isotope cluster of PI 20:4/18:0. Note that 97.25% of all

mass resolution, and isobaric ions that overlap due to either a difference in double bonds and M + 2 or M + 4 isotope clusters (e.g., PI 38:4 + $^{13}C_2$ m/z 887.55656 and PI 38:3 m/z 887.56551, $\Delta m/z = 0.00895$) or a difference in number of nitrogen atoms and M + 1 or M + 3 isotope clusters (e.g., PC 32:2 + ^{13}C m/z 731.54149 and SM 36:1;2 m/z 731.60615, $\Delta m/z = 0.06466$). Ultrahigh mass resolution analysis with a R_{FWHM} on the order of 400 000 should in principle eliminate the majority of isotope interferences. However, since the majority of contemporary mass spectrometers do not support ultrahigh mass resolution analysis, most lipidomic software features algorithms for (i) deisotoping and (ii) isotope correction [5, 14]. Deisotoping is the subtraction of isotope-specific intensity derived from M + 1, M + 2, M + 3, and M + 4 isotope clusters from interfering species (also termed type II isotope correction) [14]. Isotope correction is the adjustment of the (deisotoped) intensity of a monoisotopic peak to yield the total intensity of all lipid molecule isotope clusters (also termed type I isotope correction factor) [14]. The isotope correction is performed by multiplying the (deisotoped) intensity of the monoisotopic peak by an isotope correction factor that represents the relative molar stoichiometry of the entire isotopic cluster of the monitored lipid analyte. The isotope correction serves to correct for systematic differences in isotope distribution between lipid molecules having differences in the number of atoms. For example, the monoisotopic ions of PC 30:0 and PC 40:6 correspond to 64% and 57%, respectively, of their respective isotopic clusters. Ignoring this difference during data processing will inflict a systematic quantitative bias of 12% between these two PC species. A detailed outline of deisotoping and isotope correction of lipid data acquired on instruments with intermediate resolving powers is provided by Ejsing *et al.* [5].

We note that the above-mentioned impact of isotope effects, deisotoping, and isotope correction focused on the processing of survey MS scans where lipid species are detected as intact ions (e.g., TOF MS^1 or FT MS^1 scans). Lipid analysis is typically performed using sensitive fragmentation-based scan modes (e.g., precursor ion scanning (PIS)). Accurate deisotoping of such data requires that only a subpopulation of the isotope clusters is taken into account [5]. For example, all PC and SM molecules can be monitored by PIS for the monoisotopic phosphorylcholine fragment ion m/z 184.07 ($^{12}C_5{}^{1}C_{15}{}^{16}O_4{}^{14}N^{31}P^+$). This monoisotopic fragment

PI molecules in the first isotope cluster comprise a ^{13}C atom. PI molecules with a ^{2}H atom and a ^{17}O atom comprise 1.80% and 0.95%, respectively. (d) Zoom of the second isotope cluster of PI 20:4/18:0. Note that 16.64% of all PI molecules in the second isotope cluster comprise an ^{18}O atom. PI molecules with two ^{13}C atoms comprise 78.75% of the isotopic cluster. The remainder of the cluster is composed of combinations of ^{2}H, ^{17}O, and ^{13}C atoms. (e) Zoom of the third isotope cluster of PI 20:4/18:0. Note how the different combinations of ^{2}H, ^{17}O, ^{18}O, and ^{13}C atoms contribute to the isotopic cluster. (f) Zoom of the fourth isotope cluster of PI 20:4/18:0. Again, note how various combinations of ^{2}H, ^{17}O, ^{18}O, and ^{13}C atoms contribute to the isotopic cluster. The simulated FT MS spectrum and isotope pattern were constructed using QualBrowser 2.0.7 (Thermo Fisher Scientific Inc.) and OriginPro 8.5. For clarity, only the most pronounced isotope patterns are annotated.

ion can only be released from a fully monoisotopic PC/SM precursor ions or from $M+1$, $M+2$, and $M+3$ precursor ions having isotopes situated in the glycerol part or the two fatty acid moieties of the molecule (i.e., the part of the molecule that undergoes neutral loss). Hence, the isotope distribution of a PC or a SM molecule in a PIS *m/z* 184.07 spectrum reflects the isotope pattern of the neutral loss fragment ($[PC/SM - {}^{12}C_5{}^{1}H_{15}{}^{16}O_4{}^{14}N{}^{31}P^+]$) and not the intact lipid ion. Similar principles operate for other fragmentation-based lipidomic methods, including neutral loss scans and data-dependent acquisition. A detailed explanation of the principles underlying deisotoping of lipid fragment ion data is outlined by Ejsing *et al.* [5].

Although the above-mentioned isotope effects impose additional lipidomic data processing to minimize false-positive identification and quantification bias, stable isotopes also enable dynamic lipidomic studies of *in vivo* lipid metabolic pathway activity [15]. For example, stable isotope-labeled lipid metabolic precursors such as ${}^{2}H$-labeled choline and ethanolamine can be used to specifically monitor *de novo* PC and PE syntheses, respectively [16–19].

8.4 Data Processing

The specific workflow for processing lipidomic data depends on the data format of the applied mass analysis techniques. First of all, it is pivotal for the accurate interpretation and validation of most lipidomic data sets that the lipidomist has a firm knowledge of the molecular lipid biochemistry of the organism under investigation, mass spectrometry, and the processing of multivariate data. Depending on the number of monitored lipid classes and molecular lipid species, the mass analysis typically includes a combination of positive and negative ion mode high-resolution survey analysis and various fragmentation-based acquisition routines. For example, the global analysis of the yeast lipidome required six successive automated MS and MS/MS experiments per sample followed by dedicated software routines for processing of the recorded data [20]. Furthermore, the executed mass analysis techniques can be operated in conjunction with direct infusion (shotgun lipidomics), LC, and surface sampling where the two last approaches provide time and space, respectively, as an additional analytical dimension to account for during the data processing. A general overview of the lipidomic data processing is presented in Figure 8.5.

The workflow for processing of raw lipidomic data can be divided into *de novo* lipid identification, targeted export of lipidomic data, and data normalization. Each of these processing modules requires dedicated software routines and operations. Lipidomic software tools (Table 8.1) utilize spectral peak list data for lipid identification and export of spectral data for subsequent processing. Notably, most software tools cannot process the proprietary data formats recorded by mass spectrometers (e.g., wiff, raw) and therefore require a preprocessing module for conversion to either text files (e.g., txt, mgf) or the.mzXML data format [21].

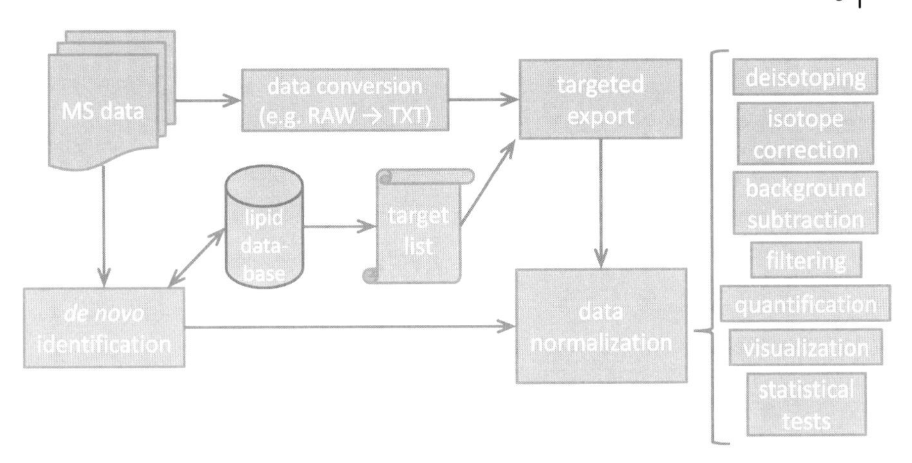

Figure 8.5 Schematic outline of lipidomic data processing. The mass spectrometry data can be analyzed either by *de novo* identification or by targeted export using a target list. A lipid database is used to guide the identification and to construct target lists for targeted processing. After identification and extraction of data from the spectra, a series of routines are carried out for data normalization that can include deisotoping, background subtraction, various filters, and absolute quantification using internal standards.

8.4.1
De Novo Lipid Identification

Lipid species can be identified and annotated using molecular composition or sum composition nomenclature. The identification of molecular lipid species requires the accurate matching of measured *m/z* values of both the intact precursor ion and structure-specific fragment ions with the true *m/z* values calculated from chemical formulas. The identification of a lipid species only by sum composition nomenclature is applicable when monitoring intact lipid species by the release of lipid class-specific fragments (e.g., PC/SM analysis by PIS *m/z* 184.07), by high-resolution FT MS analysis without fragmentation [22, 23], and by LC MS analysis combined with retention time information [24]. All lipidomic software tools are based on the premise of matching, within a user-specified *m/z* tolerance window, the measured *m/z* values with the true *m/z* values calculated from chemical formulas. The width of the *m/z* tolerance window is determined by mass resolution and mass accuracy. To assist the identification, the software tools feature a lipid database with information on the chemical composition of intact lipid species and fragment ions from which the true *m/z* values can easily be calculated and used for the matching. As an alternative to using a lipid database for identification, Herzog *et al.* recently developed *LipidXplorer* that uses a novel two-step data processing scheme [13, 25]. First, a *MasterScan* database is generated with all acquired MS and MS/MS spectra data from samples. Subsequently, lipid species identification within the *MasterScan* database is performed by writing queries in a *molecular fragmentation query language* that can easily be adapted to support *de novo* identification of any lipid

Table 8.1 Overview of different software tools for lipidomics.

	LipidQA	LIMSA	FAAT	LipID	Lipid Search	LipidXplorer	LipidView[a]	ALEX[b]
Lipid ID processing of survey MS data	Yes	Yes	Yes	Yes	Yes	Yes	Yes	Yes
Lipid ID processing of MS/MS data	Yes				Yes	Yes	Yes	Yes
Database with lipid m/z		Yes	Yes	Yes	Yes		Yes	Yes
Database with spectra	Yes							
Deisotoping and isotope correction	Yes	Yes				Yes	Yes	Yes
Cross-platform	Yes	Yes	Yes	Yes	Yes	Yes	Yes[c]	
Spectral alignment			Yes			Yes		
m/z Calibration offset correction						Yes	Yes	Yes
Lipid calculator[d]							Yes	Yes
GUI for cross-correlating lipid identification and MS data							Yes	
Spectral viewer with automated peak annotation[e]							Yes	

Adapted from Ref. [13].
a) LipidView; formerly termed Lipid Profiler [5, 6].
b) ALEX: Analysis of lipid experiments [20]; primarily handles LTQ Orbitrap data.
c) LipidView: Allows processing of all data formats recorded on AB Sciex instruments, including QSTAR, tripleTOF 5600, and 5500 and 4000 QTRAP.
d) A supplementary software tool to assist lipid identification.
e) A software tool that assists validation of lipid identifications.

species. This approach removed the requirement for a reference lipid database and makes it relatively easy to add queries to identify new lipid species.

The output of the first stage of lipid identification is a text file with annotated lipid identities and their corresponding peak m/z values and intensities for intact lipid precursor ions and/or fragment ions where relevant. In order to improve the confidence of lipid identification, several criteria in the form of Boolean expressions can be implemented. For example, stringent criteria for identification of the molecular lipid species PC 16:0–18:0 can be implemented by requiring (i) that both fatty acid-derived fragment ions m/z 255.23 (16:0) and m/z 283.26 (18:0) should be detected by MS/MS analysis of the intact precursor ion in negative ion mode [5, 26], (ii) that the lipid class-specific fragment ion m/z 184.07 should be detected by MS/MS analysis of the intact precursor ion in positive ion mode, and (iii) that only if these criteria are fulfilled, the lipid species should be displayed in the final data output. The software LipidView has a graphical user-interface that allows users to specify lipid identification criteria. Furthermore, to assist the validation of lipid identifications, LipidView features a spectral viewer that supports automated annotation of detected lipid species directly within acquired mass spectra [5].

Song *et al.* [27] recently devised a quantitative approach to monitor the confidence of lipid identification. A scoring scheme was developed based on comparing the actual number of fragment ions derived from a lipid analyte with the number of fragment ions expected based on known fragmentation pathways. Although this approach can be a valuable approach for inspecting and validating identification of lipid ions with good ion intensities, the efficacy of the technique may become limited for low-abundance lipid species with poor ion intensities and the concomitant detection of fewer fragment ions. Notably, the scoring scheme can be applied to spectral data from numerous mass spectrometers, although the accuracy of the scoring scheme is improved when using high-mass resolution data due to its better mass accuracy.

We note here that *de novo* lipid identification should always be performed when initiating lipidomic studies of a new model organism. Importantly, the existing literature on the lipid biochemistry and available lipidomic data on the organisms should be cross-referenced to validate seemingly novel lipid species that do not fit known lipid metabolic pathways. In addition, if a novel or unexpected lipid species is observed, then it should be validated by additional mass spectrometric experiments in both positive and negative ion modes and, if possible, by employing molecular biology tools to modify the organism to either produce more or less of the novel lipid species. Moreover, in all experiments, it is crucial to include appropriate blank runs in order to eliminate any spectral contaminants from the solvents and other additives that may promote false-positive identifications.

8.4.2
Targeted Export of Lipidomic Data

Targeted export of spectral data is an alternative to *de novo* lipid identification. This approach is typically applied in conjunction with multiple reaction monitoring

(MRM) or multiplexed MS/MS analysis where fragment ions of known lipid species are monitored. In addition, targeted export of lipidomic data can be done with high-resolution survey FT MS data and comprehensive MS/MS of all precursor ions as obtained by multiple precursor ion scanning analysis. The targeted export of data uses a *target list* with specific m/z values of targeted precursor lipid ions and, where relevant, fragment ion m/z values. The target list is used to specifically extract only peak intensities of targeted lipid species in a set of sample files. The advantage of this approach is that it speeds up the process of data export compared to *de novo* lipid identification, which requires spectral matching with lipid database information. Using a target list can be of interest when processing hundreds of samples. However, a limitation of this approach is that it may fail to identify novel lipid species that are not specified on the target list. As for *de novo* lipid identification, appropriate blank runs should always be included in order to eliminate potential interferences from spectral contaminants. The output of the targeted export of spectral data is similar to that of *de novo* lipid identification: a text file with annotated lipid identities and their corresponding peak m/z values and intensities for intact lipid precursor ions and fragment ions when relevant. As for *de novo* lipid identification, the specificity of the targeted export of lipid data can also be subjected to several criteria in the form of Boolean expressions as outlined in the section above.

8.4.3
Normalization of Lipidomic Data

After *de novo* lipid identification or targeted export of lipidomic data, a series of processing routines are required to convert data into a final output format. These processing routines can include quality control procedures to evaluate the quality of the exported spectral data set, filtering steps to remove, for example, low-abundance lipid precursor ions detected only in one technical replicate, and data normalization to estimate lipidomic features such as absolute abundances. The data normalization can be based on (i) absolute quantification, (ii) relative quantification, and (iii) intensity profiling. Lipidomic features include dependent variables such as pmol amounts of lipid species, lipid class composition, and double bond and acyl chain length indexes of the monitored lipid classes. The processing of lipidomic data requires data management software since the available lipidomic software tools do not support this operation (except for LipidView that allows some basic data normalization). Thus, lipidomic data are typically normalized using Microsoft Excel or more advanced programs such as SAS software.

Quality control of raw lipidomic data is pivotal for the downstream data processing. Adequate ion statistics (i.e., intensities) are of profound importance, especially for internal lipid standards since these are used for normalization of endogenous lipid species intensities. If an internal lipid standard has poor ion statistics, it may lead to an overestimate of the abundance of endogenous lipid species. An efficient approach to spectral data quality control is to plot the intensities of a few selected endogenous lipid species and internal lipid standards as a function of sample injection and/or sample name (Figure 8.6). This approach also provides

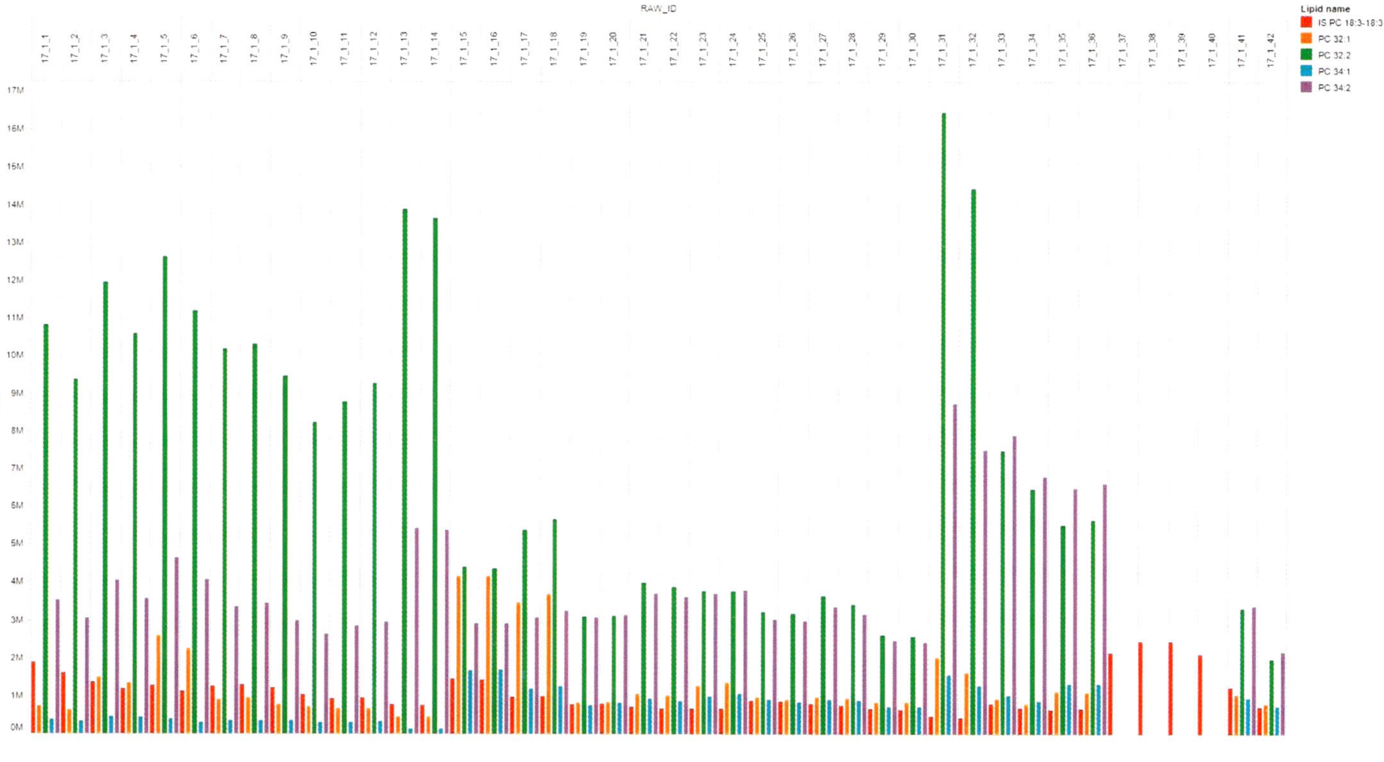

Figure 8.6 Quality control plot. Forty-two yeast lipid extracts were analyzed by positive ion mode FT MS on a LTQ Orbitrap XL equipped with a Triversa NanoMate nanoelectrospray source. FT MS spectra were processed using ALEX software. To assess the quality of the samples, the intensity of selected PC species was plotted as a function of injection number (RAW_ID). Note that only the internal standard IS PC 18:3/18:3 and not the endogenous PC species was detected in blank samples at injection 38–40. This analysis shows that the monitoring of endogenous PC species was specific. The plot was made using Tableau Desktop 6.1.

information on the analytical specificity since no endogenous lipid species should be detected in blank runs. An additional quality control approach, especially for studies with large sample sizes, is to display the number of identified lipid species in each sample as a function of their total intensity. The quality control facilitates an overview of the lipidomic data quality and the rejection of poor quality samples prior to the subsequent processing routines.

Quantitative monitoring of lipidomes is typically performed using (i) absolute quantification, (ii) relative quantification, and (iii) intensity profiling. Absolute quantification is the ultimate data format in analytical chemistry as it reflects the absolute number of molecules in a given sample matrix. One of the advantages of absolute quantification is that it provides information about molar stoichiometry, which is of interest for studies of biological membranes. Absolute quantification of a given lipid species requires at least one unique internal lipid standard of identical lipid class, that a known amount of this standard is spiked into the sample matrix prior to lipid extraction, and that the internal lipid standard is specifically detected. This internal lipid standard serves to correct for potential losses during extraction (i.e., extraction efficiency) and lipid class-specific ionization efficiency and detection efficiency if quantification is performed by MS/MS analysis. The absolute amount of an endogenous lipid species is estimated by normalizing its deisotoped intensity to the intensity of the internal standard and by multiplying the spike amount of the internal lipid standard with the isotope correction factor [5]. Absolute quantification is typically expressed in molar amount (mol) or mass (g) lipid species per unit of sample material (e.g., cells, μg total protein, μg total DNA, pmol total phosphate, and mol% of all monitored lipid species). Importantly, both the molar amount of lipid species and the unit of sample material are (stochastic) variables having an average and a variance. Thus, accurate absolute quantification of lipid species necessitates careful control of not only the mass spectrometric analysis but also the unit of sample material used for sampling, which otherwise may precipitate systematic quantification biases between samples.

Relative quantification is typically applied when an endogenous lipid species can be specifically detected, but no adequate class-specific internal lipid standard is (commercially) available. To address this analytical limitation, an internal standard that is not reminiscent of the lipid analyte is to be spiked into the sample matrix prior to or after lipid extraction. After MS analysis, the ratio between the intensities of endogenous lipid analytes and the intensity of the internal standard can be determined for all samples and compared to monitor whether a given endogenous lipid species is more or less abundant in a subset of the samples [28]. In comparison to absolute quantification, relative quantification fails to account for biased extraction, ionization, and detection efficiencies, which might jeopardize the quantification accuracy and reproducibility.

Intensity profiling refers to the quantitative monitoring of endogenous lipid species without the use of internal standards. This approach is based on the ability to simultaneously detect a multitude of endogenous lipid species from different lipid classes within the same MS or MS/MS scan and subsequent normalization of intensities of individual lipid species to the sum of all monitored lipid species (i.e., the sum = 100%). This approach can, for example, be used in conjunction with screening routines by high-mass resolution FT MS analysis, which in positive ion mode affords the simultaneous detection of PC, PE, TG, DG, ceramide, sphingomyelin, hexosylceramide, sterol ester, lysophosphatidylcholine (LPC), and lysophosphatidylethanolamine species [23]. By this strategy, the intensity% value of all monitored lipid species across a large sample set can be compared to pinpoint lipid species that display a pronounced fluctuation. Since matrix effects can affect the ionization efficiency of lipids, it is important to validate changes in lipid profiles by additional in-depth structural analysis and absolute quantification [22, 23].

The output of the above-mentioned processing routines is a data list/file with estimated abundances of monitored lipid species that constitute a set of *independent lipidomic features*. A set of *dependent lipidomic features* can be calculated from the independent lipidomic features. These features include lipid class composition (e.g., mol% of all monitored PC, PE, PI, and TG species), the lipid species double bond index (e.g., % of all PC species having zero, one, and two double bonds), and the lipid species fatty acyl chain length index (e.g., % of all PE species having an 18:1 moiety). The calculation of dependent lipidomic features serves to reduce the lipidomic data complexity and thereby assist the data interpretation and elucidation of the molecular mechanisms behind the observed lipidomic perturbations.

8.5 Lipidomic Data Mining and Visualization

In the previous sections we described identification and quantification of lipid species. The choice of statistical and data mining tools for in-depth lipidomic analysis is partly determined by the aim(s) of the lipidomic study. In this section, we summarize different approaches used for assessing and visualizing significant differences in lipidomic data sets.

8.5.1 Comparative Lipidomics

A variety of lipidomic studies aim to quantitatively characterize and compare the molecular lipid composition of distinct cell types, subcellular organelles (e.g., secretory vesicles), cell-derived particles (e.g., viruses), tissues, or biofluids [20, 29–32]. Since these types of studies typically focus on a limited number of experimental conditions, the comparative analysis is readily performed using bar plot (histograms) of lipidomic features. Error bars should be added to highlight the technical and/or biological variation between replicates. Box plot of lipid abundances is an alternative approach to compare and report characteristic lipid compositions in a given set of samples. For studies with larger numbers of sample groups (e.g., clinical cohorts), the application of univariate statistical methods can be a powerful approach to identify significant differences. Significant differences in lipidomic features between different sample groups can be determined using conventional Student's *t*-test [33] or analysis of variance for normally distributed data [29, 34] or with nonparametric analogues: Kruskal-Wallis test and Wilcoxon Rank-Sum Test [35]. Notably, the higher the number of monitored lipid features, the higher the chance of identifying false-positive significant differences. Thus, *p*-values calculated using the conventional univariate methods should be corrected to account for multiple hypothesis testing using methods based on stringent Bonferroni correction [36] or false-discovery rates [37].

8.5.2
Multivariate Data Analysis

The univariate methods described above are well suited for pinpointing lipidomic differences. To harness the more complex interrelationships between changing abundances of numerous lipid species, multivariate analyses are required. Hierarchical clustering analysis is a common method for discovering similarity trends in data sets [38]. First, the relative difference in lipid species or other lipidomic features are calculated. Subsequently, a clustering algorithm is applied for sorting lipidomic features into clusters having similar relative changes. The results are visualized as heat maps with a color-coding that accounts for the magnitude of differences in lipidomic features. In addition, tree dendograms on the side of heat maps can be used to depict the correlation between perturbed lipidomic features and sample groups.

Another powerful multivariate method is principal component analysis [22, 39, 40]. This method was developed for reducing data dimensionality by calculating (principal) components that are constructed from the variables (e.g., lipidomic features) that explain the largest proportion of the variance (differences) within a data set. Principal component score plots serve to visualize whether identical or similar samples form discrete clusters that coincide with the variance (differences) within the data set (Figure 8.7b). Loading plots serve to visualize variables that account for the biggest differences. Figure 8.7 illustrates how principal component analysis can be used to distinguish mice subjected to different experimental conditions. Importantly, to assess the efficacy of the principal component analysis, the percentage of the total variance (differences) explained by each principal component should be displayed/reported. Moreover, it is important to validate the results of principal component analysis by confirming identified differences within the raw data. Figure 8.7 displays spectral subtraction graphs that clearly demonstrate the differences between lipid species in the raw mass spectral data.

8.5.3
Lipidomics in Biomarker Research

A different type of statistical analysis is required when lipidomic data are used for biomarker diagnostics. Here, it is tested whether certain lipidomic features are able to discriminate between two or more sample groups (e.g., disease or no disease).

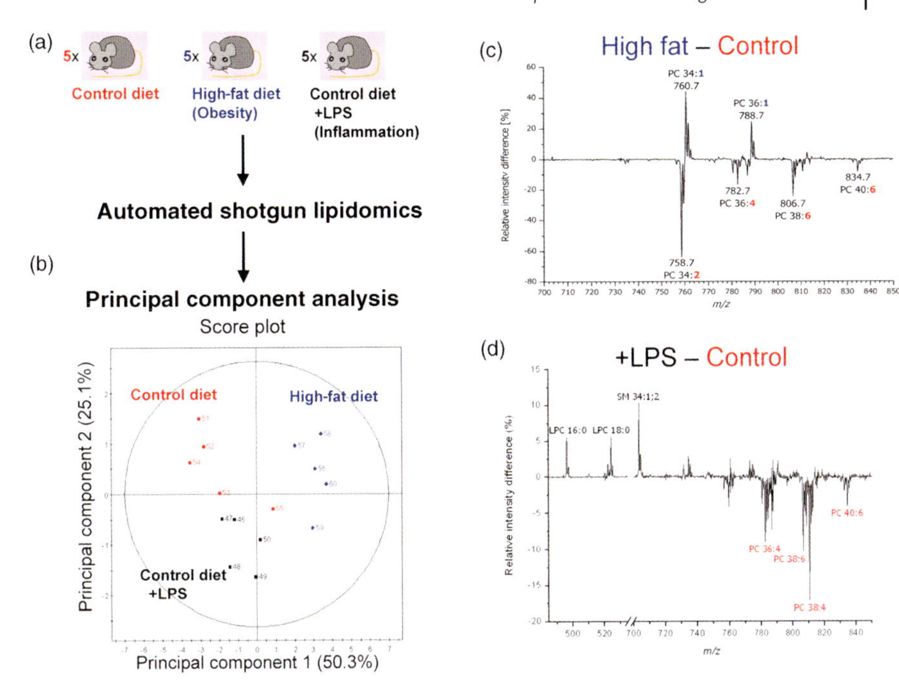

Figure 8.7 Principal component analysis of lipidomic data. (a) Mice were fed control diet, high-fat diet, or control diet and injected with lipopolysaccharide (LPS). Lipid extracts of liver tissue were analyzed by automated shotgun lipidomics using a QSTAR mass spectrometer equipped with a nanoelectrospray ion source Triversa NanoMate and data processing by LipidView [5, 41]. (b) Principal component analysis of lipidomic data. The score plot shows separate clustering of samples from each of the experimental conditions. The first and second principal components account for 50.3% and 25.1%, respectively, of the variance (differences) between lipidomic features in the samples. (c) Validation of the principal component analysis by subtraction of PIS *m/z* 184.07 spectra. This analysis demonstrates that mice fed a high-fat diet synthesize more monounsaturated and less polyunsaturated PC species compared to mice fed a control diet. (d) Spectral subtraction of PIS *m/z* 184.07 spectra. The analysis indicates that mice injected with LPS selectively hydrolyze polyunsaturated PC species to produce 20:4 and 22:6-derived signaling molecules and lysophosphatidylcholine (LPC).

Logistic regression is routinely used in medical sciences for predicting a probability of a medical condition based on cohort data from individuals. Logistic regression does not require assumption about the distribution of the variables and can be used with continuous and discrete values. The outcome variable is a categorical binary parameter being equal to 0 (nondisease) or 1 (disease). Conditional logistic regression (a special type of logistic regression for matched group analysis) was recently applied in a lipidomic cohort study to assess the risk of development of type 2 diabetes [42]. Lipid profiles of individuals who developed diabetes within a 12 year period were compared with a control group who did not develop the disease. The study discovered an increased risk associated with lipid species having a low degree of double bonds and shorter fatty acid chain moieties especially within the pool of TG species.

8.6 Lipidomic Data Integration

The integration of biological data from multiple *omics* technologies can be a powerful approach for understanding biological function. Notably, the information content of contemporary high-throughput lipidomic technologies is an emerging resource that is increasingly being combined with other *omics* data through bioinformatic strategies. These strategies also include methods for projecting lipidomic data onto the underlying lipid metabolic networks. A variety of tools are available for integration and analysis of *omics* data sets [43, 44], mining, and integration of lipidomic data sets [45, 46].

Data integration starts with the selection of a knowledge base for data mapping and interpretation of results. Information about lipid metabolic pathways can be extracted from databases of metabolic pathways using, for example, the Kyoto Encyclopedia of Genes and Genomes (KEGG) [47], Reactome [48], and MetaCyc [49]. Notably, these resources are not restricted to lipid metabolism and cover a broad range of metabolic and signaling pathways. Lipidomic data sets can be projected on lipid pathway maps. A pathway map is a diagram with connections of relevant enzymes to corresponding lipid "metabolites." Lipid metabolites that change during an experimental perturbation can be color-coded according to the magnitude and direction of change. Such representation can be produced using a graphical editor or with the help of automated pathway drawing tools such as KegArray [45] and VANTED [50]. Although knowledge about some lipids such as eicosanoids allows representation of detailed metabolic transitions of particular molecular species, the equivalent option is not yet available for individual glycerophospholipid and sphingolipid species. Distinct glycerophospholipid and sphingolipid species are projected onto simplified lipid class-centered metabolic networks despite the fact that molecular species can be differentially metabolized. To combat this limitation, the LIPID MAPS pathway browser allows differential display of glycerophospholipid and sphingolipid species depending on their acyl chain compositions (http://www.lipidmaps.org/pathways/vanted.html). Since the pathway maps also represent the interrelationship between enzymes and lipid species, the networks can also be used for projecting abundances or transcript levels of corresponding enzymes.

Several studies have been performed to survey the interrelationship of lipidomic and transcriptomic changes [45, 51–53]. The ultimate goal of such investigations is to capture system perturbations that provide evidence that changes in lipid features are related to specific expression patterns of distinct genes. Correlation analysis, such as Pearson correlation, is a useful method for mathematically confirming a significance of an association between gene transcript levels and related lipid species [31, 54].

Despite a fair understanding of lipid metabolism, our knowledge about the regulatory mechanisms that sustain lipid homeostasis is still poor. Lipid homeostasis can be controlled by multiple mechanisms (e.g., enzyme activities, mass action, transcriptional regulators, allosteric modulation, and signal transduction events). However, the complexity of lipid metabolism makes it difficult to ascertain which enzymes and regulators are responsible for a given molecular lipidome composition. Various function prediction algorithms have been developed to facilitate the functional classification of genes and proteins. These methods automatically propose gene/protein functions based on existing association data such as gene coexpression and protein–protein and genetic interactions [55, 56]. The proposed gene/protein function is based on the premise that a given gene/protein shares the biochemical function with the genes/proteins with which it interacts. Thus, a gene/protein that was not previously attributed to lipid homeostasis but has many connections to known lipid-related genes is likely to be involved in lipid metabolism.

Recently, Breslow *et al.* [57] identified ORM1 and ORM2 as conserved negative regulators of *de novo* sphingolipid metabolism. A large-scale screen for genetic interactions in the yeast *Saccharomyces cerevisiae* revealed a strong anticorrelation between ORM1 and ORM2 and the genes LCB1 and LCB2, which encode subunits of the serine palmitoyl-CoA transferase that catalyzes the first step in sphingolipid biosynthesis. The implication of ORM1 and ORM2 in sphingolipid metabolism was confirmed by lipidomic analysis that demonstrated both a reduction of long-chain bases and ceramide levels in mutants overexpressing ORM1 and ORM2 and an increase of all sphingolipid species in ORM1/2 deletion mutants. Additional experiments revealed that the regulatory activity of ORM1 and ORM2 proteins is modulated by phosphorylation. A more recent study has now identified the protein kinase Ypk1 as responsible for the phosphorylation of the ORM proteins [58].

8.7
Conclusions and Future Perspectives

For decades, lipids/fats were considered only passive structural components of biological systems. This was in part due to a series of technological developments in the 1950s and onward that enabled researchers to focus their efforts on DNA research, molecular biology, genetics, structural biology, and protein chemistry. The results of these efforts have undoubtedly shaped our modern-day society. In comparison, the progress in understanding lipid function and metabolism has been attenuated by lack of adequate technology that can cope with the structural complexity and dynamics of lipids. Notably, progress in the field of lipid research was accelerated about 10 years ago by technological developments in mass spectrometry that powered more sensitive and accurate lipid analysis.

Today we are equipped with lipidomic methodologies that afford sensitive, quantitative, and lipidome-wide analysis of molecular lipid species. The methodology has already prompted new experimental avenues for harnessing the molecular mechanisms that operate global lipid metabolism, lipid homeostasis, and the dynamics of biological membranes [20, 30, 32, 57, 59–61]. The future developments of lipidomic methodology will undoubtedly rely on future advances in mass spectrometry technology, data processing and visualization strategies, and systems biology approaches that serve to integrate lipidomic, metabolomic, proteomic, and genomic data sets.

Acknowledgments

We thank Dr. Kim Ekroos for help with depicting the lipidomic workflow (Figure 8.1). This work was supported by Lundbeckfonden (95-310-13591, CSE) and the Danish Council for Independent Research/Natural Sciences (09-072484, CSE).

References

1 van Meer, G. (2005) Cellular lipidomics. *EMBO J.*, **24**, 3159–3165.
2 Yetukuri, L., Ekroos, K., Vidal-Puig, A., and Oresic, M. (2008) Informatics and computational strategies for the study of lipids. *Mol. Biosyst.*, **4**, 121–127.
3 Shevchenko, A. and Simons, K. (2010) Lipidomics: coming to grips with lipid diversity. *Nat. Rev. Mol. Cell Biol.*, **11**, 593–598.
4 Ekroos, K., Ejsing, C.S., Bahr, U., Karas, M., Simons, K., and Shevchenko, A. (2003) Charting molecular composition of phosphatidylcholines by fatty acid scanning and ion trap MS3 fragmentation. *J. Lipid Res.*, **44**, 2181–2192.
5 Ejsing, C.S., Duchoslav, E., Sampaio, J., Simons, K., Bonner, R., Thiele, C., Ekroos, K., and Shevchenko, A. (2006) Automated identification and quantification of glycerophospholipid molecular species by multiple precursor ion scanning. *Anal. Chem.*, **78**, 6202–6214.
6 Murphy, R.C., James, P.F., McAnoy, A.M., Krank, J., Duchoslav, E., and Barkley, R.M. (2007) Detection of the abundance of diacylglycerol and triacylglycerol molecular species in cells using neutral loss mass spectrometry. *Anal. Biochem.*, **366**, 59–70.
7 Thomas, M.C., Mitchell, T.W., Harman, D.G., Deeley, J.M., Nealon, J.R., and Blanksby, S.J. (2008) Ozone-induced dissociation: elucidation of double bond position within mass-selected lipid ions. *Anal. Chem.*, **80**, 303–311.
8 Fahy, E., Subramaniam, S., Brown, H.A., Glass, C.K., Merrill, A.H., Murphy, R.C., Raetz, C.R.H., Russell, D.W., Seyama, Y., Shaw, W., Shimizu, T., Spener, F., van Meer, G., VanNieuwenhze, M.S., White, S. H., Witztum, J.L., and Dennis, E.A. (2005) A comprehensive classification system for lipids. *J. Lipid Res.*, **46**, 839–861.
9 Fahy, E., Subramaniam, S., Murphy, R.C., Nishijima, M., Raetz, C.R., Shimizu, T., Spener, F., van Meer, G., Wakelam, M.J., and Dennis, E.A. (2009) Update of the LIPID MAPS comprehensive classification system for lipids. *J. Lipid Res.*, (50 Suppl.), S9–S14.
10 Karlsson, K.A. (1970) Sphingolipid long chain bases. *Lipids*, **5**, 878–891.
11 Schuhmann, K., Almeida, R., Baumert, M., Herzog, R., Bornstein, S.R., and Shevchenko, A. (2012) Shotgun lipidomics on a LTQ Orbitrap mass spectrometer by successive switching between acquisition polarity modes. *J. Mass Spectrom.*, **47**, 96–104.
12 Olsen, J.V., de Godoy, L.M., Li, G., Macek, B., Mortensen, P., Pesch, R., Makarov, A., Lange, O., Horning, S., and Mann, M. (2005) Parts per million mass accuracy on an orbitrap mass spectrometer via lock-mass injection into a C-trap. *Mol. Cell Proteomics*, **4** (12), 2010–2021.

13 Herzog, R., Schwudke, D., Schuhmann, K., Sampaio, J.L., Bornstein, S.R., Schroeder, M., and Shevchenko, A. (2011) A novel informatics concept for high-throughput shotgun lipidomics based on the molecular fragmentation query language. *Genome Biol.*, **12**, R8.

14 Han, X. and Gross, R.W. (2005) Shotgun lipidomics: electrospray ionization mass spectrometric analysis and quantitation of cellular lipidomes directly from crude extracts of biological samples. *Mass Spectrom. Rev.*, **24**, 367–412.

15 Postle, A.D. and Hunt, A.N. (2009) Dynamic lipidomics with stable isotope labelling. *J. Chromatogr. B Analyt. Technol. Biomed. Life Sci.*, **877**, 2716–2721.

16 Pynn, C.J., Henderson, N.G., Clark, H., Koster, G., Bernhard, W., and Postle, A.D. (2011) Specificity and rate of human and mouse liver and plasma phosphatidylcholine synthesis analyzed *in vivo*. *J. Lipid Res.*, **52**, 399–407.

17 Boumann, H.A., Damen, M.J.A., Versluis, C., Heck, A.J.R., de Kruijff, B., and de Kroon, A. (2003) The two biosynthetic routes leading to phosphatidylcholine in yeast produce different sets of molecular species: evidence for lipid remodeling. *Biochemistry*, **42**, 3054–3059.

18 Boumann, H.A., de Kruijff, B., Heck, A.J., and de Kroon, A.I. (2004) The selective utilization of substrates *in vivo* by the phosphatidylethanolamine and phosphatidylcholine biosynthetic enzymes Ept1p and Cpt1p in yeast. *FEBS Lett.*, **569**, 173–177.

19 Bilgin, M., Markgraf, D.F., Duchoslav, E., Knudsen, J., Jensen, O.N., de Kroon, A.I., and Ejsing, C.S. (2011) Quantitative profiling of PE, MMPE, DMPE, and PC lipid species by multiple precursor ion scanning: a tool for monitoring PE metabolism. *Biochim. Biophys. Acta*, **1811**, 1081–1089.

20 Ejsing, C.S., Sampaio, J.L., Surendranath, V., Duchoslav, E., Ekroos, K., Klemm, R. W., Simons, K., and Shevchenko, A. (2009) Global analysis of the yeast lipidome by quantitative shotgun mass spectrometry. *Proc. Natl. Acad. Sci. USA*, **106**, 2136–2141.

21 Pedrioli, P.G., Eng, J.K., Hubley, R., Vogelzang, M., Deutsch, E.W., Raught, B., Pratt, B., Nilsson, E., Angeletti, R.H., Apweiler, R., Cheung, K., Costello, C.E., Hermjakob, H., Huang, S., Julian, R.K., Kapp, E., McComb, M.E., Oliver, S.G., Omenn, G., Paton, N.W., Simpson, R., Smith, R., Taylor, C.F., Zhu, W., and Aebersold, R. (2004) A common open representation of mass spectrometry data and its application to proteomics research. *Nat. Biotechnol.*, **22**, 1459–1466.

22 Schwudke, D., Hannich, J.T., Surendranath, V., Grimard, V., Moehring, T., Burton, L., Kurzchalia, T., and Shevchenko, A. (2007) Top-down lipidomic screens by multivariate analysis of high-resolution survey mass spectra. *Anal. Chem.*, **79**, 4083–4093.

23 Graessler, J., Schwudke, D., Schwarz, P.E., Herzog, R., Shevchenko, A., and Bornstein, S.R. (2009) Top-down lipidomics reveals ether lipid deficiency in blood plasma of hypertensive patients. *PLoS One*, **4**, e6261.

24 Hermansson, M., Uphoff, A., Kakela, R., and Somerharju, P. (2005) Automated quantitative analysis of complex lipidomes by liquid chromatography/mass spectrometry. *Anal. Chem.*, **77**, 2166–2175.

25 Herzog, R., Schuhmann, K., Schwudke, D., Sampaio, J.L., Bornstein, S.R., Schroeder, M., and Shevchenko, A. (2012) LipidXplorer: a software for consensual cross-platform lipidomics. *PLoS One*, **7**, e29851.

26 Schwudke, D., Oegema, J., Burton, L., Entchev, E., Hannich, J.T., Ejsing, C.S., Kurzchalia, T., and Shevchenko, A. (2006) Lipid profiling by multiple precursor and neutral loss scanning driven by the data-dependent acquisition. *Anal. Chem.*, **78**, 585–595.

27 Song, H., Hsu, F.F., Ladenson, J., and Turk, J. (2007) Algorithm for processing raw mass spectrometric data to identify and quantitate complex lipid molecular species in mixtures by data-dependent scanning and fragment ion database searching. *J. Am. Soc. Mass Spectrom.*, **18**, 1848–1858.

28 Guan, X.L. and Wenk, M.R. (2006) Mass spectrometry-based profiling of

phospholipids and sphingolipids in extracts from *Saccharomyces cerevisiae*. *Yeast*, **23**, 465–477.

29 Mitchell, T.W., Ekroos, K., Blanksby, S.J., Hulbert, A.J., and Else, P.L. (2007) Differences in membrane acyl phospholipid composition between an endothermic mammal and an ectothermic reptile are not limited to any phospholipid class. *J. Exp. Biol.*, **210**, 3440–3450.

30 Sampaio, J.L., Gerl, M.J., Klose, C., Ejsing, C.S., Beug, H., Simons, K., and Shevchenko, A. (2011) Membrane lipidome of an epithelial cell line. *Proc. Natl. Acad. Sci. USA*, **108**, 1903–1907.

31 Dennis, E.A., Deems, R.A., Harkewicz, R., Quehenberger, O., Brown, H.A., Milne, S.B., Myers, D.S., Glass, C.K., Hardiman, G., Reichart, D., Merrill, A.H., Sullards, M.C., Wang, E., Murphy, R.C., Raetz, C.R.H., Garrett, T.A., Guan, Z., Ryan, A.C., Russell, D.W., McDonald, J.G., Thompson, B.M., Shaw, W.A., Sud, M., Zhao, Y., Gupta, S., Maurya, M.R., Fahy, E., and Subramaniam, S. (2010) A mouse macrophage lipidome. *J. Biol. Chem.*, **285**, 39976–39985.

32 Brugger, B., Glass, B., Haberkant, P., Leibrecht, I., Wieland, F.T., and Krasslich, H.G. (2006) The HIV lipidome: a raft with an unusual composition. *Proc. Natl. Acad. Sci. USA*, **103**, 2641–2646.

33 Shui, G., Stebbins, J.W., Lam, B.D., Cheong, W.F., Lam, S.M., Gregoire, F., Kusonoki, J., and Wenk, M.R. (2011) Comparative plasma lipidome between human and cynomolgus monkey: are plasma polar lipids good biomarkers for diabetic monkeys? *PloS One*, **6**, e19731.

34 Chan, R.B., Oliveira, T.G., Cortes, E.P., Honig, L.S., Duff, K.E., Small, S.A., Wenk, M.R., Shui, G., and Di Paolo, G. (2011) Comparative lipidomic analysis of mouse and human brain with Alzheimer's disease. *J. Biol. Chem.*, **287**, 2678–2688.

35 Sergent, O., Ekroos, K., Lefeuvre-Orfila, L., Rissel, M., Forsberg, G.-B., Oscarsson, J., Andersson, T.B., and Lagadic-Gossmann, D. (2009) Ximelagatran increases membrane fluidity and changes membrane lipid composition in primary human hepatocytes. *Toxicol. In Vitro*, **23**, 1305–1310.

36 Bonferroni, C.E. (1936) Teoria statistica delle classi e calcolo delle probabilità. *Pubblicazioni del R Istituto Superiore di Scienze Economiche e Commerciali di Firenze*, **8**, 3–62.

37 Benjamini, Y.H.Y. (1995) Controlling the False Discovery Rate: A Practical and Powerful Approach to Multiple Testing. *J. R. Stat. Soc. Series B*, **57** (1), 289–300.

38 Eisen, M.B., Spellman, P.T., Brown, P.O., and Botstein, D. (1998) Cluster analysis and display of genome-wide expression patterns. *Proc. Natl. Acad. Sci. USA*, **95**, 14863–14868.

39 Jackson, J.E. (1991) *A User's Guide to Principal Components*, John Wiley & Sons, Inc., New York.

40 Ivosev, G., Burton, L., and Bonner, R. (2008) Dimensionality reduction and visualization in principal component analysis. *Anal. Chem.*, **80**, 4933–4944.

41 Stahlman, M., Ejsing, C.S., Tarasov, K., Perman, J., Boren, J., and Ekroos, K. (2009) High-throughput shotgun lipidomics by quadrupole time-of-flight mass spectrometry. *J. Chromatogr. B Analyt. Technol. Biomed. Life Sci.*, **877**, 2664–2672.

42 Rhee, E.P., Cheng, S., Larson, M.G., Walford, G.A., Lewis, G.D., McCabe, E., Yang, E., Farrell, L., Fox, C.S., O'Donnell, C.J., Carr, S.A., Vasan, R.S., Florez, J.C., Clish, C.B., Wang, T.J., and Gerszten, R.E. (2011) Lipid profiling identifies a triacylglycerol signature of insulin resistance and improves diabetes prediction in humans. *J. Clin. Invest.*, **121**, 1402–1411.

43 Gehlenborg, N., O'Donoghue, S.I., Baliga, N.S., Goesmann, A., Hibbs, M.A., Kitano, H., Kohlbacher, O., Neuweger, H., Schneider, R., Tenenbaum, D., and Gavin, A.-C. (2010) Visualization of omics data for systems biology. *Nat. Methods*, **7**, S56–S68.

44 Ng, A., Bursteinas, B., Gao, Q., Mollison, E., and Zvelebil, M. (2006) Resources for integrative systems biology: from data through databases to networks and dynamic system models. *Brief. Bioinform.*, **7**, 318–330.

45 Wheelock, C.E., Wheelock, A.M., Kawashima, S., Diez, D., Kanehisa, M.,

van Erk, M., Kleemann, R., Haeggström, J.Z., and Goto, S. (2009) Systems biology approaches and pathway tools for investigating cardiovascular disease. *Mol. Biosyst.*, **5**, 588–602.

46 Subramaniam, S., Fahy, E., Gupta, S., Sud, M., Byrnes, R.W., Cotter, D., Dinasarapu, A.R., and Maurya, M.R. (2011) Bioinformatics and systems biology of the lipidome. *Chem. Rev.*, **111**, 6452–6490.

47 Kanehisa, M., Goto, S., Sato, Y., Furumichi, M., and Tanabe, M. (2011) KEGG for integration and interpretation of large-scale molecular data sets. *Nucleic Acids Res.*, **40**, D109–D114.

48 Joshi-Tope, G., Gillespie, M., Vastrik, I., D'Eustachio, P., Schmidt, E., de Bono, B., Jassal, B., Gopinath, G.R., Wu, G.R., Matthews, L., Lewis, S., Birney, E., and Stein, L. (2005) Reactome: a knowledgebase of biological pathways. *Nucleic Acids Res.*, **33**, D428–D432.

49 Caspi, R., Altman, T., Dreher, K., Fulcher, C.A., Subhraveti, P., Keseler, I.M., Kothari, A., Krummenacker, M., Latendresse, M., Mueller, L.A., Ong, Q., Paley, S., Pujar, A., Shearer, A.G., Travers, M., Weerasinghe, D., Zhang, P., and Karp, P.D. (2011) The MetaCyc database of metabolic pathways and enzymes and the BioCyc collection of pathway/genome databases. *Nucleic Acids Res.*, **40**, D742–D753.

50 Junker, B.H., Klukas, C., and Schreiber, F. (2006) VANTED: a system for advanced data analysis and visualization in the context of biological networks. *BMC Bioinformatics*, **7**, 109.

51 Kleemann, R., Verschuren, L., van Erk, M. J., Nikolsky, Y., Cnubben, N.H.P., Verheij, E.R., Smilde, A.K., Hendriks, H.F.J., Zadelaar, S., Smith, G.J., Kaznacheev, V., Nikolskaya, T., Melnikov, A., Hurt-Camejo, E., van der Greef, J., van Ommen, B., and Kooistra, T. (2007) Atherosclerosis and liver inflammation induced by increased dietary cholesterol intake: a combined transcriptomics and metabolomics analysis. *Genome Biol.*, **8**, R200.

52 Gupta, S., Maurya, M.R., Merrill, A.H., Glass, C.K., and Subramaniam, S. (2011) Integration of lipidomics and transcriptomics data towards a systems biology model of sphingolipid metabolism. *BMC Syst. Biol.*, **5**, 26.

53 Momin, A.A., Park, H., Portz, B.J., Haynes, C.A., Shaner, R.L., Kelly, S.L., Jordan, I.K., and Merrill, A.H. (2011) A method for visualization of "omic" datasets for sphingolipid metabolism to predict potentially interesting differences. *J. Lipid Res.*, **52**, 1073–1083.

54 Gupta, S., Maurya, M.R., Stephens, D.L., Dennis, E.A., and Subramaniam, S. (2009) An integrated model of eicosanoid metabolism and signaling based on lipidomics flux analysis. *Biophys. J.*, **96**, 4542–4551.

55 Wang, P.I. and Marcotte, E.M. (2010) It's the machine that matters: predicting gene function and phenotype from protein networks. *J. Proteomics*, **73**, 2277–2289.

56 Warde-Farley, D., Donaldson, S.L., Comes, O., Zuberi, K., Badrawi, R., Chao, P., Franz, M., Grouios, C., Kazi, F., Lopes, C.T., Maitland, A., Mostafavi, S., Montojo, J., Shao, Q., Wright, G., Bader, G.D., and Morris, Q. (2010) The GeneMANIA prediction server: biological network integration for gene prioritization and predicting gene function. *Nucleic Acids Res.*, **38**, W214–W220.

57 Breslow, D.K., Collins, S.R., Bodenmiller, B., Aebersold, R., Simons, K., Shevchenko, A., Ejsing, C.S., and Weissman, J.S. (2010) Orm family proteins mediate sphingolipid homeostasis. *Nature*, **463**, 1048–1053.

58 Roelants, F.M., Breslow, D.K., Muir, A., Weissman, J.S., and Thorner, J. (2011) Protein kinase Ypk1 phosphorylates regulatory proteins Orm1 and Orm2 to control sphingolipid homeostasis in *Saccharomyces cerevisiae*. *Proc. Natl. Acad. Sci. USA*, **108**, 19222–19227.

59 Klemm, R.W., Ejsing, C.S., Surma, M.A., Kaiser, H.J., Gerl, M.J., Sampaio, J.L., de Robillard, Q., Ferguson, C., Proszynski, T.J., Shevchenko, A., and Simons, K. (2009) Segregation of sphingolipids and sterols during formation of secretory vesicles at the *trans*-Golgi network. *J. Cell Biol.*, **185**, 601–612.

60 Gijon, M.A., Riekhof, W.R., Zarini, S., Murphy, R.C., and Voelker, D.R. (2008)

Lysophospholipid acyltransferases and arachidonate recycling in human neutrophils. *J. Biol. Chem.*, **283**, 30235–30245.

61 Kurat, C.F., Wolinski, H., Petschnigg, J., Kaluarachchi, S., Andrews, B., Natter, K., and Kohlwein, S.D. (2009) Cdk1/Cdc28-dependent activation of the major triacylglycerol lipase Tgl4 in yeast links lipolysis to cell-cycle progression. *Mol. Cell*, **33**, 53–63.

9
Lipids in Human Diseases

M. Mobin Siddique and Scott A. Summers

9.1
Introduction

Lipid biosynthesis has long been recognized to be essential for the maintenance of cellular homeostasis, but recent studies reveal prominent roles in several prominent diseases. Whether this is due to specific roles as signaling intermediates triggering cellular stress responses or lipotoxic triggers of cellular dysfunction, the role of lipids in the disruption of tissue function that underlies disease is no longer debatable. Many of the recent discoveries in this area are due to advances in technologies to detect specific lipid intermediates (e.g., lipidomics) that are the focus of this book, and it is our pleasure to preface this important topic.

While most lipids are delivered to peripheral tissues on lipoproteins (as complex lipids released as triglyceride-bound fatty acids released by cellular bound lipases), they are resynthesized into complex lipids within the cell. These reactions occur predominantly in the endoplasmic reticulum (ER). Lipids produced by cells (glycerolipids (GLs), fatty sphingolipids cholesterol, and sphingolipids) are used as energy reserves, as building blocks for membranes, and as precursors for various signaling intermediates. Defects in lipid synthesis or processing give rise to a host of conditions, including obesity, diabetes, nonalcoholic fatty liver disease, cancer, and so on. The major sites of lipid synthesis or lipogenesis in the body are liver, adipose tissue, and the intestine, but the process occurs almost universally in all cell types.

Due to advances in the aforementioned mass spectroscopy-based lipidomics, which allows detection of a vast array of lipids not known to previously exist, a new complexity in the role of lipids and cell function is emerging. Indeed, the LIPID MAPS initiative has identified more than 10 000 individual lipid species. Studies resulting from the addition of exogenous lipids or from the genetic manipulations of their endogenous levels have revealed potent roles in various disease processes. Moreover, increased dietary intake of lipids has proven to have profound effects on physiology and pathology. The dyslipidemia that occurs from the irregular consumption, synthesis, packaging, and transport of lipids and lipoproteins is one of the most prominent risk factors for metabolic diseases and cancer.

Lipidomics, First Edition. Edited by Kim Ekroos.

9.2
Obesity

As the obesity epidemic sores, scientists continue to struggle to explain how excess adiposity predisposes individuals to cardiovascular disease, diabetes, and cancer. The answer clearly seems to rely upon fat. A large fraction of excess lipid is stored in the form of triglyceride, predominantly in adipose tissue, and an expanded adipose mass (i.e., obesity) is a precursor for numerous diseases. In particular, visceral adipose tissue predisposes individuals to diabetes, hypertension, atherosclerosis, cardiomyopathy, and certain forms of cancer [1]. A number of hypothetical mechanisms have been proposed to explain how increased adipose stores affect tissue function in a deleterious way.

1) One theory is that the delivery of nutrients to cells or tissues in excess of their oxidative or storage capacities leads to the creation of metabolites that antagonize tissue function (e.g., ceramides, reactive oxygen species, and acylcarnitines). In times of food overabundance, the persistent accumulation of these metabolites in peripheral tissues likely contributes to sustained tissue damage (i.e., lipotoxicity) throughout the organism [1].
2) A second theory is that the increased adiposity induces a chronic inflammatory state characterized by elevated circulating levels of cytokines produced from adipocytes or from macrophages infiltrating the fat pad [2–4]. These inflammatory mediators have direct actions of their own, but they also induce catabolic processes, thus further increasing the delivery of nutrient metabolites to insulin-responsive organs [2–4]. Both of these mechanisms likely contribute to various aspects of the tissue dysfunction associated with obesity.
3) Another major mechanism by which expanded adipose mass contributes to disease relates to its newly accepted role as an endocrine organ that produces and secretes a variety of cytokines, hormones, and other metabolic players involved in the pathogenesis of cardiovascular disorders (CVDs). The amount of lipid stored within this organ has a profound impact on the rate of adipokine secretion. Secreted hormones include peptide hormones, plasminogen activator inhibitor-1, resistin, estrogen, adiponectin, glucocorticoids, and so on, which can have either positive or negative effects on tissue function. Contributing to this is the fact that endoplasmic reticulum stress that develops in adipocytes because of increased cytokine production triggers an efflux of free fatty acids (FFA) from adipocytes into the circulation and increases lipid overload in skeletal muscle, liver, and pancreas.
4) A final hypothesis relates to the idea that expanded adipose tissue becomes undervascularized and this impairs adipocyte function [5]. In such situations, increased catabolism in mitochondria leads to oxidative stress and modulates expression of inflammation-related adipokines, inflammatory cytokines, and reactive oxygen species. The impairment in mitochondrial function further impairs metabolic disease, and the presence of dying adipocytes may lead to further recruitment of macrophages into the area.

In general, the above defects likely lead to an impairment in metabolic flexibility and also to an associated resistance to the anabolic hormone insulin, leading to an inability to increase mitochondrial respiration and/or anabolism in the catabolic or postprandial state, thus leading to an inability to control lipolysis [6].

9.3 Dyslipidemia

A major source of lipids in the body derives from dietary sources, but a number of sources are generated *de novo*. Dyslipidemia refers to a condition where there is an imbalance of high-density lipoproteins (HDLs) versus low-density lipoproteins (LDLs) or very low-density lipoproteins (VLDLs). HDLs (e.g., high-density lipoproteins or good cholesterol) are generally considered lipid scavengers that clean the body of excess fat while also initiating protective benefits of their own. In contrast, low-density or very low-density lipoproteins are enriched in saturated fats and predict an "unhealthy" lipid profile that is suggestive of disease. As an acknowledgment of the importance of lipids in disease, numerous drugs have been developed to control the lipid profile, even in the absence of specific disease symptoms. For example, statins reduce cholesterol biosynthesis when disease symptoms are not present. Similarly, metformin (by acting as a mitochondrial poison) promotes lipid oxidation, and is sometimes given to "pre"-diabetics who show signs of impaired glucose tolerance but have not yet demonstrated criteria associated with diabetes. Finally, the once widely prescribed thiazolidinediones promoted adipogenesis, leading to an enhanced accumulation of lipids where they belong (in adipose tissue) and thus to a preservation of metabolic disease.

9.4 Diabetes

The expansion of diabetes in both rich and poor nations is now almost legendary, as the popular press has made numerous attempts to demonstrate how gluttony and sloth contribute to the disease. To be honest, in the authors' opinions, we still understand very little about the obesity epidemic. Humans do not exercise enough, and increased energy expenditure delays onset of virtually all diseases. But beyond that universal truth, we know very little. Regardless of this controversial statement, there is little doubt that obesity contributes to the development of many diseases. Strangely, however, obesity rates do not correlate as strongly with diabetes as one might predict. For example, in Singapore, obesity rates are only 6%, but are far outpaced by diabetes rates. In contrast, in America obesity affects over 30% of the population, while diabetes rates are lower than those of Singaporeans. Yet the relationship between diabetes and obesity seems clear, does it not? So why this fact does not disprove this hypothesis?

Some 20+ years ago, we taught every MD/MBBS graduate that diabetes could be divided into two forms. There existed a fat, adult-onset form, dominated by insulin resistance, where the anabolic hormone insulin was incapable of promoting sufficient glucose uptake or nutrient storage. There also existed a young, thin form, characterized by β-cell dysfunction that led to impaired insulin secretion. This line of thinking was wrong back then, however, sometimes it is still perpetuated by physicians today. While it is true that thin, young diabetics tend to display more severe β-cell dysfunction than the obese ones, impaired β-cell function is absolutely required for diabetes, while insulin resistance (the ability to respond to a maximal dose of insulin with optimal anabolic responses) is not.

In β-cells and many other cells, glucose and lipid metabolic pathways converge into a glycerolipid/free fatty acid cycle, which is driven by the substrates glycerol-3-phosphate and fatty acyl-CoA that are derived from glucose and fatty acids, respectively. Chronically elevated plasma lipid concentrations, particularly when combined with high glucose, elicit a plethora of effects that cause the progressive deterioration of insulin sensitivity and ultimately cellular malfunction or β-cell death, a process known as glucolipotoxicity. Ohtsubo *et al.* [7] reported a novel mechanism by which high-fat diet induces type 2 diabetes. Pancreatic β-cell surface express the Glut 2 glucose transporter that stimulates insulin secretion and is essential for glucose sensing in the liver and β-cells. This alteration may lead to hyperglycemia, impaired glucose tolerance, hyperinsulinemia, hepatic steatosis, and diminished insulin action in muscle and adipose tissues [7].

Several different sphingolipid metabolites have emerged as potentially important regulators of β-cell survival, proliferation, and function. Ceramide, which can be produced in response to inflammatory cytokines (e.g., tumor necrosis factor-α or interleukin-1), inhibits insulin gene expression, blocks β-cell proliferation, and induces β-cell apoptosis [8–12]. Gangliosides, which are glycosylated derivatives of ceramide, are antigens implicated in the onset of the autoimmune response [13, 14]. Conversely, sphingosine-1-phosphate (S1P), a bioactive metabolite of ceramide that is a ligand for a family of G-protein-coupled receptors, promotes β-cell growth and survival and negates ceramide-induced apoptosis. Using pharmacological strategies to slow down the rates of ceramide synthesis in rodents, the Unger group demonstrated that this prevented diabetes, though it seems to be limited to those lacking an intact leptin signaling system [11, 15]. We later confirmed that inhibition of ceramide synthesis was protective using highly specific reagents [16]. Insulin stimulates postprandial glucose uptake into skeletal muscle, and this phenomenon is impaired by fatty acids. While the condition (i.e., insulin resistance) is not in itself a disease, it promotes a hyperinsulinemic state that predisposes individuals to diabetes, cardiovascular disease, and cancer. Experimental elevation of nonesterified fatty acids causes insulin resistance in muscle and liver and inhibits insulin secretion. Lipid accumulation in nonadipose tissues not suited for fat storage leads to the buildup of bioactive lipids that inhibit insulin signaling and metabolism. Epidemiological studies show that saturated fat significantly worsen insulin resistance, while monounsaturated and polyunsaturated fatty acids improve it through modifications in the composition of cell membranes explained at least in part by dietary fat composition. A number of mechanisms have been proposed to explain how increased fat utilization slows down the rates of glucose entry and utilization. One mechanism is that fatty acid metabolites such as ceramide, diacyglycerol, and phosphatidic acid impair insulin signaling [17]. Another mechanism is that fatty acids impair mitochondrial function, leading to the buildup of mitochondrial metabolites that impair insulin action [17].

9.5 Cardiovascular Disorders

The majority of diabetics do not die from diabetes, but rather from cardiovascular complications of the disease. The mechanisms for this relationship have always been unclear, but a common comment from physicians is that "the easiest way to identify a diabetic is when they show up at the hospital for their first heart attack." Increased insulin resistance, glycogen, cholesterol, saturated fats, and so on have all been implicated in the cardiovascular complications of diabetes, but the specific cause is yet to be resolved. Likewise, lipotoxicity associated with abnormal accumulation or metabolism (e.g., oxidative stress) of fat, inflammation, and so on are all implicated with these disorders. Once again, the aforementioned mechanisms are implicated. Regardless, interventional studies have left little doubt that excessive fat delivery to peripheral tissues is problematic [18].

Heart failure results from progressive contractile dysfunction coupled with cardiomyocyte apoptosis [19]. Increased delivery of fat to cardiomyocytes *in vivo* leads to impaired mitochondrial function and cardiomyopathy. For example, treating mice with inhibitors of fatty acid oxidation (i.e., 2-tetradecylglycidic acid [20] or oxfenicine [21]) induces lipid accumulation in the heart, uncouples mitochondria, and impairs diastolic function. In the Zucker diabetic fatty rat, lipid accumulation in the heart is associated with signs of increased apoptosis and impaired heart function [22]. Impressively, cardiac overexpression of long-chain acyl-CoA synthetase [23] or fatty acid transfer protein-1 [24] markedly increases lipid accrual and causes severe cardiac dysfunction.

One of the lipids that has emerged as a lipotoxic metabolite in the heart is again ceramide, which is a favored lipotoxic intermediate by these authors. The Goldberg group found that inhibitors of ceramide biosynthesis ameliorated symptoms of cardiomyopathy, improving systolic function and increasing mouse survival [25]. These data are consistent with *in vitro* reports demonstrating that FFAs that induce ceramide synthesis (i.e., palmitate and stearate), but not those that failed to produce the sphingolipid (i.e., myristate, oleate, or palmitoleate), failed to induce cardiomyocyte damage [26–30]. Moreover, Dyntar *et al.* [29] found that membrane-permeable ceramide analogues mimicked the effects of palmitate on cardiomyocyte apoptosis and myofibril degeneration and that inhibitors of ceramide synthase impaired this palmitate effect. Surprisingly, recent studies suggest that excessive lipid delivery to the heart impairs heart function through a PKC-mediated mechanism [31].

Dyslipidemia also increases the risk of atherosclerosis, which is initiated by the aggregation of low-density lipoproteins within the arterial wall. In 1959, Stamler *et al.* [32] described the association of lipid and atherogenesis in cockerels where they have observed that hypercholesterolemia led to atherogenesis. Atherosclerosis can be considered the most detrimental terminal effect of excess lipid in the body, where LDL level becomes high, enters to the circulatory system, causes lesion in the smooth muscle cells on the inner arterial wall, and finally leads to the thickening of the vessel and interruption of the circulation. Lipid lowering agents "statins" are effective, widely used drugs against atherosclerosis. The process of atherogenesis starts with the formation of plaque due to accumulation of lipids and monocyte-derived foam cells beneath the endothelial layer of the inner arterial wall. Over time, vascular smooth muscle cells form a collagenous cap over this plaque of lipid core and inflammatory cells. Oxidation of the lipids in the plaque aggravates local inflammation and recruitment of smooth muscle cells leading to the thickening of the arterial wall by enlarging the plaque. Damage or rupture of the plaque cap causes release of the thrombogenic collagenous matrix and lipid core to the circulation. These ultimately block the vessel due to fibrin deposition and thrombus formation. Hyperlipidemia, obesity, type 2 diabetes, and insulin resistance have been associated with atherosclerosis, suggesting the role of insulin in this process. Long before we knew the insulin signaling in cells, Stamler *et al.* [33] suggested that insulin might be an intensifier of atherogenesis based on the study done by Duff and Payne [34] on diabetic rabbits. Indeed, they were correct in their hypothesis as insulin signaling is also responsible for this process. Impairment in insulin signaling affects production of nitric oxide by downregulating nitric oxide synthase. Nitric oxide deficiency is reported to induce atherogenesis by multiple pathways. In the event of insulin resistance, hyperactivation of the MAPK pathway causes vascular smooth muscle cell proliferation, increased collagen formation, and excessive production of inflammatory cytokines. It is evident that atherosclerosis occurs due to the presence of higher inflammatory cells in the vessel, producing an influx of inflammatory cytokines. Adipocytes produce several proinflammatory cytokines or adipokines such as TNFα, IL-1β, IL-6, monocyte chemotactic protein-1 (MCP-1), and so on [35–38]. All these events augment the process of atherogenesis. Patients with obesity, insulin resistance, and type 2 diabetes have elevated levels of triglycerides, apolipoprotein B, small dense LDL cholesterol, and low levels of HDL cholesterol. As such, it is also speculated that HDL may have protective role against atherosclerosis and other CVDs.

The mechanisms underlying this abnormal aggregation remain elusive. Some studies have demonstrated that ceramide or other sphingolipids may play a critical role in the progression of the disease. First, the inclusion of excess sphingolipids in the diet of LDL receptor knockout mice significantly influences the formation of atherosclerotic lesions [39]. Second, treating apolipoprotein E knockout mice, which are susceptible to atherosclerosis, with the SPT inhibitor myriocin dramatically reduces lesion area formation [40, 41]. The latter is particularly exciting, as it demonstrates that blocking ceramide formation *in vivo* ameliorates symptoms related to the metabolic syndrome. Several mechanisms have been proposed by which ceramide may contribute to the formation of atherosclerotic lesions. Due to its hydrophobicity and capacity to undergo extensive hydrogen bonding, ceramides have a pronounced tendency for self-aggregation [42, 43]. Thus, several groups have proposed that ceramide generation in LDLs contributes to aggregation of LDLs, which drives the formation of atherosclerotic plaques. Support for this hypothesis includes the following: the ceramide content of aggregated LDLs is 10–50-fold higher than that of plasma LDLs [44]; and, exposing LDL particles to a bacterial sphingomyelinase promotes LDL aggregation [44]. Alternatively, sphingomyelin has been proposed to block uptake of LDLs by blocking access to apolipoprotein E and lipoprotein lipase [45]. Moreover, ceramide has been shown to induce apoptosis of certain cells lining the vascular wall, a process implicated in plaque erosion and thrombosis [46, 47]. And finally, the ceramide metabolite sphingosine-1-phosphate stimulates the proliferation of endothelial and smooth muscle cells in vascular walls [48, 49], which would promote thickening and favor plaque stabilization. *Trans*-fatty acids contribute to atherosclerosis by increasing total cholesterol levels and LDL cholesterol levels, and at the same time they can reduce HDL cholesterol levels. These factors increase cholesterol transport into arteriolar smooth muscle cells, leading to atherogenesis and CVD.

Stroke is an emerging major health problem that often results in death or disability. Hyperlipidemia, high blood pressure, and diabetes are again well-established risk factors. Endothelial dysfunction associated with these risk factors underlies pathological processes leading to atherogenesis and cerebral ischemic injury. While the mechanisms of disease are complex, endothelial dysfunction involves decreased nitric oxide and elevated levels of reactive oxygen species (ROS). At physiological levels, ROS participate in regulation of cellular metabolism. However, when ROS increase to toxic levels through an imbalance of production and neutralization by antioxidant enzymes, they cause cellular injury in the form of lipid peroxidation, protein oxidation, and DNA damage. Central nervous system cells are more vulnerable to ROS toxicity due to their inherent higher oxidative metabolism and less antioxidant enzymes, as well as higher content of membranous fatty acids. During ischemic stroke, ROS concentration rises from normal low levels to a peak point during reperfusion underlying apoptosis or cellular necrosis [50].

9.6 Hereditary Sensory Neuropathy

Hereditary sensory and autonomic neuropathy type 1 (HSAN1) is an inherited neuropathy associated with severe sensory loss, muscle wasting, and ulcers predominantly affecting the lower limbs [51–53]. The condition has long been known to be associated with mutations in the SPTLC1 subunit of serine palmitoyltransferase (SPT) [53, 54], which condenses serine and palmitoyl-CoA to produce a sphingolipid precursor that is essential for the production of all sphingolipids. However, the relative impact of the mutation on SPTLC1 activity has always been controversial, and it was unclear whether a deficit in sphingolipid synthesis was the primary mechanism by which the mutations induced pathogenesis [53]. Thorsten Hornemann from the University Hospital Zurich, Switzerland, presented a definitive piece of work that provides resolution of the issue, and it represents an example of the interaction of human physiology, molecular biology, and lipidomic research to make a substantial impact on human health. Hornemann applied lipidomic strategies to reveal that cells expressing the mutant enzyme produce two atypical deoxysphingoid bases: 1-deoxy-sphinganine and 1-deoxymethyl-sphinganine in patients' plasma samples [55, 56]. The mechanistic explanation for this is that the mutations shift the substrate specificity of SPTLC1, allowing it to utilize as substrates alanine and glycine in addition to serine. These metabolites show neurotoxic effects and are highly elevated in mice and patients expressing the mutant protein. The accumulation of these toxic metabolites affects the sensory neurons and develop the disease [57], and supplementing the diet with excess serine, which leads to the formation of "normal" sphingolipids, appears to be protective [58, 59]. These studies provide a clear indication of mechanism of disease onset and suggest immediately applicable therapeutic interventions that could prevent this debilitating condition.

9.7 Neurodegeneration

Shinitzky [60] from the Weismann Institute of Science published an interesting review on the status of lipids (cholesterol and sphingomyelin) in the brain with aging. Shinitzky suggested in this review that in the aging tissue, subtle imbalances in lipid species occur due to accumulation of cholesterol and other rigidifying lipids such as sphingomyelin in the membranes. This was further confirmed by several groups who showed that peripheral cholesterol and sphingolipids increased progressively with aging and were associated with a range of age-related diseases. Neuronal membranes contain phospholipid pools that are reservoirs for the synthesis of specific lipid messengers upon neuronal stimulation or injury. These messengers in turn participate in signaling cascades that can either promote neuronal injury or neuroprotection. Proinflammatory lipid mediator prostaglandins are synthesized by the release of the substrate fatty acids, such as arachidonic acid, from the cellular phospholipids. Cyclooxygenase 2 (COX-2) catalyzes the conversion of arachidonic acid to prostaglandins that are involved in a number of physiological processes, including inflammation and regulating functions in brain. Inflammatory responses cause activation of the immune system in the brain and result in chronic inflammation that leads to neuronal cell death. This suggests that changes in lipid profile may play significant role in the progression of these disorders. A number of neurodegenerative diseases are reported to be the downstream effects of abnormal lipid metabolites, such as Parkinson's disease (PD) and Alzheimer's disease (AD).

Elevated lipid peroxidation has been reported in PD that may account for cell death in the substantia nigra pars compacta (SNpc) of the brain [61], the region that refines the movement pattern of the body. In the presence of ROS, lipid peroxidation leads to the formation of lipid radicals that cause cellular damage. The membrane phospholipids containing polyunsaturated fatty acids are predominantly susceptible to lipid peroxidation. This can greatly alter the physicochemical properties of membrane lipid bilayers, resulting in severe cellular dysfunction. Lipid peroxidation during oxidative stress in tissues produces 4-hydroxynonenal, isoprostanes, isofurans, isoketals, neuroprostanes, and neurofurans that can induce cell death by inducing ROS and DNA damage.

Two commonly mutated genes in PD, α-synuclein and LRRK2, have been proposed to regulate sphingolipid metabolism in cells. Mutated or truncated α-synuclein is responsible for forming Lewy bodies, which is a hallmark of PD. Lewy bodies are aggregates of α-synuclein and lipids. It has also been suggested that α-synuclein interacts with free PUFA and forms insoluble aggregates. Increases in specific ceramide species and other sphingolipids may alter intracellular signaling and contribute to neuronal dysfunction in PD [62]. Similarly, alterations in DAG and glycerophospholipid metabolism may also modulate neuronal function in the PD [63]. Cholesterol metabolism through the oxysterol pathway is upregulated in the PD brain. These changes are associated with selective changes in the expression of genes responsible for the control of lipid biosynthesis and with increased expression of antioxidant genes [64].

Several findings also document abnormalities in sphingolipid and cholesterol metabolism in the brain of Alzheimer's disease patients. The deposition of β-amyloid peptides in extracellular amyloid plaques within the human brain is a hallmark of AD. The perturbed membrane lipid metabolism in AD may result, at least in part, from this increased β-amyloid production, and the increased ceramide production resulting from the oxidative stress induced by β-amyloid formation may trigger the death of neurons [65]. Lipid rafts that contain sphingolipids and cholesterol within membranes are involved in regulating β-amyloid peptide generation. Sphingolipid–protein interactions are responsible for the misfolding events that cause amyloidogenic (release of β-amyloid peptides) processing of amyloid precursor proteins (APPs). Accumulation of sphingolipids decreases the capacity of cells to clear potentially amyloidogenic fragments of the APP during autophagy. APP is a type I membrane protein and could undergo sequential proteolytic processing by β- and γ-secretase resulting in the generation of the β-amyloid peptide. Genetic, molecular, and biochemical evidence indicates that the accumulation of toxic β-amyloid peptide aggregates plays a critical role in the degeneration of neurons during the pathogenesis of Alzheimer's disease [66].

In the sphingomyelin pathway of ceramide synthesis, sphingomyelinase converts sphingomyelin to ceramide and phosphocholine. In Niemann–Pick disease (types A and B), deficiency of this enzyme causes accumulation of fat and cholesterol in several organs including the brain. The other two variants of this disease (Niemann–Pick diseases C and D) occur due to lack of two lipid-transporting proteins NPC1 and NPC2, which facilitate the intracellular transport of lipids from the lysosome to other cellular sites. These are also known as lysosomal storage disorder and are discussed later.

9.8 Cancer

The diversity in types and causes of cancers makes it difficult to summarize in a succinct fashion the role of lipids in disease progression, and we will only summarize it briefly here. But it is clear that abnormalities in lipid metabolism are key components of a number of cell immortalization and carcinogenic events. Changes or abnormalities in lipid metabolism can result in alterations in cell survival, endocrine balance, eicosanoid production, and immune system function. Numerous studies have demonstrated the role of high-fat diet in increasing the risk of cancer. In particular, altered sphingolipid metabolism has been observed in many carcinogenic conditions. This is consistent with the well-established role of ceramide as an inducer of apoptosis [67–70].

Liver is the central organ in lipogenesis, gluconeogenesis, and cholesterol metabolism. An increase in the prevalence of obesity and lipid metabolic syndrome promotes physiological changes such as alterations in the insulin response, β-oxidation, autophagy, and lipid storage and transport. These may lead to nonalcoholic fatty liver disease (NAFLD), which is affecting 20–30% of adults in industrialized countries and its prevalence is increasing worldwide. NAFLD increases the susceptibility of the liver to acute liver injury and may lead to nonalcoholic steatohepatitis (NASH). NASH is characterized by fatty liver inflammation and is believed to cause fibrosis and cirrhosis. The latter is a known liver cancer risk factor. Inflammatory progression to NASH is proposed to occur by hepatic fat accumulation due to increased hepatic *de novo* lipogenesis, inhibition of fatty acid β-oxidation, impaired triglyceride clearance, and decreased VLDL export. These alterations might lead to inflammation caused by oxidative stress associated with lipid peroxidation, cytokine activation, nitric oxide, and reactive oxygen species formation [71]. Higher levels of LDL and TG were significantly associated with increased breast cancer risk. Elevated plasma LDL-C concentration, which is more susceptible to oxidation, may result in higher lipid peroxidation in breast cancer patients that may cause oxidative stress leading to malignant transformations [72, 73]. During the lipid peroxidation process, the decomposition of lipid hydroperoxides leads to the generation of many compounds known to be reactive intermediates. Reactive oxygen and nitrogen species are the most common electrophiles formed during lipid peroxidation and lead to the formation of both stable and unstable lipid peroxidation products (LPPs). Of the LPPs formed, highly reactive aldehydes are a well-recognized causative factor in aging and age-associated diseases, including cardiovascular disease and diabetes. Mitochondria are both a primary source and a target of LPPs. LPPs are involved in transcriptionally regulating endogenous antioxidant systems. Lipid peroxidation generates lipid hydroperoxides and reactive aldehyde derivatives are capable of modifying proteins, both *in vivo* and *in vitro*, which can react with DNA bases causing massive DNA damage and leading to carcinogenesis [74–77].

Elevated GL/FFA cycle favors tumor cell growth in environments poor in nutrients and growth factors and inhibits apoptosis. Monounsaturated FFA can activate proto-oncogene AKT/PKB as well as phosphoinositide 3-kinase in breast cancer cells [78]. This pro-oncogenic cancer cell growth occurs through the FFA-activated GPCR GPR40 via secretion of FFA released by lipolysis. Studies indicate that the majority of fatty acids in cancer cells are derived from *de novo* synthesis. In cancer, decreased serum cholesterol concentrations are probably related to the increased consumption of cholesterol by the tumor cells, as cholesterol levels in cancer tissues are doubled compared to the normal tissues [79]. In cancer patients with solid tumors, the continuous weight loss and depletion of tissue fat stores have been attributed in part to elevate GL/FFA cycling.

The sphingolipid metabolite, ceramide is one of the most widely studied molecules implicated in several types of cancers. Ceramide acts as tumor-suppressing lipid because of its ability to induce apoptosis and halt cell growth. Diverse cytokine receptors and environmental stresses utilize ceramide to activate apoptotic cascades. C18 ceramide is significantly lower in approximately 70% of the tumor tissues of head and neck squamous cell carcinomas (HNSCCs) patients and decreased C18 ceramide significantly correlates with lymphovascular invasion and nodal metastasis in HNSCC patients [80]. SphK metabolizes ceramide to S1P, which is a tumor-promoting molecule. Elevated levels of S1P or the enzyme that produces it, sphingosine kinase, have been observed in different cancer and tumor tissues. Kawamori *et al.* [81] reported that 89% (42 out of 47) of human colon cancers had higher expression of SphK 1 compared to adjacent normal colon mucosa.

Sphingosine-1-phosphate is a bioactive sphingolipid that promotes cell proliferation, differentiation, angiogenesis, chemotaxis, and migration [82–85]. Many of the activities of secreted S1P molecules are mediated through five closely related G-protein-coupled receptors of the sphingosine-1-phosphate receptor family (S1PR) that play a crucial role in sphingolipid action. Neuroblastoma (NB), the most common extracranial solid tumor of childhood, is highly angiogenic and often displays poor prognosis. NB expresses high levels of sphingosine kinase (SphK2), which is essential for the formation of sphingolipid metabolite sphingosine-1-phosphate. Expression of vascular endothelial growth factor (VEGF) that stimulates vasculogenesis and angiogenesis can be induced S1P. Elevated levels of S1P in NB profoundly influence the tumor microenvironment by inducing VEGF expression. S1P also activates NF-κB by signaling through G-protein-coupled S1P receptors. *In vitro*, S1P protects cells from Fas-mediated apoptosis in a NF-κB-dependent manner [86].

Some sphingolipids like ceramide or gangliosides have been described to promote autophagy or apoptosis in several cancer cell lines. Autophagy is an evolutionary conserved process by which cells recycle intracellular materials to maintain homeostasis in different cellular contexts. Under basal conditions, it prevents accumulation of damaged proteins and organelles; during starvation, autophagy provides cells with sufficient nutrients to survive. The role of autophagy in cancer may differ depending on tumor type or context; however, this plays a significant role in the progression of certain cancers where autophagy provides a protective function to limit tumor necrosis and inflammation.

Fatty acid derivatives like prostaglandins promote tumor cell proliferation and survival via multiple signaling pathways. In hypoxia, prostaglandins enhance ANGPTL4 expression and promote colorectal cancer growth [87]. ANGPTL4 has also been reported as a mediator of vascular metastasis in hypoxic breast cancer cells [88] and human hepatocellular carcinoma [89].

9.9 Lysosomal Storage Disorders

Lipidoses that refer to lysosomal lipid storage disorders occur due to accumulation of lipids in lysosomes, which lead to a series of destructive events that impair cell function. Complex lipids, such as glycosphingolipids, are constitutively degraded within the endolysosomal system by soluble hydrolytic enzymes with the help of lipid binding proteins in a sequential manner. Because of a functionally impaired hydrolases or auxiliary proteins, their lipid substrates cannot be degraded and therefore accumulate in the lysosome, slowly spreading to other intracellular membranes.

Wolman's disease, also known as acid lipase deficiency or Gaucher's disease, is a severe lipid storage disorder resulting from the deficiency of acid β-glucosidase, the enzyme responsible for the lysosomal hydrolysis of the sphingolipid, glucosylceramide, to glucose and ceramide.

Fabry disease, also known as α-galactosidase-A deficiency, causes a buildup of fatty material in the autonomic nervous system, eyes, kidneys, and cardiovascular system. Fabry disease is the only X-linked lipid storage disease occurring due to mutations in the α-galactosidase gene. Deficiency of α-galactosidase, which acts on glycoproteins and glycolipids, impairs the lysosomal hydrolysis of their substrates leading to the development of this disease.

Niemann–Pick disease is caused by the accumulation of fat and cholesterol in the cells of the liver, spleen, bone marrow, lungs, and brain. Niemann–Pick disease is currently subdivided into four categories. Deficiency of acid sphingomyelinase activity accounts for NPD types A and B. Acid sphingomyelinase converts sphingomyelin to ceramide. Defect in acid sphingomyelinase leads to the accumulation of sphingomyelin, cholesterol, glycosphingolipids, and bis(monoacylglycero)-phosphate in the visceral organs such as liver, spleen, and lung [90]. Niemann–Pick disease types C and D are caused by lack of the NPC1 or NPC2 proteins. The NPC proteins, responsible for trafficking low-density lipoproteins to late endosomes and lysosomes, are mutated resulting in the accumulation of cholesterol and glycosphingolipid in late endosomal compartments. This can impair lysosomal function, such as delivery of nutrients through the endolysosomal system, leading to a state of cellular starvation. Excessive cholesterol buildup has also been implicated in clinical manifestations associated with a number of genetically unrelated diseases including cystic fibrosis (CF).

Farber's disease is caused by a deficiency of the enzyme called acid ceramidase in which ceramide accumulates in the lysosomes. Ceramidases are key enzymes that attenuate ceramide-mediated effects and regulate ceramide/sphingosine/sphingosine-1-phosphate levels in cells. These are lysosomal hydrolases encoded by the ASAH1 gene, which hydrolyzes ceramide to sphingosine [91]. Sphingosine kinases then subsequently phosphorylate sphingosine to sphingosine-1-phosphate.

GM1 gangliosidosis and Morquio B are autosomal recessive storage disorders caused by the deficiency of β-galactosidase (GLB1), a lysosomal enzyme that hydrolyzes the terminal β-galactose from gangliosides substrates, keratan sulfate, and other glycoconjugates. Intracellular accumulation of gangliosides causes GM1 gangliosidosis, while accumulation of keratan sulfate and chondroitin-6-sulfate causes Morquio B disorder.

GM2 gangliosidoses are caused by an inherited deficiency of lysosomal β-hexosaminidase that catalyzes the degradation of the GM2 gangliosides [92]. This results in ganglioside accumulation in the brain. Deficiency of hexosaminidase-α causes Tay–Sachs disease, also known as GM2 gangliosidosis variant B. The accumulation of GM2 gangliosides in lysosomes in neurons interrupts other cellular processes and eventually destroys the neurons [93].

9.10 Cystic Fibrosis

Erich Gulbins of the University of Duisburg-Essen in Germany has published compelling data suggesting that ceramide accumulation was involved in the pathogenesis of pulmonary inflammation associated with cystic fibrosis [94, 95]. To summarize briefly, Dorothy H. Andersen (1938) described the disease with its clinical features and she was the pioneer in using pancreatic enzymes to treat patients with CF, as this condition affects the mucous layer of the pancreatic duct and reduces pancreatic enzyme secretion [96, 97]. CF results from mutations in a gene (CFTR) encoding an epithelial chloride channel [98–100]. The life span of these patients depends on the extent of pulmonary inflammation and infection with several bacterial strains [101]. Deficiency in the CFTR protein results in a number of gastrointestinal and pulmonary problems, and a great number of those afflicted suffer from recurrent *Pseudomonas aeruginosa* pneumonia [102, 103]. Gulbins has shown that mice deficient in the CFTR protein have altered sphingolipid metabolism, and accrue high concentration of ceramide in lung tissues [104]. This is problematic, as ceramides contribute to inflammatory processes by clustering cytokine receptors and inducing oxidative stress. In a compelling piece of work, Gulbins showed that heterozygosity of acid sphingomyelinase (ASM), which hydrolyzes sphingomyelin to produce ceramide, or inhalation of specific ASM-inhibitors normalizes ceramide levels and reduces markers of pulmonary inflammation in mice lacking CFTR. Moreover, Gulbins has demonstrated efficacy in a patient with cystic fibrosis, noting improved lung function and few systemic side effects associated with inhalation of an ASM inhibitor [105]. These data suggest novel therapeutic approaches that hold enormous promise as a means for improving the health of cystic fibrosis patients.

9.11 Anti-Inflammatory Lipid Mediators

Lipids have also been shown to have both pro- and anti-inflammatory properties. In a recent meeting of the International Conference on the Bioscience of Lipids, a couple of new types of anti-inflammatory lipids were identified. Charles Serhan from the Brigham and Women's Hospital and the Harvard Medical School in Boston has identified a novel group of lipid-derived anti-inflammatory molecules termed resolvins, protectins, and maresins [106, 107]. Using a systems approach to dissect the molecular mechanism(s) underlying the ways in which mammalian systems control the extent of inflammation, his group identified factors derived from essential omega-3 fatty acid precursors that are capable of inducing an extensive and diverse, yet specific, series of anti-inflammatory events [108, 109]. Moreover, they do so at exceedingly low concentrations. These data provide important insight into the evolution of the resolution of inflammation, challenging the dogma that it is essentially a passive process. Moreover, they suggest novel therapeutic strategies, currently being tested in the clinics, for reducing inflammation [110, 111].

Dennis Voelker of the National Jewish Health Center in Denver, Colorado, has identified another lipid with potentially important anti-inflammatory properties. Voelker determined that palmitoyl-oleoyl-phosphatidylglycerol (POPG), which is the major molecular PG species in surfactant, is an antagonist of proinflammatory cytokine production and suggested that it served an evolutionary conserved role for preventing lung inflammation in response to inhaled bacteria and viruses [112, 113]. These findings suggest a potential therapeutic role, should a mechanism for introducing the lipid be devised, as a means for suppressing pulmonary infections.

9.12 Conclusions

A complete analysis of the role of lipids in cell growth and function is beyond the scope of this relatively short introduction. For example, we have spent little time discussing polyphosphoinositides, which are perhaps the best characterized set of lipids involved in signal transduction, cell proliferation, and calcium homeostasis. What is becoming apparent from efforts such as the LIPID MAPS initiative, however, is that the diversity in the lipid pool has purpose and each of the distinct lipid molecules produced by cells has distinct functions in cellular homeostasis and signal transduction. As a result, the lipidomic technologies described herein are essential for understanding the complex ways in which macromolecules interact to govern cell function. More importantly, they provide opportunities for therapeutic intervention in a wide array of disease arenas.

References

1 Ferris, W.F. and Crowther, N.J. (2011) Once fat was fat and that was that: our changing perspectives on adipose tissue. *Cardiovasc. J. Afr.*, **22**, 147–154.

2 Dandona, P., Aljada, A., and Bandyopadhyay, A. (2004) Inflammation: the link between insulin resistance, obesity and diabetes. *Trends Immunol.*, **25**, 4–7.

3 Fantuzzi, G. (2005) Adipose tissue, adipokines, and inflammation. *J. Allergy Clin. Immunol.*, **115**, 911–919, quiz 920.

4 Hotamisligil, G.S. (2003) Inflammatory pathways and insulin action. *Int. J. Obes. Relat. Metab. Disord.*, (Suppl. 3), S53–S55.

5 Ye, J. (2011) Adipose tissue vascularization: its role in chronic inflammation. *Curr. Diab. Rep.*, **11**, 203–210.

6 Boyle, K.E., Zheng, D., Anderson, E.J., Neufer, P.D., and Houmard, J.A. (2011) Mitochondrial lipid oxidation is impaired in cultured myotubes from obese humans. *Int. J. Obes. (Lond.)*. doi: 10.1038/ijo.2011.201.

7 Ohtsubo, K., Chen, M.Z., Olefsky, J.M., and Marth, J.D. (2011) Pathway to diabetes through attenuation of pancreatic beta cell glycosylation and glucose transport. *Nat. Med.*, **17**, 1067–1075.

8 Kelpe, C.L., Moore, P.C., Parazzoli, S.D., Wicksteed, B., Rhodes, C.J., and Poitout, V. (2003) Palmitate inhibition of insulin gene expression is mediated at the transcriptional level via ceramide synthesis. *J. Biol. Chem.*, **278**, 30015–30021.

9 Maedler, K., Oberholzer, J., Bucher, P., Spinas, G.A., and Donath, M.Y. (2003) Monounsaturated fatty acids prevent the deleterious effects of palmitate and high glucose on human pancreatic beta-cell turnover and function. *Diabetes*, **52**, 726–733.

10 Ishizuka, N., Yagui, K., Tokuyama, Y., Yamada, K., Suzuki, Y., Miyazaki, J., Hashimoto, N., Makino, H., Saito, Y., and Kanatsuka, A. (1999) Tumor necrosis factor alpha signaling pathway and apoptosis in pancreatic beta cells. *Metabolism*, **48**, 1485–1492.

11 Shimabukuro, M., Higa, M., Zhou, Y.T., Wang, M.Y., Newgard, C.B., and Unger, R.H. (1998) Lipoapoptosis in beta-cells of obese prediabetic fa/fa rats. Role of serine palmitoyltransferase overexpression. *J. Biol. Chem.*, **273**, 32487–32490.

12 Sjoholm, A. (1995) Ceramide inhibits pancreatic beta-cell insulin production and mitogenesis and mimics the actions of interleukin-1 beta. *FEBS Lett.*, **367**, 283–286.

13 Dionisi, S., Dotta, F., Diaz-Horta, O., Carabba, B., Viglietta, V., and Di Mario, U. (1997) Target antigens in autoimmune diabetes: pancreatic gangliosides. *Ann. Ist. Super Sanita*, **33**, 433–435.

14 Misasi, R., Dionisi, S., Farilla, L., Carabba, B., Lenti, L., Di Mario, U., and Dotta, F. (1997) Gangliosides and autoimmune diabetes. *Diabetes Metab. Rev.*, **13**, 163–179.

15 Shimabukuro, M., Zhou, Y.T., Levi, M., and Unger, R.H. (1998) Fatty acid-induced beta cell apoptosis: a link between obesity and diabetes. *Proc. Natl. Acad. Sci. USA*, **95**, 2498–2502.

16 Holland, W.L., Brozinick, J.T., Wang, L. P., Hawkins, E.D., Sargent, K.M., Liu, Y., Narra, K., Hoehn, K.L., Knotts, T.A., Siesky, A., Nelson, D.H., Karathanasis, S. K., Fontenot, G.K., Birnbaum, M.J., and Summers, S.A. (2007) Inhibition of ceramide synthesis ameliorates glucocorticoid-, saturated-fat-, and

obesity-induced insulin resistance. *Cell Metab.*, **5**, 167–179.

17 Chavez, J.A. and Summers, S.A. (2010) Lipid oversupply, selective insulin resistance, and lipotoxicity: molecular mechanisms. *Biochim. Biophys. Acta*, **1801**, 252–265.

18 Wende, A.R. and Abel, E.D. (2010) Lipotoxicity in the heart. *Biochim. Biophys. Acta*, **1801**, 311–319.

19 Chen, Q.M. and Tu, V.C. (2002) Apoptosis and heart failure: mechanisms and therapeutic implications. *Am. J. Cardiovasc. Drugs*, **2**, 43–57.

20 Litwin, S.E., Raya, T.E., Gay, R.G., Bedotto, J.B., Bahl, J.J., Anderson, P.G., Goldman, S., and Bressler, R. (1990) Chronic inhibition of fatty acid oxidation: new model of diastolic dysfunction. *Am. J. Physiol.*, **258**, H51–H56.

21 Bachmann, E. and Weber, E. (1988) Biochemical mechanisms of oxfenicine cardiotoxicity. *Pharmacology*, **36**, 238–248.

22 Zhou, Y.T., Grayburn, P., Karim, A., Shimabukuro, M., Higa, M., Baetens, D., Orci, L., and Unger, R.H. (2000) Lipotoxic heart disease in obese rats: implications for human obesity. *Proc. Natl. Acad. Sci. USA*, **97**, 1784–1789.

23 Chiu, H.C., Kovacs, A., Ford, D.A., Hsu, F.F., Garcia, R., Herrero, P., Saffitz, J.E., and Schaffer, J.E. (2001) A novel mouse model of lipotoxic cardiomyopathy. *J. Clin. Invest.*, **107**, 813–822.

24 Chiu, H.C., Kovacs, A., Blanton, R.M., Han, X., Courtois, M., Weinheimer, C.J., Yamada, K.A., Brunet, S., Xu, H., Nerbonne, J.M., Welch, M.J., Fettig, N.M., Sharp, T.L., Sambandam, N., Olson, K.M., Ory, D.S., and Schaffer, J.E. (2005) Transgenic expression of fatty acid transport protein 1 in the heart causes lipotoxic cardiomyopathy. *Circ. Res.*, **96**, 225–233.

25 Park, T.S., Hu, Y., Noh, H.L., Drosatos, K., Okajima, K., Buchanan, J., Tuinei, J., Homma, S., Jiang, X.C., Abel, E.D., and Goldberg, I.J. (2008) Ceramide is a cardiotoxin in lipotoxic cardiomyopathy. *J. Lipid Res.*, **49**, 2101–2112.

26 Paumen, M.B., Ishida, Y., Muramatsu, M., Yamamoto, M., and Honjo, T. (1997) Inhibition of carnitine palmitoyltransferase I augments sphingolipid synthesis and palmitate-induced apoptosis. *J. Biol. Chem.*, **272**, 3324–3329.

27 de Vries, J.E., Vork, M.M., Roemen, T.H., de Jong, Y.F., Cleutjens, J.P., van der Vusse, G.J., and van Bilsen, M. (1997) Saturated but not mono-unsaturated fatty acids induce apoptotic cell death in neonatal rat ventricular myocytes. *J. Lipid Res.*, **38**, 1384–1394.

28 Hickson-Bick, D.L., Buja, M.L., and McMillin, J.B. (2000) Palmitate-mediated alterations in the fatty acid metabolism of rat neonatal cardiac myocytes. *J. Mol. Cell Cardiol.*, **32**, 511–519.

29 Dyntar, D., Eppenberger-Eberhardt, M., Maedler, K., Pruschy, M., Eppenberger, H.M., Spinas, G.A., and Donath, M.Y. (2001) Glucose and palmitic acid induce degeneration of myofibrils and modulate apoptosis in rat adult cardiomyocytes. *Diabetes*, **50**, 2105–2113.

30 Sparagna, G.C., Hickson-Bick, D.L., Buja, L.M., and McMillin, J.B. (2000) A metabolic role for mitochondria in palmitate-induced cardiac myocyte apoptosis. *Am. J. Physiol. Heart Circ. Physiol.*, **279**, H2124–H2132.

31 Drosatos, K., Bharadwaj, K.G., Lymperopoulos, A., Ikeda, S., Khan, R., Hu, Y., Agarwal, R., Yu, S., Jiang, H., Steinberg, S.F., Blaner, W.S., Koch, W.J., and Goldberg, I.J. (2011) Cardiomyocyte lipids impair beta-adrenergic receptor function via PKC activation. *Am. J. Physiol. Endocrinol. Metab.*, **300**, E489–E499.

32 Stamler, J., Pick, R., and Katz, L.N. (1959) Saturated and unsaturated fats: effects on cholesterolemia and atherogenesis in chicks on high-cholesterol diets. *Circ. Res.*, **7**, 398–402.

33 Stamler, J., Pick, R., and Katz, L.N. (1960) Effect of insulin in the induction and regression of atherosclerosis in the chick. *Circ. Res.*, **8**, 572–576.

34 Duff, G.L. and Payne, T.P. (1950) The effect of alloxan diabetes on experimental cholesterol atherosclerosis in the rabbit: III. The mechanism of the inhibition of experimental cholesterol atherosclerosis

in alloxan-diabetic rabbits. *J. Exp. Med.*, **92**, 299–317.

35 Weisberg, S.P., McCann, D., Desai, M., Rosenbaum, M., Leibel, R.L., and Ferrante, A.W., Jr. (2003) Obesity is associated with macrophage accumulation in adipose tissue. *J. Clin. Invest.*, **112**, 1796–1808.

36 Rajala, M.W. and Scherer, P.E. (2003) Minireview: the adipocyte – at the crossroads of energy homeostasis, inflammation, and atherosclerosis. *Endocrinology*, **144**, 3765–3773.

37 Kern, P.A., Ranganathan, S., Li, C., Wood, L., and Ranganathan, G. (2001) Adipose tissue tumor necrosis factor and interleukin-6 expression in human obesity and insulin resistance. *Am. J. Physiol. Endocrinol. Metab.*, **280**, E745–E751.

38 Takahashi, K., Mizuarai, S., Araki, H., Mashiko, S., Ishihara, A., Kanatani, A., Itadani, H., and Kotani, H. (2003) Adiposity elevates plasma MCP-1 levels leading to the increased CD11b-positive monocytes in mice. *J. Biol. Chem.*, **278**, 46654–46660.

39 Li, Z., Basterr, M.J., Hailemariam, T.K., Hojjati, M.R., Lu, S., Liu, J., Liu, R., Zhou, H., and Jiang, X.C. (2005) The effect of dietary sphingolipids on plasma sphingomyelin metabolism and atherosclerosis. *Biochim. Biophys. Acta*, **1735**, 130–134.

40 Hojjati, M., Li, Z., Zhou, H., Tang, S., Huan, C., Ooi, E., Lu, S., and Jiang, X.C. (2005) Effect of myriocin on plasma sphingolipid metabolism and atherosclerosis in apoE-deficient mice. *J. Biol. Chem*, **280** (11), 10284–10289.

41 Park, T.S., Panek, R.L., Mueller, S.B., Hanselman, J.C., Rosebury, W.S., Robertson, A.W., Kindt, E.K., Homan, R., Karathanasis, S.K., and Rekhter, M.D. (2004) Inhibition of sphingomyelin synthesis reduces atherogenesis in apolipoprotein E-knockout mice. *Circulation*, **110**, 3465–3471.

42 Holopainen, J.M., Lehtonen, J.Y., and Kinnunen, P.K. (1997) Lipid microdomains in dimyristoylphosphatidylcholine-ceramide liposomes. *Chem. Phys. Lipids*, **88**, 1–13.

43 Holopainen, J.M., Lemmich, J., Richter, F., Mouritsen, O.G., Rapp, G., and Kinnunen, P.K. (2000) Dimyristoylphosphatidylcholine/C16:0-ceramide binary liposomes studied by differential scanning calorimetry and wide- and small-angle X-ray scattering. *Biophys. J.*, **78**, 2459–2469.

44 Schissel, S.L., Tweedie-Hardman, J., Rapp, J.H., Graham, G., Williams, K.J., and Tabas, I. (1996) Rabbit aorta and human atherosclerotic lesions hydrolyze the sphingomyelin of retained low-density lipoprotein: proposed role for arterial-wall sphingomyelinase in subendothelial retention and aggregation of atherogenic lipoproteins. *J. Clin. Invest.*, **98**, 1455–1464.

45 Morita, S.Y., Okuhira, K., Tsuchimoto, N., Vertut-Doi, A., Saito, H., Nakano, M., and Handa, T. (2003) Effects of sphingomyelin on apolipoprotein E- and lipoprotein lipase-mediated cell uptake of lipid particles. *Biochim. Biophys. Acta*, **1631**, 169–176.

46 Mallat, Z. and Tedgui, A. (2000) Apoptosis in the vasculature: mechanisms and functional importance. *Br. J. Pharmacol.*, **130**, 947–962.

47 Mallat, Z. and Tedgui, A. (2001) Current perspective on the role of apoptosis in atherothrombotic disease. *Circ. Res.*, **88**, 998–1003.

48 Yatomi, Y., Ohmori, T., Rile, G., Kazama, F., Okamoto, H., Sano, T., Satoh, K., Kume, S., Tigyi, G., Igarashi, Y., and Ozaki, Y. (2000) Sphingosine 1-phosphate as a major bioactive lysophospholipid that is released from platelets and interacts with endothelial cells. *Blood*, **96**, 3431–3438.

49 Auge, N., Nikolova-Karakashian, M., Carpentier, S., Parthasarathy, S., Negre-Salvayre, A., Salvayre, R., Merrill, A.H., Jr., and Levade, T. (1999) Role of sphingosine 1-phosphate in the mitogenesis induced by oxidized low density lipoprotein in smooth muscle cells via activation of sphingomyelinase, ceramidase, and sphingosine kinase. *J. Biol. Chem.*, **274**, 21533–21538.

50 Ölmez, I. and Özyurt, H. (2012) Reactive oxygen species and ischemic cerebrovascular disease. *Neurochem. Int.*, **60**, 208–212.

51 Auer-Grumbach, M., De Jonghe, P., Verhoeven, K., Timmerman, V., Wagner, K., Hartung, H.P., and Nicholson, G.A. (2003) Autosomal dominant inherited neuropathies with prominent sensory loss and mutilations: a review. *Arch. Neurol.*, **60**, 329–334.

52 Nicholson, G.A. (2006) The dominantly inherited motor and sensory neuropathies: clinical and molecular advances. *Muscle Nerve*, **33**, 589–597.

53 Hornemann, T., Penno, A., Richard, S., Timmerman, V., and von Eckardstein, A. (2009) A systematic comparison of all mutations in hereditary sensory neuropathy type I (HSAN I) reveals that the G387A mutation is not disease associated. *Neurogenetics*, **10**, 135–143.

54 McCampbell, A., Truong, D., Broom, D. C., Allchorne, A., Gable, K., Cutler, R.G., Mattson, M.P., Woolf, C.J., Frosch, M.P., Harmon, J.M., Dunn, T.M., and Brown, R.H., Jr. (2005) Mutant SPTLC1 dominantly inhibits serine palmitoyltransferase activity *in vivo* and confers an age-dependent neuropathy. *Hum. Mol. Genet.*, **14**, 3507–3521.

55 Rotthier, A., Penno, A., Rautenstrauss, B., Auer-Grumbach, M., Stettner, G.M., Asselbergh, B., Van Hoof, K., Sticht, H., Levy, N., Timmerman, V., Hornemann, T., and Janssens, K. (2011) Characterization of two mutations in the SPTLC1 subunit of serine palmitoyl-transferase associated with hereditary sensory and autonomic neuropathy type I. *Hum. Mutat.*, **32**, E2211–E2225.

56 Penno, A., Reilly, M.M., Houlden, H., Laura, M., Rentsch, K., Niederkofler, V., Stoeckli, E.T., Nicholson, G., Eichler, F., Brown, R.H., Jr., von Eckardstein, A., and Hornemann, T. (2010) Hereditary sensory neuropathy type 1 is caused by the accumulation of two neurotoxic sphingolipids. *J. Biol. Chem.*, **285**, 11178–11187.

57 Eichler, F.S., Hornemann, T., McCampbell, A., Kuljis, D., Penno, A., Vardeh, D., Tamrazian, E., Garofalo, K., Lee, H.J., Kini, L., Selig, M., Frosch, M., Gable, K., von Eckardstein, A., Woolf, C. J., Guan, G., Harmon, J.M., Dunn, T.M., and Brown, R.H., Jr. (2009) Overexpression of the wild-type SPT1 subunit lowers desoxysphingolipid levels and rescues the phenotype of HSAN1. *J. Neurosci.*, **29**, 14646–14651.

58 Garofalo, K., Penno, A., Schmidt, B.P., Lee, H.J., Frosch, M.P., von Eckardstein, A., Brown, R.H., Hornemann, T., and Eichler, F.S. (2011) Oral L-serine supplementation reduces production of neurotoxic deoxysphingolipids in mice and humans with hereditary sensory autonomic neuropathy type 1. *J. Clin. Invest.*, **121**, 4735–4745.

59 Scherer, S.S. (2011) The debut of a rational treatment for an inherited neuropathy? *J. Clin. Invest.*, **121**, 4624–4627.

60 Shinitzky, M. (1987) Patterns of lipid changes in membranes of the aged brain. *Gerontology*, **33**, 149–154.

61 Dexter, D.T., Carter, C.J., Wells, F.R., Javoy-Agid, F., Agid, Y., Lees, A., Jenner, P., and Marsden, C.D. (1989) Basal lipid peroxidation in substantia nigra is increased in Parkinson's disease. *J. Neurochem.*, **52**, 381–389.

62 Bras, J., Singleton, A., Cookson, M.R., and Hardy, J. (2008) Emerging pathways in genetic Parkinson's disease: potential role of ceramide metabolism in Lewy body disease. *FEBS J.*, **275**, 5767–5773.

63 Farooqui, A.A., Horrocks, L.A., and Farooqui, T. (2007) Interactions between neural membrane glycerophospholipid and sphingolipid mediators: a recipe for neural cell survival or suicide. *J. Neurosci. Res.*, **85**, 1834–1850.

64 Cheng, D., Jenner, A.M., Shui, G., Cheong, W.F., Mitchell, T.W., Nealon, J. R., Kim, W.S., McCann, H., Wenk, M.R., Halliday, G.M., and Garner, B. (2011) Lipid pathway alterations in Parkinson's disease primary visual cortex. *PLoS One*, **6**, e17299.

65 Cutler, R.G., Kelly, J., Storie, K., Pedersen, W.A., Tammara, A., Hatanpaa, K., Troncoso, J.C., and Mattson, M.P. (2004) Involvement of oxidative stress-induced abnormalities in ceramide and cholesterol metabolism in brain aging and Alzheimer's disease. *Proc. Natl. Acad. Sci. USA*, **101**, 2070–2075.

66 Tamboli, I.Y., Hampel, H., Tien, N.T., Tolksdorf, K., Breiden, B., Mathews, P. M., Saftig, P., Sandhoff, K., and Walter, J. (2011) Sphingolipid storage affects autophagic metabolism of the amyloid precursor protein and promotes Abeta generation. *J. Neurosci.*, **31**, 1837–1849.

67 Van Brocklyn, J.R. and Williams, J.B. (2012) The control of the balance between ceramide and sphingosine-1-phosphate by sphingosine kinase: oxidative stress and the seesaw of cell survival and death. *Comp. Biochem. Physiol. B Biochem. Mol. Biol.*, **163**, 26–36.

68 Sassa, T., Suto, S., Okayasu, Y., and Kihara, A. (2012) A shift in sphingolipid composition from C24 to C16 increases susceptibility to apoptosis in HeLa cells. *Biochim. Biophys. Acta*, **1821**, 1031–1037.

69 Meyers-Needham, M., Lewis, J.A., Gencer, S., Sentelle, R.D., Saddoughi, S.A., Clarke, C.J., Hannun, Y.A., Norell, H., da Palma, T.M., Nishimura, M., Kraveka, J.M., Khavandgar, Z., Murshed, M., Cevik, M.O., and Ogretmen, B. (2012) Off-target function of the sonic hedgehog inhibitor cyclopamine in mediating apoptosis via nitric oxide-dependent neutral sphingomyelinase 2/ceramide induction. *Mol. Cancer Ther.*, **11**, 1092–1102.

70 Aflaki, E., Doddapattar, P., Radovic, B., Povoden, S., Kolb, D., Vujic, N., Wegscheider, M., Koefeler, H., Hornemann, T., Graier, W.F., Malli, R., Madeo, F., and Kratky, D. (2012) C16 ceramide is crucial for triacylglycerol-induced apoptosis in macrophages. *Cell Death Dis.*, **3**, e280.

71 Nomura, K. and Yamanouchi, T. (2012) The role of fructose-enriched diets in mechanisms of nonalcoholic fatty liver disease. *J. Nutr. Biochem.*, **23**, 203–208.

72 Shah, F.D., Shukla, S.N., Shah, P.M., Patel, H.R.H., and Patel, P.S. (2008) Significance of alterations in plasma lipid profile levels in breast cancer. *Integr. Cancer Ther.*, **7**, 33–41.

73 Ray, G. and Husain, S.A. (2001) Role of lipids, lipoproteins and vitamins in women with breast cancer. *Clin. Biochem.*, **34**, 71–76.

74 Winczura, A., Zdzalik, D., and Tudek, B. (2012) Damage of DNA and proteins by major lipid peroxidation products in genome stability. *Free Radic. Res.*, **46**, 442–459.

75 Nair, U., Bartsch, H., and Nair, J. (2007) Lipid peroxidation-induced DNA damage in cancer-prone inflammatory diseases: a review of published adduct types and levels in humans. *Free Radic. Biol. Med.*, **43**, 1109–1120.

76 Bartsch, H. and Nair, J. (2004) Oxidative stress and lipid peroxidation-derived DNA-lesions in inflammation driven carcinogenesis. *Cancer Detect. Prev.*, **28**, 385–391.

77 Fraga, C.G. and Tappel, A.L. (1988) Damage to DNA concurrent with lipid peroxidation in rat liver slices. *Biochem. J.*, **252**, 893–896.

78 Hardy, S., Langelier, Y., and Prentki, M. (2000) Oleate activates phosphatidylinositol 3-kinase and promotes proliferation and reduces apoptosis of MDA-MB-231 breast cancer cells, whereas palmitate has opposite effects. *Cancer Res.*, **60**, 6353–6358.

79 Eggens, I., Ekstrom, T.J., and Aberg, F. (1990) Studies on the biosynthesis of polyisoprenols, cholesterol and ubiquinone in highly differentiated human hepatomas. *J. Exp. Pathol.*, **71**, 219–232.

80 Karahatay, S., Thomas, K., Koybasi, S., Senkal, C.E., Elojeimy, S., Liu, X., Bielawski, J., Day, T.A., Gillespie, M.B., Sinha, D., Norris, J.S., Hannun, Y.A., and Ogretmen, B. (2007) Clinical relevance of ceramide metabolism in the pathogenesis of human head and neck squamous cell carcinoma (HNSCC): attenuation of C(18)-ceramide in HNSCC tumors correlates with lymphovascular invasion and nodal metastasis. *Cancer Lett.*, **256**, 101–111.

81 Kawamori, T., Kaneshiro, T., Okumura, M., Maalouf, S., Uflacker, A., Bielawski, J., Hannun, Y.A., and Obeid, L.M. (2009)

Role for sphingosine kinase 1 in colon carcinogenesis. *FASEB J.*, **23**, 405–414.

82 Meacci, E., Nuti, F., Donati, C., Cencetti, F., Farnararo, M., and Bruni, P. (2008) Sand sphingosine-1 activity is required for myogenic differentiation of C2C12 myoblasts. *J. Cell Physiol.*, **214**, 210–220.

83 Skoura, A., Sanchez, T., Claffey, K., Mandala, S.M., Proia, R.L., and Hla, T. (2007) Essential role of sphingosine 1-phosphate receptor 2 in pathological angiogenesis of the mouse retina. *J. Clin. Invest.*, **117**, 2506–2516.

84 Takuwa, Y., Okamoto, Y., Yoshioka, K., and Takuwa, N. (2008) Sand sphing-1-phosphate signaling and biological activities in the cardiovascular system. *Biochim. Biophys. Acta*, **1781**, 483–488.

85 Heo, K., Park, K.A., Kim, Y.H., Kim, S. H., Oh, Y.S., Kim, I.H., Ryu, S.H., and Suh, P.G. (2009) Sand sphing 1-phosphate induces vascular endothelial growth factor expression in endothelial cells. *BMB Rep.*, **42**, 685–690.

86 Blom, T., Bergelin, N., Meinander, A., Lof, C., Slotte, J.P., Eriksson, J.E., and Tornquist, K. (2010) An autocrine sand sphing-1-phosphate signaling loop enhances NF-kappaB-activation and survival. *BMC Cell Biol.*, **11**, 45.

87 Kim, S.H., Park, Y.Y., Kim, S.W., Lee, J. S., Wang, D., and DuBois, R.N. (2011) ANGPTL4 induction by prostaglandin E2 under hypoxic conditions promotes colorectal cancer progression. *Cancer Res.*, **71**, 7010–7020.

88 Zhang, H., Wong, C.C., Wei, H., Gilkes, D.M., Korangath, P., Chaturvedi, P., Schito, L., Chen, J., Krishnamachary, B., Winnard, P.T., Jr., Raman, V., Zhen, L., Mitzner, W.A., Sukumar, S., and Semenza, G.L. (2011) HIF-1-dependent expression of angiopoietin-like 4 and L1CAM mediates vascular metastasis of hypoxic breast cancer cells to the lungs. *Oncogene*, **31** (14), 1757–1770.

89 Li, H., Ge, C., Zhao, F., Yan, M., Hu, C., Jia, D., Tian, H., Zhu, M., Chen, T., Jiang, G., Xie, H., Cui, Y., Gu, J., Tu, H., He, X., Yao, M., Liu, Y., and Li, J. (2011) Hypoxia-inducible factor 1 alpha-activated angiopoietin-like protein 4 contributes to tumor metastasis via vascular cell adhesion molecule-1/integrin beta1 signaling in human hepatocellular carcinoma. *Hepatology*, **54**, 910–919.

90 Schuchman, E.H. (2010) Acid sphingomyelinase, cell membranes and human disease: lessons from Niemann–Pick disease. *FEBS Lett.*, **584**, 1895–1900.

91 Park, J.H. and Schuchman, E.H. (2006) Acid ceramidase and human disease. *Biochim. Biophys. Acta*, **1758**, 2133–2138.

92 Paw, B.H., Moskowitz, S.M., Uhrhammer, N., Wright, N., Kaback, M.M., and Neufeld, E.F. (1990) Juvenile GM2 gangliosidosis caused by substitution of histidine for arginine at position 499 or 504 of the alpha-subunit of beta-hexosaminidase. *J. Biol. Chem.*, **265**, 9452–9457.

93 Huang, J.Q., Trasler, J.M., Igdoura, S., Michaud, J., Hanal, N., and Gravel, R.A. (1997) Apoptotic cell death in mouse models of GM2 gangliosidosis and observations on human Tay–Sachs and Sandhoff diseases. *Hum. Mol. Genet.*, **6**, 1879–1885.

94 Becker, K.A., Grassme, H., Zhang, Y., and Gulbins, E. (2010) Ceramide in *Pseudomonas aeruginosa* infections and cystic fibrosis. *Cell Physiol. Biochem.*, **26**, 57–66.

95 Becker, K.A., Tummler, B., Gulbins, E., and Grassme, H. (2010) Accumulation of ceramide in the trachea and intestine of cystic fibrosis mice causes inflammation and cell death. *Biochem. Biophys. Res. Commun.*, **403**, 368–374.

96 Andersen, D.H. and Hodges, R.G. (1946) Celiac syndrome; genetics of cystic fibrosis of the pancreas, with a consideration of etiology. *Am. J. Dis. Child*, **72**, 62–80.

97 Andersen, D.H. (1958) Cystic fibrosis of the pancreas. *J. Chronic. Dis.*, **7**, 58–90.

98 Doring, G. and Gulbins, E. (2009) Cystic fibrosis and innate immunity: how chloride channel mutations provoke lung disease. *Cell Microbiol.*, **11**, 208–216.

99 Kerem, B., Rommens, J.M., Buchanan, J.A., Markiewicz, D., Cox, T.K., Chakravarti, A., Buchwald, M., and Tsui, L.C. (1989) Identification of the cystic fibrosis gene: genetic analysis. *Science*, **245**, 1073–1080.

100 Riordan, J.R., Rommens, J.M., Kerem, B., Alon, N., Rozmahel, R., Grzelczak, Z., Zielenski, J., Lok, S., Plavsic, N., Chou, J.L. *et al.* (1989) Identification of the cystic fibrosis gene: cloning and characterization of complementary DNA. *Science*, **245**, 1066–1073.

101 Heijerman, H. (2005) Infection and inflammation in cystic fibrosis: a short review. *J. Cyst. Fibros*, **4** (Suppl. 2), 3–5.

102 Del Porto, P., Cifani, N., Guarnieri, S., Di Domenico, E.G., Mariggio, M.A., Spadaro, F., Guglietta, S., Anile, M., Venuta, F., Quattrucci, S., and Ascenzioni, F. (2011) Dysfunctional CFTR alters the bactericidal activity of human macrophages against *Pseudomonas aeruginosa*. *PLoS One*, **6**, e19970.

103 Grassme, H., Becker, K.A., Zhang, Y., and Gulbins, E. (2010) CFTR-dependent susceptibility of the cystic fibrosis-host to *Pseudomonas aeruginosa*. *Int. J. Med. Microbiol.*, **300**, 578–583.

104 Teichgraber, V., Ulrich, M., Endlich, N., Riethmuller, J., Wilker, B., De Oliveira-Munding, C.C., van Heeckeren, A.M., Barr, M.L., von Kurthy, G., Schmid, K.W., Weller, M., Tummler, B., Lang, F., Grassme, H., Doring, G., and Gulbins, E. (2008) Ceramide accumulation mediates inflammation, cell death and infection susceptibility in cystic fibrosis. *Nat. Med.*, **14**, 382–391.

105 Becker, K.A., Riethmuller, J., Luth, A., Doring, G., Kleuser, B., and Gulbins, E. (2010) Acid sand sphingomyelinase inhibitors normalize pulmonary ceramide and inflammation in cystic fibrosis. *Am. J. Respir. Cell Mol. Biol.*, **42**, 716–724.

106 Ponnappan, R.K., Serhan, H., Zarda, B., Patel, R., Albert, T., and Vaccaro, A.R. (2009) Biomechanical evaluation and comparison of polyetheretherketone rod system to traditional titanium rod fixation. *Spine J.*, **9**, 263–267.

107 Serhan, C.N., Chiang, N., and Van Dyke, T.E. (2008) Resolving inflammation: dual anti-inflammatory and pro-resolution lipid mediators. *Nat. Rev. Immunol.*, **8**, 349–361.

108 Serhan, C.N., Gotlinger, K., Hong, S., and Arita, M. (2004) Resolvins, docosatrienes, and neuroprotectins, novel omega-3-derived mediators, and their aspirin-triggered endogenous epimers: an overview of their protective roles in catabasis. *Prostaglandins Other Lipid Mediat.*, **73**, 155–172.

109 Serhan, C.N., Clish, C.B., Brannon, J., Colgan, S.P., Chiang, N., and Gronert, K. (2000) Novel functional sets of lipid-derived mediators with antiinflammatory actions generated from omega-3 fatty acids via cyclooxygenase 2-nonsteroidal antiinflammatory drugs and transcellular processing. *J. Exp. Med.*, **192**, 1197–1204.

110 Ji, R.R., Xu, Z.Z., Strichartz, G., and Serhan, C.N. (2011) Emerging roles of resolvins in the resolution of inflammation and pain. *Trends Neurosci.*, **34**, 599–609.

111 Serhan, C.N., Krishnamoorthy, S., Recchiuti, A., and Chiang, N. (2011) Novel anti-inflammatory–pro-resolving mediators and their receptors. *Curr. Top. Med. Chem.*, **11**, 629–647.

112 Numata, M., Chu, H.W., Dakhama, A., and Voelker, D.R. (2010) Pulmonary surfactant phosphatidylglycerol inhibits respiratory syncytial virus-induced inflammation and infection. *Proc. Natl. Acad. Sci. USA*, **107**, 320–325.

113 Numata, M., Kandasamy, P., Nagashima, Y., Posey, J., Hartshorn, K., Woodland, D., and Voelker, D.R. (2012) Phosphatidylglycerol suppresses influenza A virus infection. *Am. J. Respir. Cell Mol. Biol.*, **46**, 479–487.

10
Lipidomics in Lipoprotein Biology

Marie C. Lhomme, Laurent Camont, M. John Chapman, and Anatol Kontush

10.1
Introduction

Due to the central role of lipoproteins in the development of dyslipidemia and cardiovascular disease, their lipid and protein composition has been extensively studied. For many years, lipoproteins were typically characterized in terms of their content of major lipid classes, that is, free cholesterol (FC), cholesteryl ester (CE), triacylglycerol (TG), and phospholipid (PL). Recent technological advances in mass spectrometry (MS) have enabled applications of this powerful technique to provide a detailed quantification of individual molecular species of lipids in a framework of the field known as lipidomics. This technique not only resolves the issues of detection sensitivity, which has been a typical drawback of traditional techniques such as enzymatic assays, thin-layer chromatography, or high-pressure liquid chromatography (HPLC), but also offers new dimensions in the lipid identification. It is indeed possible by current lipidomic methodologies to identify individual molecular species of lipids with more than 200 molecules detected in plasma-derived lipoprotein samples [1–3].

Following the emergence of genomics, proteomics, and metabolomics, lipidomics has been the latest "omics" technique developed over the last decade. Lipidomics has been extensively employed in the past 3–5 years to characterize the plasma lipidome in health and disease as recently reviewed [4]. In the framework of the LIPID MAPS project, Dennis, Brown, Merrill and others have published pioneering data on the plasma lipidome produced with newly developed protocols for extraction, separation, and detection of major lipid classes [1, 5–9]. On the contrary, lipidomic studies of isolated lipoproteins, referred to as lipoproteinomics in this chapter, are much less abundant. Such studies are however essential to understand molecular mechanisms underlying development of atherosclerosis and other lipoprotein-related disorders. Furthermore, lipidomics can prove useful for the identification of biomarkers of disease risk.

Importantly, lipid composition of lipoproteins using classical methods together with the lipidome of plasma in normo- and dyslipidemia have already been extensively studied [10–21]. Therefore, this chapter will essentially focus on

Lipidomics, First Edition. Edited by Kim Ekroos.

lipoproteinomics in normo- and dyslipidemic subjects characterized by lipidomic methodologies.

10.2 Metabolism of Lipoproteins

Cholesterol is essential for animal life as it is required for the synthesis of cell membranes, bile acids, and steroid hormones [22]. Since cholesterol is insoluble in blood, it is transported in the circulatory system within lipoproteins [23]. In addition to providing a soluble means for transporting cholesterol and other lipids through the blood, lipoproteins possess cell-targeting signals that direct the lipids they carry to certain tissues [23]. This biological function is ensured by apolipoproteins (apo). ApoB is the predominant protein component of proatherogenic, cholesterol-rich low-density lipoproteins (LDLs), triglyceride-rich very low-density lipoproteins (VLDLs), VLDL remnants, and intermediate density lipoproteins (IDLs), whereas apoA-I is the major protein component of antiatherogenic high-density lipoproteins (HDLs).

Lipoproteins are plurimolecular, quasi-spherical, and pseudomicellar complexes that ensure lipid transport in biological fluids. They are divided into five major classes in the order of increasing hydrated density and decreasing size, notably into chylomicrons, VLDL, IDL, LDL, and HDL. In addition, lipoprotein (a) [Lp(a)] forms a separate subclass distinguished by the presence of apolipoprotein (a). Chylomicrons, VLDL, IDL, and LDL are generally regarded as light, large, and lipid-rich lipoproteins, whereas HDL belongs to dense, small, and protein-rich particles.

The four major lipid classes present in all lipoproteins are PL, FC, CE, and TG. The percent of each lipid class varies across the lipoprotein types. From a structural viewpoint, all lipoproteins are formed of the surface lipid monolayer, which contains embedded proteins, and hydrophobic lipid core. PL and free cholesterol build the surface monolayer, whereas CE and TG ensure the existence of the hydrophobic core.

Chylomicrons, which are the largest lipid-transporting particles, are assembled with apoB-48 as nascent particles. In the bloodstream, HDL particles donate apoC-II and apoE to nascent chylomicrons in intestinal enterocytes. They carry dietary lipids from the intestine to the muscle and adipose tissues for energy production and storage and are predominantly composed of TG. Through apoC-II, mature chylomicrons activate lipoprotein lipase (LPL), which hydrolyzes TG, resulting in the release of glycerol and free fatty acids (FA). Hydrolyzed chylomicrons, termed chylomicron remnants, are subsequently taken up by the liver via apoE receptors [24]. The liver subsequently loads lipids onto apoB-100 and secretes them as VLDL particles that mainly contain endogenous TG [25]. Like nascent chylomicrons, newly released VLDLs acquire apoCs and apoE from circulating HDLs. ApoC-II activates LPL, causing hydrolysis of VLDL TGs and the release of glycerol and free fatty acids that can be absorbed by adipose tissue and muscle [26].

The action of LPL coupled to the loss of apoCs converts VLDLs to IDLs. The predominant remaining apolipoproteins are apoB-100 and apoE. IDL particles can be taken up by the liver via apoE receptors, or they can be further hydrolyzed by hepatic lipase (HL), releasing glycerol and free FA and converting IDLs to LDLs. As an intermediate product of lipoprotein metabolism, IDLs are not always detectable in the blood [26].

The exclusive apolipoprotein of LDL is apoB-100 [27]. The uptake of LDL occurs predominantly in the liver (75%), adrenals, and adipose tissue via LDL receptor-mediated endocytosis. The endocytosed membrane vesicles (endosomes) fuse with lysosomes, in which apoB-100 is degraded and CE is hydrolyzed into FC. The latter is incorporated into cellular membranes or re-esterified by acyl-CoA-cholesterol acyltransferase (ACAT) for intracellular storage. LDL can be modified (e.g., by oxidation) and subsequently taken up via scavenger receptors A1 (SR-AI) and CD36 by macrophages, which become lipid loaded with formation of foam cells, the early hallmark of the atherosclerotic plaque formation [28].

HDLs are distinguished from other lipoprotein classes by their small size, high density, and enrichment in protein (30–60 wt%). HDLs form a heterogeneous population of functionally distinct particles that differ in density, size, electrophoretic mobility, protein content, and lipid composition [29, 30]. The prominent heterogeneity of HDL particles primarily results from the highly dynamic structure of apoA-I that can adopt distinct conformations, forming discoid and spherical particles as a function of the amount of bound lipid [31]. Discoid HDLs, also called pre-β HDLs, are small and lipid-poor (lipid content $\leq$30%) particles made up of mainly apoA-I with small amounts of lipid (mainly PL and FC). Spherical HDL particles are larger and in addition contain a hydrophobic core of CE and TG. Two major HDL subpopulations that can be distinguished by density are light, large, lipid-rich HDL2 and dense, small, lipid-poor HDL3 [30].

Nascent HDLs are synthesized in the liver and small intestine as protein-rich discoid particles that primarily contain apoA-I, PL, and FC [32]. Nascent HDLs display pre-β mobility, are unstable, and readily acquire lipids via the ATP-binding cassette transporter (ABC) A1 (ABCA1)-mediated efflux of cholesterol and PL from cells [33]. In addition, small, dense pre-β HDL can be generated from PL released by lipolysis from both chylomicrons and VLDL under the action of LPL [33].

Free cholesterol in nascent HDLs is esterified to CE by the enzyme lecithin: cholesterol acyltransferase (LCAT) [33]. This enzyme transfers a fatty acid from *sn*-2 position of phosphatidylcholine (PC) to the hydroxyl group of cholesterol, resulting in the formation of CE and lysophosphatidylcholine (lysoPC). This latter is largely removed from HDL by albumin, while CEs are internalized into the hydrophobic core of HDL particle to generate larger and spherical HDL3 and subsequently HDL2 particles. Such HDLs can undergo further remodeling via particle fusion and surface remnant transfer mediated by phospholipid transfer protein (PLTP) [33].

HDL3 particles are further remodeled via acquisition of cholesterol during cellular efflux through ABCA1, ABCG1, and SR-BI. In turn, HDL2 promotes cellular cholesterol efflux via ABCG1 and SR-BI [30]. In addition, CE in HDL can be transferred to VLDL and LDL by the action of cholesteryl ester transfer protein (CETP) [34]. Such transfer involves exchange of TGs from VLDLs and LDLs to HDLs. TG-enriched HDLs, produced as a result of the action of CETP, are targets for HL that produces progressively smaller and unstable particles that release apoA-I [35, 36]. This newly released apoA-I can interact with ABCA1 in the next lipidation cycle [37].

HDL lipids can be catabolized in the liver either separately from HDL proteins through selective uptake by the receptor SR-BI or following HDL holoparticle uptake via an HDL holoparticle receptor. ApoE-containing HDLs can be targeted to LDL receptors as well as to hitherto unidentified receptors in the form of holoparticles.

The return of cholesterol from peripheral tissues back to the liver is referred to as reverse cholesterol transport (RCT). The role of HDLs in RCT is thought to represent the major atheroprotective function of this lipoprotein. In addition to its function in RCT, HDLs exert other atheroprotective activities, including antioxidative, anti-inflammatory, cytoprotective, vasodilatory, antithrombotic, and anti-infectious actions. Across the HDL subpopulation spectrum, small, dense, protein-rich HDLs display potent atheroprotective properties, which can be attributed to specific clusters of proteins and lipids yet to be fully identified [30].

10.3 Lipoproteinomics in Normolipidemic Subjects

Early data obtained using spectrophotometric techniques in the 1960s and 1970s have extensively documented the fact that chemical composition of isolated plasma lipoproteins largely reflects differences in their biology and metabolism.

Chylomicrons and VLDL, the large and very light particles that ensure transport of TG through the circulation, are enriched in this lipid class. Indeed, chylomicrons are composed of TG by 80–95 wt%, with PL (3–6 wt%), FC (2–7 wt%), and protein (0.5–1.0 wt%) accounting for the rest. VLDL equally contains mainly TG (55–65 wt%), together with CE (12–14 wt%), FC (6–8 wt%), PL (12–18 wt%), and protein (5–10 wt%).

LDLs, the large, light particles primarily destined to carry cholesterol, are accordingly enriched in this lipid class. Thus, LDLs reveal elevated cholesterol content (35–40 wt% of CE and 5–10 wt% of FC), while PL (20–25 wt%), TG (8–12 wt%), and protein (20–24 wt%) account for the rest.

HDLs represent small, protein-rich lipoproteins displaying multiple biological activities and are enriched in total protein (30–60 wt%) and in surface lipid components relative to the core. Indeed, PLs present at 20–30 wt% constitute the major lipid class in HDL. The rest of the HDL lipidome is represented by FC (3–5 wt%), CE (14–18 wt%), and TG (3–6 wt%), together with minor lipid classes.

Lipidomic data on individual molecular species of lipids present in major lipoproteins obtained from healthy normolipidemic subjects have been rapidly accumulating over the last years. The pioneering study of Wiesner *et al.* published in 2009 [2] provided reference values for VLDL, LDL, and HDL obtained from healthy normolipidemic controls. In this study, lipoproteins were isolated from blood serum of 21 donors by fast-performance liquid chromatography (FPLC), and molecular species of PC, lysoPC, sphingomyelin (SM), ceramide, phosphatidylethanolamine (PE), PE-based plasmalogens, and CEs were quantified by electrospray ionization tandem mass spectrometry. In less comprehensive studies, other researchers were successful in applying lipidomic techniques to the determination of particular lipid classes and characterization of specific lipoproteins [38].

Consistent with earlier data obtained by other methodologies, lipoproteinomic analyses reveal that each lipoprotein class displays a specific pattern of major lipid classes (Table 10.1). VLDLs are mainly composed of TG that account for approximately 52 mol% of total lipids. Other lipids are present in VLDL at lesser amounts, that is, phospholipids at 20 mol%, CE at 17 mol%, and sterols at 11 mol%. In LDL, CEs predominate (48 mol%), with phospholipids (24 mol%), sterols (17 mol%), and TG (10 mol%) building the rest of the lipidome. In HDL, phospholipids account for the majority of lipids (37–54 mol%), whereas sterols, CE, and TG are less abundant

Table 10.1 Lipoproteinomics from healthy normolipidemic human subjects.

Lipid class	VLDL content	LDL content	HDL content
	Molar % of total lipids		
Phospholipids	19.4–20.3	24.4–36.6	37–54
Phosphatidylcholine	15	17–23	31–37
Sphingomyelin	3.0–3.7	6–12	4.5–6
Lysophosphatidylcholine	0.50–0.75	1–1.2	1–8
Phosphatidylethanolamine	0.6	0.2	0.5–1.5
PE plasmalogens	0.3	0.25	0.5
Phosphatidylinositol	ND	ND (18:0/20:3)[a]	0.9 (18:0/20:4)[a]
Sterols	11–21	8–17[b]	9–20[b]
Cholesteryl ester	17–21	48–52	30–40
Triacylglycerides	37–52	3–10[b]	5–12[b]
Diacylglycerides	0.9	0.14	ND
Minor lipids			
Ceramide d18:1	0.12	0.14	0.05
Cardiolipin	0.08[c]	0.19[c]	0.2[c]
Phosphatidylserine	0.02[c]	0.09[c]	0.03[c]
Lactosyl ceramides	0.02	0.04	0.04
S1P (d18:1 and d18:0)	0.002	0.001	0.02
SPC d18:1	0.001	0.0006	0.0005
Isoprostane-containing PC	ND	ND	ND (IPGE2/D2-PC (36:4))[a]
Phosphatidylglycerol	ND	ND (18:3/20:3)[a]	ND (18:1/20:2)[a]
Phosphatidic acid	ND	ND (20:4/20:2)[a]	ND (20:4/20:0)[a]

According to Refs [2, 3, 38, 48]. ND: not determined; SPC: sphingosyl-phosphorylcholine; S1P: sphingosine-1 phosphate; IPGE2: isoprostaglandin E2.

a) Main species detected.

b) Determined by enzymatic assay.

c) Determined by HPLC.

at 9–20, 30–40, and 5–12 mol%, respectively. Importantly, lipidomics reveal that most lipid species in the circulation are exclusively associated with lipoproteins; lysolipids and platelet-activating factor (PAF) represent rare exceptions, which are in part carried by albumin [2, 39].

The real power of the lipidomic technologies however results from its ability to provide quantitative data on individual molecular species of lipids and on low-abundance lipid molecules rather than on major lipid classes and as such data are becoming increasingly available.

10.3.1 Phospholipids

Phosphatidylcholine predominates as major molecular class of phospholipids in plasma lipoproteins. In addition, lipoproteins contain significant amounts of phosphatidylinositol (PI), lysoPC, PE, and PC- and PE-derived plasmalogens [2, 11, 40]. Minor phospholipids are represented by phosphatidylglycerol (PG), phosphatidylserine (PS), phosphatidic acid (PA), and cardiolipin [41, 42].

Phospholipids are unequally distributed in the circulation across serum lipoproteins. HDL, which is enriched in phospholipids, represents a major carrier of PC, PE, PE-derived plasmalogens, and lysoPC [2]. Specifically, HDLs contain more than 50% of total circulating PC, PE, plasmalogens, and lysoPC.

10.3.1.1 Phosphatidylcholine

PC is the principal plasma phospholipid that accounts for 70 mol% of phospholipid in HDL, 69 mol% in LDL, and 77 mol% in VLDL [2]. Major molecular species of PC are represented by the 16:0/18:2, 18:0/18:2, and 16:0/20:4 species [2, 3, 38, 41, 42]. Compared to other lipoproteins, HDL is enriched in PC containing polyunsaturated fatty acid (PUFA) moieties [2]. Phosphatidylcholine in lipoproteins can be of both hepatic (via formation of nascent lipoprotein particles) and extrahepatic (via the actions of PLTP and CETP) origins.

10.3.1.2 Lysophosphatidylcholine

LysoPC, which is the product of the LCAT reaction, constitutes a minor subclass of phospholipids in apoB-containing lipoproteins, which ranges from 3 mol% in VLDL to 4 mol% LDL [2, 3]. The content of lysoPC is however greater in HDL (up to 17 mol%), consistent with the predominant association of the LCAT reaction with this lipoprotein [11]. As considerable amounts of serum lysoPC are also associated with albumin [2], HDL contamination by the latter as typically occurring upon FPLC isolation can significantly contribute to the enrichment of this lipid in the total HDL fraction isolated by this approach.

Major molecular species of serum lysoPC contain saturated fatty acid moieties of predominantly 16 and 18 carbon atoms [2, 3], reflecting LCAT preferencing 16 and 18 carbon atom long PCs [43]. Interestingly, the 16:0 species is predominantly associated with HDL, whereas the 18:0 lysoPC is enriched in VLDL and LDL, potentially reflecting distinct metabolic pathways for individual lysoPC molecules.

10.3.1.3 Phosphatidylethanolamine

PE is another plasma phospholipid that accounts for 1 mol% of phospholipid in HDL, 1 mol% in LDL, and 3 mol% in VLDL [2]. Principal molecular species of PE are represented by the 36:2 and 38:4 species that are evenly distributed across major lipoprotein classes [11].

10.3.1.4 Phosphatidylethanolamine Plasmalogens

PE plasmalogens are minor plasma phospholipids with antioxidative properties [44, 45] and are slightly higher in VLDL (2 mol%) than in LDL (1 mol%) and HDL (1 mol%) [2]. Species containing arachidonic acid residues predominate in all lipoprotein classes. Interestingly, HDL is enriched in the 20:4 species relative to VLDL and LDL, in parallel to the depletion of species containing 18:1 and 18:2 residues.

10.3.1.5 Phosphatidylinositol, Phosphatidylserine, Phosphatidylglycerol, and Phosphatidic Acid

PI, PS, PG, and PA are negatively charged minor phospholipids that are present in lipoproteins in trace amounts (Table 10.1). These lipids may significantly impact the net surface charge of lipoproteins [41, 46, 47]. The content of these lipids can thereby modulate lipoprotein interactions with lipases, extracellular matrix, and other protein components, which are largely charge dependent. Major molecular species of PI in LDL include the 18:0/20:3 and 18:0/20:4 species, whereas those of phosphatidylglycerol contain 18:3/20:3, 18:2/18:1, and 18:1/20:1 moieties [41].

10.3.1.6 Cardiolipin

CL is a minor anionic phospholipid with potent anticoagulant properties and is equally present in trace amounts across all lipoprotein classes (Table 10.1). This lipid may contribute to the effects of lipoproteins on coagulation and platelet aggregation [48].

10.3.1.7 Isoprostane-Containing PC

Isoprostane-containing PCs represent stable products of nonenzymatic oxidation of PCs containing polyunsaturated fatty acid moieties. These lipids are well established as biomarkers of oxidative stress [49], are present in the circulation at very low (nanomolar) concentrations, and are largely associated with HDL [50]. Major molecular species of isoprostane-containing PCs include 5,6-epoxy isoprostaglandine A2-PC (EIPGA2-PC) (36:3), 5,6 EIPGE2-PC (36:4), IPGE2/D2-PC (36:4), IPGF-PC (36:4), IPGE2/D2-PC (38:4), and IPGF-PC (38:4) [38].

10.3.2 Sphingolipids

Contrary to phospholipids, the largest amounts of SM (50 mol%) and ceramide (60 mol%) are carried in the circulation by LDL [2].

10.3.2.1 **Sphingomyelin**

SM is the second major phospholipid in circulating lipoproteins, which largely originates from triglyceride-rich lipoproteins and only to a minor extent from nascent HDL [51]. SM accounts for approximately 10 mol% of phospholipid in HDL, 25 mol% in LDL, and 14 mol% in VLDL. Major molecular species of sphingomyelin are the 34:1 and 42:2 species [2, 3, 11]. Interestingly, some molecular species of sphingomyelin reveal markedly heterogeneous distribution across lipoprotein classes. As an example, the 38:1 molecule is highly enriched in LDL [2]. Sphingomyelin content constitutes a critical factor in determining surface pressure in lipid membranes and lipoproteins, enhancing rigidity and thereby influencing activity of embedded proteins [52, 53].

10.3.2.2 **Lysosphingolipids**

Among lysosphingolipids, sphingosine 1-phosphate (S1P) is particularly interesting as this bioactive lipid plays key roles in vascular biology and can function as a ligand for the family of G-protein-coupled S1P receptors present on endothelial and smooth muscle cells, which regulate cell proliferation, motility, apoptosis, angiogenesis, wound healing, and immune response [54]. HDL is the major carrier of S1P in the circulation, which ensures its bioavailability [54]. Indeed, more than 90% of sphingoid base phosphates are found in HDL and albumin-containing fractions by LC-MS analysis (Table 10.1) [55]. S1P is produced by phosphorylation of sphingosine by sphingosine kinases, which are expressed in platelets, erythrocytes, neutrophils, and mononuclear cells. Erythrocytes appear to represent the primary source of S1P in plasma followed by platelets [56, 57]; S1P release from erythrocytes can be triggered by HDL and serum albumin [58]. Other biologically active lysosphingolipids carried by HDL are represented by lysoshpingomyelin and lysosulfatide [59].

10.3.2.3 **Ceramide**

Ceramides (Table 10.1) are preferentially carried by LDL (60 mol% of total plasma ceramides) compared to HDL (25 mol%) and VLDL (15 mol%) [2]. Furthermore, the pattern of ceramide species is distinct in VLDL and LDL relative to HDL (Table 10.1). Thus, the percentage of the 16:0 species is almost doubled and the proportion of the 24:0 species is reduced in HDL compared to that in VLDL and LDL [2, 3]. Significantly, ceramides play a key role as signaling molecules involved in cellular survival, growth, and differentiation.

10.3.2.4 **Minor Sphingolipids**

Truly, lipidomic data on glycosphingolipids, gangliosides, and sulfatides are scarce [60, 61].

Interestingly, HDL is equally depleted in glycosphingolipids and gangliosides compared to LDL [60]. Hexosyl and lactosyl species constitute the major glycosphingolipids in plasma lipoproteins (Table 10.1) [55]. Gangliosides determine interactions with protein receptors and signal transduction; the physiological relevance of their presence in lipoproteins remains indeterminate.

10.3.3 Sterols

Together with phospholipids, sterols are located in the surface lipid monolayer of lipoprotein particles and regulate its fluidity. Lipoprotein sterols are dominated by cholesterol, reflecting the key role of lipoproteins in cholesterol transport through the body. Other sterols are present in lipoproteins at much lower levels as exemplified by minor amounts of lathosterol, ergosterol, phytosterols (β-sitosterol, campesterol), oxysterols (27-hydroxycholesterol, 24-hydroxycholesterol, cholesterol-5,6-beta-epoxide, 7-ketocholesterol) [62], and estrogens (largely circulating as esters) [63]. Plasma levels of exogenous versus endogenous steroids can provide valuable information on cholesterol metabolism and absorption and on key processes regulating plasma levels of cholesterol. Wide application of lipidomic techniques assessing the determination of lipoprotein steroids is still awaited.

10.3.4 Cholesteryl Esters

CEs are largely (up to 80%) formed in plasma HDL as a result of transesterification of PL and cholesterol catalyzed by LCAT. These highly hydrophobic lipids form the lipid core of HDL and are subsequently transferred to apoB-containing lipoproteins by CETP in exchange for TG. Most of lipoprotein CE is accounted by cholesteryl linoleate, although this pattern differs to a minor degree between lipoprotein classes [2, 3, 11].

10.3.5 Triacylglycerides

Circulating TGs are dominated by species containing oleic, palmitic, and linoleic acid moieties [3, 42]. High molecular diversity of plasma TG has recently been reported [9]. Furthermore, the presence of multiple molecular species of TG was documented in chylomicron remnants by MALDI/TOF, reflecting high diversity of their dietary sources [42]. Distribution of individual molecular species of triglycerides across VLDL and LDL was determined by Stahlman *et al.* and showed similar profiles. Both VLDL and LDL were preferentially enriched in TAGs 16:0/18:1/18:1 and 16:0/18:1/18:2 [3].

10.3.6 Minor Lipids

Minor bioactive lipids present in lipoproteins include diacylglycerides (DAG), monoacylglycerides (MAG), and free FA [2, 3, 10, 64]. Recently published first detailed characterization of DAG molecular species in VLDL and LDL reveals 18:1/18:1, 16:0/18:1, and 18:1/18:2 as main species [3]. Lipidomic characterization of the distribution of MAGs and free FAs across lipoprotein classes remains to be performed.

The currently available study results reveal that lipidomic techniques have already provided basic characterization of three major FPLC-isolated lipoprotein classes – VLDL, LDL, and HDL – that would need to be verified on ultracentrifugation-isolated lipoproteins. In view of high internal heterogeneity of each major plasma lipoprotein class, the next relevant question therefore involves detailed molecular characterization of lipids carried by key lipoprotein subspecies, such as VLDL1, VLDL2, LDL1 + 2, LDL3, LDL4 + 5, HDL2, HDL3, and others [65–67]. Inherent heterogeneity of the HDL lipidome was addressed in an HPLC-based study of major molecular species of lipids present in five subclasses of human HDL [11]. Progressive reduction in HDL particle size with increase in hydrated density was associated with progressive elevation in S1P content and reduction in sphingomyelin content, consistent with distinct metabolic origins and potent biological activities of small, dense HDL [11, 68]. This example indicates the power of lipoproteinomics to obtain critical information regarding the metabolism and function of lipoproteins relevant for the development of cardiovascular disease, which can in turn provide novel biomarkers of cardiovascular risk.

10.4 Altered Lipoproteinomics in Dyslipidemia

In order to understand molecular mechanisms underlying altered lipid metabolism in dyslipidemic states, detailed lipidomic characterization of isolated lipoproteins is required. An important number of studies have already been carried out on the plasma lipidome in order to characterize alterations of plasma lipids in various pathologies, including obesity [12], metabolic syndrome [13, 14], coronary heart disease [15], insulin resistance [16], nonalcoholic fatty liver disease (NAFLD), nonalcoholic steatohepatitis (NASH) [18], myocardial infarction [19], hypertension [20], and cystic fibrosis [21].

However, limited information about alteration of the lipidome of isolated plasma lipoproteins in dyslipidemia and other pathological conditions is available at present as reviewed below.

10.4.1 Phospholipids

10.4.1.1 Phosphatidylcholine

PC content was shown to be altered in HDL isolated from subjects with Niemann–Pick disease type B (NPD B), an autosomal disorder characterized by the lack of sphingomyelinase (SMase). This deficiency leads to the accumulation of sphingomyelin and cholesterol in multiple organs such as the liver and the spleen [69]. HDL isolated from NPD B subjects showed a +95% increase in total PC compared

Table 10.2 Alterations in the lipids of lipoproteins isolated from dyslipidemic subjects.

Molecular lipid classes	Pathology and lipoprotein studied					
	Ischemic heart disease	Type 2 diabetes			NPD B	Acute phase
	HDL [76]	LDL [71]	apoC-III-enriched LDL [74]	HDL [80]	HDL [69]	HDL [38]
PC					$\uparrow^{a}$	
SM			$\downarrow$		$\uparrow^{b}$	$\downarrow^{c}$
LysoPC		$\uparrow^{d}\downarrow^{e}$				$\uparrow$
PE						$\downarrow$
PI						$\downarrow$
Isoprostane-containing PC						$\downarrow^{f}$
FC			$\downarrow$			$\uparrow$
CE		$\uparrow$	$\downarrow$			$\downarrow$
TAG			$\uparrow$			$\uparrow$
Ceramide	$\downarrow^{g}$		$\downarrow$			
S1P and dihydro S1P	$\downarrow$					
NEFA						$\uparrow^{h}$
GM1			$\downarrow$			
Oxidized fatty acids				$\uparrow^{i}$		

a) PC 34:2, 36:4, 34:1, and 36:2.
b) SM (d18:1/16:0).
c) SM 33:1 and SM 38:1.
d) LPC 16:0.
e) LPC decreased after simvastatin treatment.
f) 5;6 EIPGA2-PC (36:3); 5;6 EIPGE2-PC (36:4); IPGE2/D2-PC (36:4); IPGF-PC (36:4); IPGE2/D2-PC (38:4), and IPGF-PC (38:4).
g) Ceramide 24:1.
h) Palmitic acid.
i) 5-HETE; PC: phosphatidylcholine; PE: phosphatidylethanolamine; PI: phosphatidylinositol; SM: sphingomyelin; CE: choelsteryl ester; FC: free cholesterol; TAG: triacylglycerides; S1P: sphingosine-1 phosphate; NEFA: not esterified fatty acids; HETE: 5-Hydroxyeicosatetraenoic; GM1: ganglioside 1.

to HDL from healthy controls (Table 10.2). Major PC species enriched in NPD B HDL were identified as PC 34:2, 36:4, 34:1, and 36:2 [69].

10.4.1.2 **Lysophosphatidylcholine**

LysoPC displays potent proatherogenic activities such as proinflammatory and proapoptotic actions involving induction of monocyte adhesion [70, 71]. Since lysoPC is the product of PC degradation, it can be used as both a biomarker of enzymatic activity of PC-degrading enzymes and a proatherogenic biomarker. Three studies have investigated changes in lysoPC molecular species in disease state such as during the acute-phase response [38] and in patients with type 2 diabetes [71, 72]. Interestingly, lysoPC content measured in HDL during the acute-phase response was elevated compared to that in HDL isolated from normolipidemic subjects (Table 10.2) [38]. In this study, six lysoPC molecular species were affected, specifically lysoPC 18:0, 18:2, 20:4, 20:5, and 22:6. Similarly, lysoPC 16:0 content measured in LDL from patients with type 2 diabetes was increased relative to healthy controls (Table 10.2) [71]. Interestingly, simvastatin [71] lowered the lysoPC contents of LDL in patients with type 2 diabetes (Table 10.2), consistent with the proatherogenic properties of lysoPC.

10.4.1.3 **Phosphatidylethanolamine**

To our knowledge, only one study has to date reported lipidomic data on PE content in isolated lipoproteins under pathological conditions. This study showed that isolated HDLs were significantly depleted in diacyl PE following remodeling of normal HDL into acute-phase HDL (Table 10.2) [38].

10.4.1.4 **Phosphatidylethanolamine Plasmalogens**

In the FIELD (fibrate intervention and event lowering in diabetes) study of the effects of fenofibrate in patients with type 2 diabetes, fenofibrate failed to significantly decrease cardiovascular events. Remarkably, treated subjects displayed elevated levels of atherogenic homocysteine [73]. Interestingly, patients with elevated plasma levels of homocysteine displayed HDL depleted in PE plasmalogens relative to patients with reduced homocysteine levels. As plasmalogens display antioxidative properties, it is interesting to note that patients responding to fenofibrate treatment by increasing their plasma homocysteine levels were less protected against oxidative stress than patients responding by lowering homocysteine levels. Consistent with these data, HDL isolated during the acute-phase response, which features elevated oxidative stress, showed a significant decrease in ether-linked PE (Table 10.2) [38].

10.4.1.5 **Phosphatidylinositol**

Despite the fact that PI may influence the interaction of HDL with enzymes and lipid transfer proteins due to its negative charge, this lipid has rarely been measured in isolated lipoproteins. The lack of data on PI lipoprotein content can be attributed to its elevated charge and highly dynamic nature, which together with increased propensity to phosphorylation render this lipid difficult to isolate from biological samples. Moreover, PI is almost exclusively measured in negative ion mode where detection limit is at lowest. When PI was measured using LC/MS in isolated acute-phase HDL, it was shown to be depleted compared to normal HDL (Table 10.2) [38].

10.4.1.6 **Isoprostane-Containing PC**

Isoprostane-containing PCs were detected in acute-phase HDL and their content compared to that of normolipidemic HDL [38]. The detected species were the following: 5,6-EIPGA2-PC (36:3), 5,6-EIPGE2-PC (36:4), IPGE2/D2-PC (36:4),

IPGF-PC (36:4), IPGE2/D2-PC (38:4), and IPGF-PC (38:4). It was shown that isoprostanes were 10-fold depleted in acute-phase HDL relative to control HDL (Table 10.2).

10.4.2 Sphingolipids

10.4.2.1 Sphingomyelin

The content of SM species in HDL can be modified in NPD B patients [69], in fenofibrate-treated patients with type 2 diabetes [73], in low-HDL-C subjects [72], and during the acute-phase response (Table 10.2) [38]. SM levels were also modified in LDL isolated from patients with type 2 diabetes [74]. As expected, total SM was increased by +28% in HDL isolated from NPD B patients [69]. Moreover, the SM (d18:1/16:0) species were elevated by +95%. The SM/PC ratio was however not significantly different between NPD B and control HDLs as PC was increased in parallel to SM. In contrary, nascent HDLs prepared *in vitro* from NPD B fibroblasts showed a twofold increase in the SM/PC ratio [69]. This modification deleteriously affects LCAT activity and HDL maturation that largely depend on LCAT-mediated cholesterol esterification. Finally, the acute-phase HDL isolated by ultracentrifugation was depleted in SM (Table 10.2), and particularly of SM 33:1 and SM 38:1, compared to normolipidemic HDL [38]. Since apoA-I preferentially interacts with SM [75], such HDL depletion in SM could reflect the replacement of apoA-I by serum amyloid A. In addition, activation of endogenous SMase could further contribute to the preferential loss of SM in the acute phase [38].

LDL content of SM isolated from patients with type 2 diabetes was inversely correlated with that of apoC-III [74]. ApoC-III is an inhibitor of LPL and its elevated plasma levels can be considered as a risk factor of cardiovascular disease. Using shotgun lipidomics, it was shown that LDL particles enriched in apoC-III were depleted in SM (Table 10.2) and more susceptible to SMase-induced hydrolysis and aggregation [74]. Formation of LDL aggregates via this pathway could contribute to the proatherogenic effect of apoC-III.

10.4.2.2 Lysosphingolipids: S1P and Dihydro S1P

Sphingoid bases were measured by LC-MS in HDL-containing fraction obtained from patients presenting with cardiac ischemia (Table 10.2) [76]. Four different groups were compared, subjects with ischemia presenting high or low HDL-C concentrations and subjects with no ischemia displaying high or low HDL-C. The results showed an inverse correlation between HDL content of S1P and dihyrdo S1P and the occurrence of ischemic heart disease. Interestingly, such correlation was not observed when S1P was measured in total serum. Moreover, S1P levels in HDL were positively correlated with the capacity of HDL to induce endothelial cell barrier signaling. This study not only reflects the atheroprotective role of S1P but also emphasizes the importance of lipoproteinomics rather than plasma lipidomics on the former and provides greater insight into the multiple biological roles of plasma lipids.

10.4.2.3 Ceramide

Ceramide levels were measured by LC-MS in HDL isolated from patients with ischemic heart disease [76] and in apoC-III-enriched LDL from patients with type 2 diabetes [74]. In HDL-isolated apoB-depleted serum, abundance of ceramide 24:1 was inversely correlated with the occurrence of ischemic heart disease [76]. Similarly, ceramide depletion was observed in apoC-III-enriched LDL from patients with type 2 diabetes (Table 10.2) [74]. Since ceramides induce a more rigid surface lipid monolayer, the partial loss of SM and ceramides from apoC-III-enriched LDL could increase the surface monolayer fluidity and favor conformational changes of apoB that enhance its affinity for arterial wall proteoglycans. On the contrary however, LDLs isolated from human atherosclerosis lesions were enriched in ceramides (Table 10.2) [77]. Whether SMase-generated ceramides bind differently with circulating plasma LDL compared to LDL derived from atherosclerotic lesions remains a plausible explanation for this discrepancy.

10.4.3 Free Cholesterol

FC was quantified by evaporative light scattering in apoC-III-enriched LDL from patients with type 2 diabetes and found to be depleted relative to LDL from normolipidemic subjects (Table 10.2) [74]. As a consequence, the surface lipid monolayer poor in free cholesterol should exhibit an elevated fluidity and favor the transfer of LDL core lipids toward the surface, making the hydrophobic lipid core accessible to hydrolytic enzymes in patients type 2 diabetes. On the contrary, unesterified cholesterol was elevated in acute-phase HDL compared to that in normal HDL.

10.4.4 Cholesteryl Esters

CEs were measured by LC/MS and shotgun MS in HDL isolated during the acute-phase response and from apoC-III-enriched LDL from patients with type 2 diabetes [38, 74]. In both studies, CE was depleted relative to corresponding lipoproteins from normal controls (Table 10.2).

10.4.5 Triacylglycerides

Unlike esterified cholesterol, TGs were enriched in acute-phase HDL (Table 10.2) [38], suggesting that CE was replaced by TG in the core of these particles. Similarly, TGs were elevated in apoC-III-enriched LDL (Table 10.2) [74]. As TGs decrease the affinity of LDL for proteoglycans [78], enhanced affinity of apoC-III-enriched LDL toward proteoglycans cannot be attributed to TG enrichment.

10.4.6 Minor Lipids

10.4.6.1 Nonesterified Fatty Acids

Free FAs measured by GC/MS were significantly enriched in acute-phase HDL (Table 10.2) [38], with the major affected species being palmitic acid. Such elevation in a hydrolytic product is consistent with the upregulation of HDL hydrolysis during the acute phase.

10.4.6.2 Ganglioside GM1

The acidic glycosphingolipid, ganglioside GM1, was reduced in LDL from patients with type 2 diabetes compared to normolipidemic LDL (Table 10.2) [74]. Similarly, GM1 was diminished in apoC-III-enriched LDL (Table 10.2) [74]. It is of interest to note that GM1 is an inhibitor of SMase [79] and that apoC-III-enriched LDL possesses an elevated susceptibility to hydrolysis by this enzyme, consistent with enhanced proatherogenic properties of LDL in patients type 2 diabetes.

10.4.6.3 Oxidized Lipids

Quantification of oxidized FA in HDL isolated from patients with type 2 diabetes using LC-MS revealed a 60-fold increase in 5-hydroxyeicosatetraenoic compared to that in control HDL (Table 10.2) [80]. These data suggest that HDL from patients with type 2 diabetes are exposed to elevated oxidative stress *in vivo* and that elevated concentrations of oxidized lipids might reflect enhanced production of reactive oxygen species secondary to hyperglycemia.

10.5 Conclusions

Newly developed lipidomic methodologies have provided first insights into lipid species profiles of major lipoprotein classes. Such studies were, however, systematically performed in lipoproteins obtained only from healthy normolipidemic subjects. Thus, results of comprehensive lipoproteinomics of disease states are eagerly awaited.

The results of the first studies devoted to detailed characterization of isolated lipoproteins suggest that lipidomics can deliver biomarkers of lipoprotein functionality, which may in turn prove useful as biomarkers of cardiovascular risk. Systematic lipoproteinomics performed in normo- and dyslipidemic subjects could not only provide such biomarkers but also deepen our understanding of molecular mechanisms involved in the atheroprotection characteristic of some lipoproteins such as HDL and in proatherogenic properties of other lipoproteins such as LDL. For novel mechanistic insights, lipoproteinomics appear to be essential.

However, for this to happen, numerous technical issues remain to be thoroughly addressed. One of them is the standardization of lipidomic analyses. Indeed, although a systematic study characterizing the effect of fatty acid chain length and level of unsaturation on the MS response factor of individual PL species was recently published by Genest and coworkers [69], it was limited to the quantification of PC and SM. Moreover, extraction procedures and internal standards vary from one study to another and matrix effects are not always taken into account by both LC-MS and shotgun analyses. Other technical issues that remain to be resolved include chromatographical separation of isomeric lipids differing in the position of fatty acid residues (*sn*-1 versus *sn*-2 versus *sn*-3) and identification of double bond positions in fatty acid chains. This issue was recently addressed in the laboratory of Jan Boren who was able to show the enrichment of vaccenic acid in the TAG fraction isolated from VLDL in diabetic dyslipidemia [3]. In the same study, elevated amounts of hexadecenoic (16:1) and eicosatrienoic (20:3) species in VLDL PC and CE and increased amounts of palmitic acid in VLDL TAG were also measured [3].

Finally, the major technical challenge yet to be addressed is the handling of the vast amount of data produced in lipidomics. This issue remains indeed the bottleneck of the analysis and includes peak picking, quantification, statistical analysis, metabolic pathway investigation, and molecular dynamic simulation as recently reviewed by Brown *et al.* and by Oresic *et al.* [7, 81]. Despite all these limitations, lipoproteinomics is on the way to become a valuable research tool in the field of cardiovascular and metabolic diseases.

References

1 Jung, H.R., Sylvanne, T., Koistinen, K.M., Tarasov, K., Kauhanen, D., and Ekroos, K. (2011) High throughput quantitative molecular lipidomics. *Biochim. Biophys. Acta*, **1811**, 925–934.

2 Wiesner, P., Leidl, K., Boettcher, A., Schmitz, G., and Liebisch, G. (2009) Lipid profiling of FPLC-separated lipoprotein fractions by electrospray ionization tandem mass spectrometry. *J. Lipid Res.*, **50**, 574–585.

3 Stahlman, M., Pham, H.T., Adiels, M., Mitchell, T.W., Blanksby, S.J., Fagerberg, B., Ekroos, K., and Boren, J. (2012) Clinical dyslipidaemia is associated with changes in the lipid composition and inflammatory properties of apolipoprotein-B-containing lipoproteins from women with type 2 diabetes. *Diabetologia*, **55** (4), 1156–1166.

4 Kontush, A. and Chapman, M.J. (2010) Lipidomics as a tool for the study of lipoprotein metabolism. *Curr. Atheroscler Rep.*, **12**, 194–201.

5 Ivanova, P.T., Milne, S.B., Byrne, M.O., Xiang, Y., and Brown, H.A. (2007) Glycerophospholipid identification and quantitation by electrospray ionization mass spectrometry. *Methods Enzymol.*, **432**, 21–57.

6 Harkewicz, R. and Dennis, E.A. (2011) Applications of mass spectrometry to lipids and membranes. *Annu. Rev. Biochem.*, **80**, 301–325.

7 Myers, D.S., Ivanova, P.T., Milne, S.B., and Brown, H.A. (2011) Quantitative analysis of glycerophospholipids by LC-MS: acquisition, data handling, and interpretation. *Biochim. Biophys. Acta*, **1811** (11): 748–757.

8 Hutchins, P.M., Barkley, R.M., and Murphy, R.C. (2008) Separation of cellular nonpolar neutral lipids by normal-phase chromatography and analysis by electrospray ionization mass spectrometry. *J. Lipid Res.*, **49**, 804–813.

9 Quehenberger, O., Armando, A.M., Brown, A.H., Milne, S.B., Myers, D.S.,

Merrill, A.H., Bandyopadhyay, S., Jones, K.N., Kelly, S., Shaner, R.L., Sullards, C. M., Wang, E., Murphy, R.C., Barkley, R. M., Leiker, T.J., Raetz, C.R.H., Guan, Z., Laird, G.M., Six, D.A., Russell, D.W., McDonald, J.G., Subramaniam, S., Fahy, E., and Dennis, E.A. (2010) Lipidomics reveals a remarkable diversity of lipids in human plasma. *J. Lipid Res.*, **51**, 3299–3305.

10 Skipski, V.P., Barclay, M., Barclay, R.K., Fetzer, V.A., Good, J.J., and Archibald, F. M. (1967) Lipid composition of human serum lipoproteins. *Biochem. J.*, **104**, 340–352.

11 Kontush, A., Therond, P., Zerrad, A., Couturier, M., Negre-Salvayre, A., de Souza, J.A., Chantepie, S., and Chapman, M.J. (2007) Preferential sphingosine-1-phosphate enrichment and sphingomyelin depletion are key features of small dense HDL3 particles: relevance to antiapoptotic and antioxidative activities. *Arterioscler. Thromb. Vasc. Biol.*, **27**, 1843–1849.

12 Pietilainen, K.H., Sysi-Aho, M., Rissanen, A., Seppanen-Laakso, T., Yki-Jarvinen, H., Kaprio, J., and Oresic, M. (2007) Acquired obesity is associated with changes in the serum lipidomic profile independent of genetic effects: a monozygotic twin study. *PLoS One*, **2**, e218.

13 Schwab, U., Seppanen-Laakso, T., Yetukuri, L., Agren, J., Kolehmainen, M., Laaksonen, D.E., Ruskeepaa, A.L., Gylling, H., Uusitupa, M., and Oresic, M. (2008) Triacylglycerol fatty acid composition in diet-induced weight loss in subjects with abnormal glucose metabolism: the GENOBIN study. *PLoS One*, **3**, e2630.

14 Lankinen, M., Schwab, U., Gopalacharyulu, P.V., Seppanen-Laakso, T., Yetukuri, L., Sysi-Aho, M., Kallio, P., Suortti, T., Laaksonen, D.E., Gylling, H., Poutanen, K., Kolehmainen, M., and Oresic, M. (2009) Dietary carbohydrate modification alters serum metabolic profiles in individuals with the metabolic syndrome. *Nutr. Metab. Cardiovasc. Dis*, **20** (4): 249–257.

15 de Mello, V.D., Lankinen, M., Schwab, U., Kolehmainen, M., Lehto, S., Seppanen-Laakso, T., Oresic, M., Pulkkinen, L., Uusitupa, M., and Erkkila, A.T. (2009) Link between plasma ceramides, inflammation and insulin resistance: association with serum IL-6 concentration in patients with coronary heart disease. *Diabetologia*, **52**, 2612–2615.

16 Kotronen, A., Velagapudi, V.R., Yetukuri, L., Westerbacka, J., Bergholm, R., Ekroos, K., Makkonen, J., Taskinen, M.R., Oresic, M., and Yki-Jarvinen, H. (2009) Serum saturated fatty acids containing triacylglycerols are better markers of insulin resistance than total serum triacylglycerol concentrations. *Diabetologia*, **52**, 684–690.

17 Aalto-Setala, K., Laitinen, K., Erkkila, L., Leinonen, M., Jauhiainen, M., Ehnholm, C., Tamminen, M., Puolakkainen, M., Penttila, I., and Saikku, P. (2001) *Chlamydia pneumoniae* does not increase atherosclerosis in the aortic root of apolipoprotein E-deficient mice. *Arterioscler. Thromb. Vasc. Biol.*, **21**, 578–584.

18 Puri, P., Wiest, M.M., Cheung, O., Mirshahi, F., Sargeant, C., Min, H.K., Contos, M.J., Sterling, R.K., Fuchs, M., Zhou, H., Watkins, S.M., and Sanyal, A.J. (2009) The plasma lipidomic signature of nonalcoholic steatohepatitis. *Hepatology*, **50**, 1827–1838.

19 Lankinen, M., Schwab, U., Erkkila, A., Seppanen-Laakso, T., Hannila, M.L., Mussalo, H., Lehto, S., Uusitupa, M., Gylling, H., and Oresic, M. (2009) Fatty fish intake decreases lipids related to inflammation and insulin signaling: a lipidomics approach. *PLoS One*, **4**, e5258.

20 Graessler, J., Schwudke, D., Schwarz, P.E., Herzog, R., Shevchenko, A., and Bornstein, S.R. (2009) Top-down lipidomics reveals ether lipid deficiency in blood plasma of hypertensive patients. *PLoS One*, **4**, e6261.

21 Guerrera, I.C., Astarita, G., Jais, J.P., Sands, D., Nowakowska, A., Colas, J., Sermet-Gaudelus, I., Schuerenberg, M., Piomelli, D., Edelman, A., and Ollero, M. (2009) A novel lipidomic strategy reveals plasma phospholipid signatures associated with respiratory disease severity in cystic fibrosis patients. *PLoS One*, **4**, e7735.

22 Ikonen, E. (2008) Cellular cholesterol trafficking and compartmentalization. *Nat. Rev. Mol. Cell Biol.*, **9**, 125–138.
23 Olson, R.E. (1998) Discovery of the lipoproteins, their role in fat transport and their significance as risk factors. *J. Nutr.*, **128**, 439S–443S.
24 Mansbach, C.M. and Siddiqi, S.A. (2010) The biogenesis of chylomicrons. *Annu. Rev. Physiol.*, **72**, 315–333.
25 Gibbons, G.F., Wiggins, D., Brown, A.M., and Hebbachi, A.M. (2004) Synthesis and function of hepatic very-low-density lipoprotein. *Biochem. Soc. Trans.*, **32**, 59–64.
26 Berneis, K.K. and Krauss, R.M. (2002) Metabolic origins and clinical significance of LDL heterogeneity. *J. Lipid Res.*, **43**, 1363–1379.
27 Esterbauer, H., Gebicki, J., Puhl, H., and Jurgens, G. (1992) The role of lipid peroxidation and antioxidants in oxidative modification of LDL. *Free Radic. Biol. Med.*, **13**, 341–390.
28 Itabe, H. (2003) Oxidized low-density lipoproteins: what is understood and what remains to be clarified. *Biol. Pharm. Bull.*, **26**, 1–9.
29 Rosenson, R.S., Brewer, H.B., Jr., Chapman, M.J., Fazio, S., Hussain, M.M., Kontush, A., Krauss, R.M., Otvos, J.D., Remaley, A.T., and Schaefer, E.J. (2011) HDL measures, particle heterogeneity, proposed nomenclature, and relation to atherosclerotic cardiovascular events. *Clin. Chem.*, **57**, 392–410.
30 Camont, L., Chapman, M.J., and Kontush, A. (2011) Biological activities of HDL subpopulations and their relevance to cardiovascular disease. *Trends Mol. Med.*, **17** (10), 594–603.
31 Huang, R., Silva, R.A., Jerome, W.G., Kontush, A., Chapman, M.J., Curtiss, L.K., Hodges, T.J., and Davidson, W.S. (2011) Apolipoprotein A-I structural organization in high-density lipoproteins isolated from human plasma. *Nat. Struct. Mol. Biol.*, **18**, 416–422.
32 Rye, K.A. and Barter, P.J. (2004) Formation and metabolism of prebeta-migrating, lipid-poor apolipoprotein A-I. *Arterioscler. Thromb. Vasc. Biol.*, **24**, 421–428.
33 Rothblat, G.H. and Phillips, M.C. (2010) High-density lipoprotein heterogeneity and function in reverse cholesterol transport. *Curr. Opin. Lipidol.*, **21**, 229–238.
34 von Eckardstein, A., Nofer, J.R., and Assmann, G. (2001) High density lipoproteins and arteriosclerosis: role of cholesterol efflux and reverse cholesterol transport. *Arterioscler. Thromb. Vasc. Biol.*, **21**, 13–27.
35 Chapman, M.J., Le Goff, W., Guerin, M., and Kontush, A. (2010) Cholesteryl ester transfer protein: at the heart of the action of lipid-modulating therapy with statins, fibrates, niacin, and cholesteryl ester transfer protein inhibitors. *Eur. Heart J.*, **31**, 149–164.
36 Santamarina-Fojo, S., Gonzalez-Navarro, H., Freeman, L., Wagner, E., and Nong, R. (2004) Hepatic lipase, lipoprotein metabolism, and atherogenesis. *Arterioscler. Thromb. Vasc. Biol*, **24**, 1750–1754.
37 Curtiss, L.K., Valenta, D.T., Hime, N.J., and Rye, K.A. (2006) What is so special about apolipoprotein AI in reverse cholesterol transport? *Arterioscler. Thromb. Vasc. Biol.*, **26**, 12–19.
38 Pruzanski, W., Stefanski, E., de Beer, F.C., de Beer, M.C., Ravandi, A., and Kuksis, A. (2000) Comparative analysis of lipid composition of normal and acute-phase high density lipoproteins. *J. Lipid Res.*, **41**, 1035–1047.
39 Ojala, P.J., Hermansson, M., Tolvanen, M., Polvinen, K., Hirvonen, T., Impola, U., Jauhiainen, M., Somerharju, P., and Parkkinen, J. (2006) Identification of alpha-1 acid glycoprotein as a lysophospholipid binding protein: a complementary role to albumin in the scavenging of lysophosphatidylcholine. *Biochemistry*, **45**, 14021–14031.
40 Scherer, M., Bottcher, A., Schmitz, G., and Liebisch, G. (2011) Sphingolipid profiling of human plasma and FPLC-separated lipoprotein fractions by hydrophilic interaction chromatography tandem mass spectrometry. *Biochim. Biophys. Acta*, **1811**, 68–75.
41 Lee, J.Y., Min, H.K., Choi, D., and Moon, M.H. (2010) Profiling of phospholipids in

lipoproteins by multiplexed hollow fiber flow field-flow fractionation and nanoflow liquid chromatography-tandem mass spectrometry. *J. Chromatogr. A*, **1217**, 1660–1666.

42 Hidaka, H., Hanyu, N., Sugano, M., Kawasaki, K., Yamauchi, K., and Katsuyama, T. (2007) Analysis of human serum lipoprotein lipid composition using MALDI-TOF mass spectrometry. *Ann. Clin. Lab Sci.*, **37**, 213–221.

43 Subbaiah, P.V. and Liu, M. (1996) Comparative studies on the substrate specificity of lecithin:cholesterol acyltransferase towards the molecular species of phosphatidylcholine in the plasma of 14 vertebrates. *J. Lipid Res.*, **37**, 113–122.

44 Maeba, R. and Ueta, N. (2003) Ethanolamine plasmalogen and cholesterol reduce the total membrane oxidizability measured by the oxygen uptake method. *Biochem. Biophys. Res. Commun.*, **302**, 265–270.

45 Maeba, R. and Ueta, N. (2003) Ethanolamine plasmalogens prevent the oxidation of cholesterol by reducing the oxidizability of cholesterol in phospholipid bilayers. *J. Lipid Res.*, **44**, 164–171.

46 Davidson, W.S., Sparks, D.L., Lund-Katz, S., and Phillips, M.C. (1994) The molecular basis for the difference in charge between pre-beta- and alpha-migrating high density lipoproteins. *J. Biol. Chem.*, **269**, 8959–8965.

47 Boucher, J.G., Nguyen, T., and Sparks, D. L. (2007) Lipoprotein electrostatic properties regulate hepatic lipase association and activity. *Biochem. Cell Biol.*, **85**, 696–708.

48 Deguchi, H., Fernandez, J.A., Hackeng, T. M., Banka, C.L., and Griffin, J.H. (2000) Cardiolipin is a normal component of human plasma lipoproteins. *Proc. Natl. Acad. Sci. USA*, **97**, 1743–1748.

49 Morrow, J.D. (2005) Quantification of isoprostanes as indices of oxidant stress and the risk of atherosclerosis in humans. *Arterioscler. Thromb. Vasc. Biol.*, **25**, 279–286.

50 Proudfoot, J.M., Barden, A.E., Loke, W.M., Croft, K.D., Puddey, I.B., and Mori, T.A. (2009) HDL is the major lipoprotein carrier of plasma F2-isoprostanes. *J. Lipid Res.*, **50**, 716–722.

51 Nilsson, A. and Duan, R.D. (2006) Absorption and lipoprotein transport of sphingomyelin. *J. Lipid Res.*, **47**, 154–171.

52 Rye, K.A., Hime, N.J., and Barter, P.J. (1996) The influence of sphingomyelin on the structure and function of reconstituted high density lipoproteins. *J. Biol. Chem.*, **271**, 4243–4250.

53 Saito, H., Arimoto, I., Tanaka, M., Sasaki, T., Tanimoto, T., Okada, S., and Handa, T. (2000) Inhibition of lipoprotein lipase activity by sphingomyelin: role of membrane surface structure. *Biochim. Biophys. Acta*, **1486**, 312–320.

54 Lucke, S. and Levkau, B. (2010) Endothelial functions of sphingosine-1-phosphate. *Cell Physiol. Biochem.*, **26**, 87–96.

55 Scherer, M., Bottcher, A., Schmitz, G., and Liebisch, G. (2010) Sphingolipid profiling of human plasma and FPLC-separated lipoprotein fractions by hydrophilic interaction chromatography tandem mass spectrometry. *Biochim. Biophys. Acta*, **1181** (2): 68–75.

56 Pappu, R., Schwab, S.R., Cornelissen, I., Pereira, J.P., Regard, J.B., Xu, Y., Camerer, E., Zheng, Y.W., Huang, Y., Cyster, J.G., and Coughlin, S.R. (2007) Promotion of lymphocyte egress into blood and lymph by distinct sources of sphingosine-1-phosphate. *Science*, **316**, 295–298.

57 Yatomi, Y., Ruan, F., Hakomori, S., and Igarashi, Y. (1995) Sphingosine-1-phosphate: a platelet-activating sphingolipid released from agonist-stimulated human platelets. *Blood*, **86**, 193–202.

58 Bode, C., Sensken, S.C., Peest, U., Beutel, G., Thol, F., Levkau, B., Li, Z., Bittman, R., Huang, T., Tolle, M., van der Giet, M., and Graler, M.H. (2010) Erythrocytes serve as a reservoir for cellular and extracellular sphingosine 1-phosphate. *J. Cell Biochem.*, **109**, 1232–1243.

59 Nofer, J.-R. and Assmann, G. (2005) Atheroprotective effects of high-density lipoprotein-associated lysosphingolipids. *Trends Cardiovasc. Med.*, **15**, 265–271.

60 Dawson, G., Kruski, A.W., and Scanu, A.M. (1976) Distribution of glycosphingolipids

in the serum lipoproteins of normal human subjects and patients with hypo- and hyperlipidemias. *J. Lipid Res.*, **17**, 125–131.

61 Senn, H.J., Orth, M., Fitzke, E., Wieland, H., and Gerok, W. (1989) Gangliosides in normal human serum: concentration, pattern and transport by lipoproteins. *Eur. J. Biochem.*, **181**, 657–662.

62 Burkard, I., von Eckardstein, A., Waeber, G., Vollenweider, P., and Rentsch, K.M. (2007) Lipoprotein distribution and biological variation of 24S- and 27-hydroxycholesterol in healthy volunteers. *Atherosclerosis*, **194**, 71–78.

63 Tikkanen, M.J., Vihma, V., Jauhiainen, M., Hockerstedt, A., Helisten, H., and Kaamanen, M. (2002) Lipoprotein-associated estrogens. *Cardiovasc Res.*, **56**, 184–188.

64 Lalanne, F., Pruneta, V., Bernard, S., and Ponsin, G. (1999) Distribution of diacylglycerols among plasma lipoproteins in control subjects and in patients with non-insulin-dependent diabetes. *Eur. J. Clin. Invest.*, **29**, 139–144.

65 Havel, R.J., Eder, H.A., and Bragdon, J.H. (1955) The distribution and chemical composition of ultracentrifugally separated lipoproteins in human serum. *J. Clin. Invest.*, **34**, 1345–1353.

66 Chapman, M.J., Goldstein, S., Lagrange, D., and Laplaud, P.M. (1981) A density gradient ultracentrifugal procedure for the isolation of the major lipoprotein classes from human serum. *J. Lipid Res.*, **22**, 339–358.

67 Guerin, M., Lassel, T.S., Le Goff, W., Farnier, M., and Chapman, M.J. (2000) Action of atorvastatin in combined hyperlipidemia: preferential reduction of cholesteryl ester transfer from HDL to VLDL1 particles. *Arterioscler. Thromb. Vasc. Biol.*, **20**, 189–197.

68 Kontush, A. and Chapman, M.J. (2006) Antiatherogenic small, dense HDL: guardian angel of the arterial wall? *Nat. Clin. Pract. Cardiovasc. Med.*, **3**, 144–153.

69 Lee, C.Y., Lesimple, A., Larsen, A., Mamer, O., and Genest, J. (2005) ESI-MS quantitation of increased sphingomyelin in Niemann–Pick disease type B HDL. *J. Lipid Res.*, **46**, 1213–1228.

70 Glass, C.K. and Witztum, J.L. (2001) Atherosclerosis. the road ahead. *Cell*, **104**, 503–516.

71 Iwase, M., Sonoki, K., Sasaki, N., Ohdo, S., Higuchi, S., Hattori, H., and Iida, M. (2008) Lysophosphatidylcholine contents in plasma LDL in patients with type 2 diabetes mellitus: relation with lipoprotein-associated phospholipase A2 and effects of simvastatin treatment. *Atherosclerosis*, **196**, 931–936.

72 Yetukuri, L., Soderlund, S., Koivuniemi, A., Seppanen-Laakso, T., Niemela, P.S., Hyvonen, M., Taskinen, M.-R., Vattulainen, I., Jauhiainen, M., and Oresic, M. (2010) Composition and lipid spatial distribution of HDL particles in subjects with low and high HDL-cholesterol. *J. Lipid Res.*, **51**, 2341–2351.

73 Yetukuri, L., Huopaniemi, I., Koivuniemi, A., Maranghi, M., Hiukka, A., Nygren, H., Kaski, S., Taskinen, M.R., Vattulainen, I., Jauhiainen, M., and Oresic, M. (2011) High density lipoprotein structural changes and drug response in lipidomic profiles following the long-term fenofibrate therapy in the FIELD substudy. *PLoS One*, **6**, e23589.

74 Hiukka, A., Stahlman, M., Pettersson, C., Levin, M., Adiels, M., Teneberg, S., Leinonen, E.S., Hulten, L.M., Wiklund, O., Oresic, M., Olofsson, S.O., Taskinen, M.R., Ekroos, K., and Boren, J. (2009) ApoCIII-enriched LDL in type 2 diabetes displays altered lipid composition, increased susceptibility for sphingomyelinase, and increased binding to biglycan. *Diabetes*, **58**, 2018–2026.

75 Forte, T.M., Bielicki, J.K., Goth-Goldstein, R., Selmek, J., and McCall, M.R. (1995) Recruitment of cell phospholipids and cholesterol by apolipoproteins A-II and A-I: formation of nascent apolipoprotein-specific HDL that differ in size, phospholipid composition, and reactivity with LCAT. *J. Lipid Res.*, **36**, 148–157.

76 Argraves, K.M., Sethi, A.A., Gazzolo, P.J., Wilkerson, B.A., Remaley, A.T., Tybjaerg-Hansen, A., Nordestgaard, B.G., Yeatts, S.D., Nicholas, K.S., Barth, J.L., and Argraves, W.S. (2011) S1P, dihydro-S1P and C24:1-ceramide levels in the HDL-containing fraction of serum inversely

correlate with occurrence of ischemic heart disease. *Lipids Health Dis.*, **10**, 70.

77 Schissel, S.L., Tweedie-Hardman, J., Rapp, J.H., Graham, G., Williams, K.J., and Tabas, I. (1996) Rabbit aorta and human atherosclerotic lesions hydrolyze the sphingomyelin of retained low-density lipoprotein: proposed role for arterial-wall sphingomyelinase in subendothelial retention and aggregation of atherogenic lipoproteins. *J. Clin. Invest.*, **98**, 1455–1464.

78 Flood, C., Gustafsson, M., Pitas, R.E., Arnaboldi, L., Walzem, R.L., and Boren, J. (2004) Molecular mechanism for changes in proteoglycan binding on compositional changes of the core and the surface of low-density lipoprotein-containing human apolipoprotein B100. *Arterioscler. Thromb. Vasc. Biol.*, **24**, 564–570.

79 Fanani, M.L. and Maggio, B. (1997) Mutual modulation of sphingomyelinase and phospholipase A2 activities against mixed lipid monolayers by their lipid intermediates and glycosphingolipids. *Mol. Membr. Biol.*, **14**, 25–29.

80 Morgantini, C., Natali, A., Boldrini, B., Imaizumi, S., Navab, M., Fogelman, A.M., Ferrannini, E., and Reddy, S.T. (2011) Anti-inflammatory and antioxidant properties of HDLs are impaired in type 2 diabetes. *Diabetes*, **60**, 2617–2623.

81 Oresic, M. (2011) Informatics and computational strategies for the study of lipids. *Biochim. Biophys. Acta*, **4** (2), 121–127.

11
Mediator Lipidomics in Inflammation Research

Makoto Arita, Ryo Iwamoto, and Yosuke Isobe

11.1
Introduction

Polyunsaturated fatty acids (PUFAs) exhibit a range of biological effects, many of which are mediated through the formation and actions of lipid mediators such as prostaglandins (PGs), leukotrienes (LTs), lipoxins (LX), resolvins, and protectins. These lipid mediators are potent endogenous regulators of inflammation and related diseases. To better understand the molecular and cellular mechanisms underlying the coordinated processes of inflammation and resolution, it is important to know when, where, and how much of those lipid mediators are formed in the inflammatory sites. Mediator lipidomics in general, when combined with information on metabolic pathways such as Kyoto Encyclopedia of Genes and Genomes (KEGG), have broad applications in understanding the role of bioactive lipid mediators under certain physiological and/or pathological conditions. The potential challenges of lipid mediator analyses are low abundance, labile, and overlapping parent masses and products. In this chapter, we will introduce LC-ESI-MS/MS-based lipidomics, which is suitable for quantifying lipid mediators in complex samples because this system can provide superior sensitivity, selectivity, and rapid analysis.

11.2
PUFA-Derived Lipid Mediators: Formation and Action

PUFA-derived lipid mediators are formed by enzymatic oxidation through the action of cyclooxygenases (COX), lipoxygenases (LOX), and cytochrome P450 monooxygenases (CYP). Arachidonic acid (20:4 *n*-6) is a common precursor of many eicosanoids, which are bioactive lipid mediators that control inflammatory responses. When cells are stimulated, arachidonic acid is released from membrane phospholipids by phospholipase A_2 (PLA_2), which hydrolyzes the acyl ester bond. This is the first step in the arachidonic acid cascade, and is the overall rate-determining step in the generation of eicosanoids. Mammals have three types of PLA_2, which are classified as secretory, cytoplasmic, and calcium-independent

Lipidomics, First Edition. Edited by Kim Ekroos.

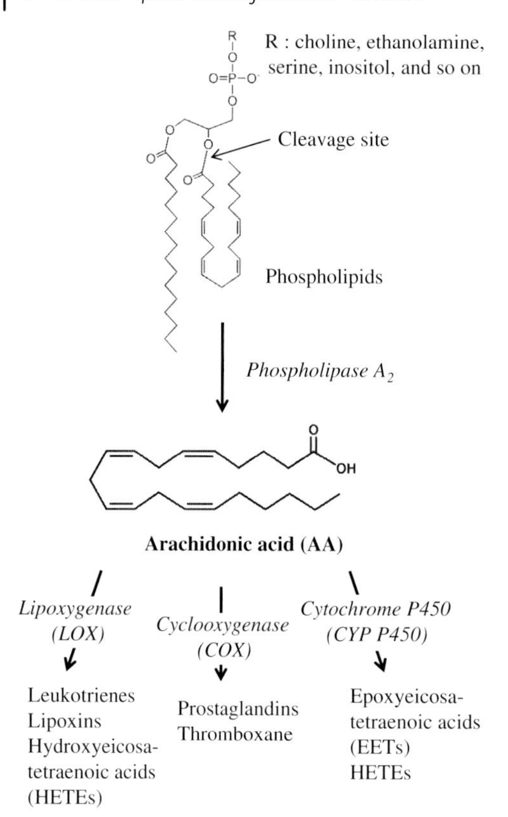

Figure 11.1 Eicosanoid production from arachidonic acid. Phospholipase A_2 cleaves the ester bond of phospholipids marked by the arrow to release arachidonic acid. Free arachidonic acid is used as a substrate for the cyclooxygenase (COX), lipoxygenase (LOX), and cytochrome P450 monooxygenase (CYP) pathways. The COX pathway produces prostaglandins and thromboxane. The LOX pathway produces leukotrienes, lipoxins, and hydroxyeicosatetraenoic acids (HETEs). The CYP pathway produces epoxyeicosatrienoic acids (EETs) and HETEs.

PLA_2s ($sPLA_2$s, $cPLA_2$s, $iPLA_2$s), respectively [1]. The different types of PLA_2 are regulated differently and are expressed in different tissues. Unesterified intracellular arachidonic acid is immediately metabolized by COX, LOX, or CYP. The COX pathway leads to the formation of prostaglandins and thromboxanes (TX); the LOX pathway leads to leukotrienes, lipoxins, and hydroeicosatetraenoic acids (HETE); and the CYP pathway leads to HETEs and epoxyeicosatrienoic acids (EET) (Figure 11.1). As a class, these molecules act as autacoids that are rapidly synthesized in response to specific stimuli, act quickly at the immediate locality, and remain active for only a short time before degradation.

Arachidonic acid-derived eicosanoids are important in many physiological processes. Under nondisease conditions, PGs contribute to the maintenance of homeostasis, for example, they have cytoprotective roles in the gastric mucosa, respiratory tract, and renal parenchyma. PGs are also involved in proinflammatory processes and are responsible for many of the hallmark signs of inflammation such as heat, redness, swelling, and pain [2]. COX occurs in two isoforms, called the constitutive isoform (COX-1) and the inducible isoform (COX-2) [3]. These enzymes may contribute to the production of different sets of eicosanoids at different locations at different times. The LOX pathway represents another major pathway to produce LTs and LXs. Mammals have at least three LOXs, 5-, 12-, and 15-LOX present in mammalian systems [4, 5]. 5-LOX-derived LTs (LTB_4, cysteinyl LTs) are involved in proinflammatory processes such as neutrophil infiltration, in creased vascular permeability, and smooth muscle contraction [6]. In contrast, 5- and 15-LOX-derived LXA_4 counterregulate the proinflammatory processes and may be important in the resolution of inflammation [7]. An imbalance in lipoxin–leukotriene homeostasis may be a key factor in the pathogenesis of inflammatory disease s. The epoxygenation of arachidonic acid by CYP generates EETs, which may have roles in the regulation of smooth muscle cells and vascular tone [8]. Many of the eicosanoids signal via seven-transmembrane G-protein-coupled receptors [9].

Eicosapentaenoic acid (EPA) (20:5 *n*-3) and docosahexaenoic acid (DHA) (22:6 *n*-3) are *n*-3 PUFAs that are abundant in fish oils. EPA-derived mediators include 3-series PGs, 5-series LTs and LXs, hydroxyeicosapentaenoic acids (HEPE), and epoxyeicosatetraenoic acids (EpETE) (Figure 11.2). DHA is also converted to

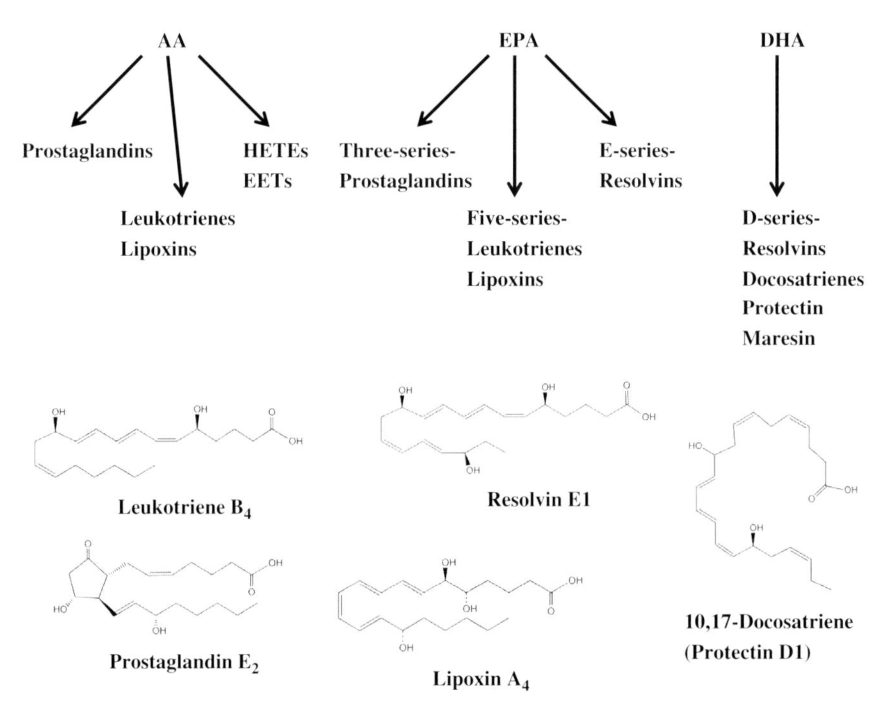

Figure 11.2 Scheme of eicosanoid, docosanoid, and hydroxy-fatty acid production from arachidonic acid (AA), eicosapentaenoic acid (EPA), and docosahexaenoic acid (DHA).

hydroxydocosahexaenoic acids (HDoHE) and epoxy-fatty acids by enzymatic oxidation (Figure 11.2). Dietary supplementation of *n*-3 PUFA has beneficial effects in many inflammatory disorders, including cardiovascular disease, arthritis, colitis, and metabolic syndrome [10]. Also, elevation in *n*-3 PUFA levels in *n*-3 desaturase (*fat-1*) transgenic mice protected against inflammatory disease models [11]. *n*-3 PUFAs are thought to act via several mechanisms. One role is to serve as an altern ative substrate for COX or LOX, resulting in the production of less potent products [12]. Another role is to be converted to potent anti-inflammatory and protective mediators. For example, E-series resolvins are produced from EPA, and D-series re solvins, protectin, and maresin are produced from DHA (Figure 11.2) [13]. These metabolites may have some roles in the beneficial actions of *n*-3 PUFAs in controlling inflammation and related diseases.

11.3 LC-ESI-MS/MS-Based Lipidomics

A powerful approach for the analysis of mono- and polyhydroxylated fatty acids is liquid chromatography tandem mass spectrometry (LC-ESI-MS/MS). The development of electrospray ionization (ESI) technology provided an ideal interface between LC and MS, which paved the way for the analysis of hydroxy-fatty acids without derivatization and decomposition. ESI is a soft ionization technology used to form either positive or negative ions through the addition of a proton to form $[M+H]^+$ or through the removal of a proton to form $[M-H]^-$. In case of fatty acid-derived mediators, ESI results in $[M-H]^-$ carboxylate ions that can be detected with relatively high sensitivity. A triple quadrupole mass spectrometer is capable of carrying out an MS/MS method called multiple reaction monitoring (MRM). A specified precursor ion is selected according to its mass-to-charge ratio in the first quadrupole mass filter and is fragmented into product ions in the second chamber by collision-induced dissociation (CID). Then, the third quadrupole mass filter is locked on its specified product ion. This MRM mode leads to further improvement of the detection and quantification limits when combined with high-resolution LC separations. We have developed a comprehensive LC-ESI-MS/MS method that can simultaneously detect and quantify more than 250 PUFA metabolites, including PGs, LTs, LXs, resolvins, protectins, and other AA-, EPA-, DHA-derived products (Figure 11.3).

11.3.1 Sample Preparation

Liquid samples are adjusted to 67% methanol (v/v) and kept at $-20\,^{\circ}C$. Frozen tissue samples are homogenized in ice-cold methanol using Precellys 24 biological sample grinder (Bertin Technologies). The samples are centrifuged to remove precipitated proteins. The supernatants are diluted with ice-cold water and adjusted to 10% (v/v) methanol. Internal standards (1 ng of PGE_2-d4, LTB_4-d4, 15-HETE-d8,

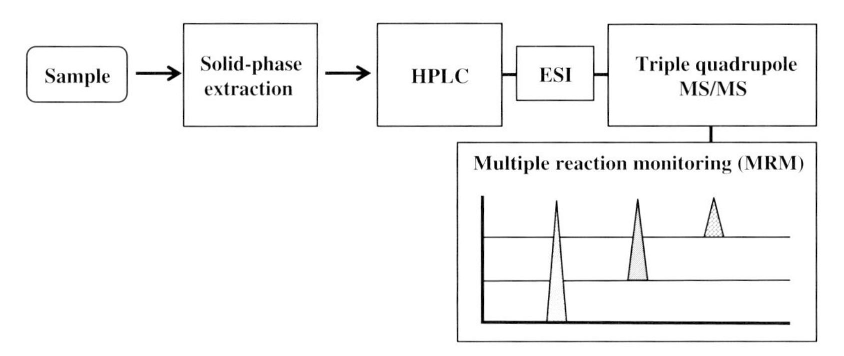

Figure 11.3 Flow chart depicting the system of LC-ESI-MS/MS-based lipidomics. After solid-phase extraction, samples are separated by HPLC and fatty acid metabolites are detected and quantified by multiple reaction monitoring (MRM) using triple quadrupole MS/MS.

and arachidonic acid-d8) are added to each sample, and the clear supernatants acidified to pH 4.0 are immediately applied to preconditioned solid-phase extraction cartridges (Sep-Pak C18, Waters) to extract the lipid mediators. Sep-Pak cartridges are first washed with 20 ml water and then with 20 ml hexane, and finally the lipid mediators (hydroxy-fatty acids) are eluted with 10 ml methyl formate. The lipid extracts are dried under a gentle stream of nitrogen gas, dissolved in 60% methanol, and stored at −20 °C before analysis.

11.3.2
LC-ESI-MS/MS Analysis

LC-MS/MS-based lipidomic analyses are performed on an Acquity UPLC BEH C_{18} column (1.0 mm × 150 mm × 1.7 μm) using an Acquity UltraPerformance LC system (UPLC; Waters) coupled to an electrospray (ESI) triple quadrupole mass spectrometer (5500 QTRAP; AB SCIEX). Instrument control and data acquisition are performed using Analyst 1.5.1 (AB SCIEX). Samples are eluted with a mobile phase consisting of water/acetate (100 : 0.1, v/v) and acetonitrile/methanol (4 : 1, v/v) (73 : 27) for 5 min and ramped to 30 : 70 after 15 min, to 20 : 80 after 25 min and held for 8 min, to 0 : 100 after 35 min and held for 10 min with flow rates of 50 μl/min (0–30 min), 80 μl/min (30–33 min), and 100 μl/min (33–45 min). MS/MS analyses are conducted in negative ion mode, and fatty acid metabolites are detected and quantified by scheduled MRM with dwell times of ≥10 ms and interchannel delay of 3 ms, giving a total cycle time of 1.3 s. Conditions for the detection of each compound by MRM are represented in Table 11.1. Identification of the target compounds is based on the LC retention time of the analyte compared to that of a standard and on the ratio of abundance of two specific MRM transitions. For quantification, composite standard solutions ranging from 0.1 to 100 pg/μl are prepared, and calibration lines are calculated by the least-squares linear regression method, as represented in Figure 11.4.

Table 11.1 Multiple reaction monitoring (MRM) transitions and LC retention times of lipid mediators and hydroxy-fatty acids.

Compound	MRM transition (*m*/*z*)	Retention time (min)
PGE2	351 > 271	14.9
PGD2	351 > 271	15.3
15-Keto-PGE2	349 > 235	15.4
PGE2-20-OH	367 > 287	6.4
15-Deoxy-PGJ2	315 > 271	20.9
PGF2α	353 > 193	14.6
6-Keto-PGF1α	369 > 163	12.3
TXB2	369 > 195	13.9
LTB4	335 > 195	18.5
LTB4-20-OH	351 > 195	12.6
HXB3	335 > 183	20.6
LXA4	351 > 217	15.9
LXB4	351 > 221	15.0
5-HETE	319 > 115	24.4
5,6-EET	319 > 191	26.7
5,6-DHT	337 > 145	21.4
8-HETE	319 > 155	23.5
9-HETE	319 > 123	24.0
8,9-EET	319 > 155	26.3
8,9-DHT	337 > 127	20.6
11-HETE	319 > 167	23.0
12-HETE	319 > 179	23.4
11,12-EET	319 > 167	25.9
11,12-DHT	337 > 167	20.1
15-HETE	319 > 219	22.4
14,15-EET	319 > 219	25.0
14,15-DHT	337 > 207	19.5
16-HETE	319 > 189	21.6
17-HETE	319 > 247	21.4
18-HETE	319 > 261	21.2
19-HETE	319 > 275	20.8
20-HETE	319 > 289	21.0
5,6-Di-HETE	335 > 219	20.8
5,15-Di-HETE	335 > 201	18.1
8,15-Di-HETE	335 > 235	17.7
5-Oxo-ETE	317 > 203	25.8
12-Oxo-ETE	317 > 179	23.9
15-Oxo-ETE	317 > 113	22.9
AA-d8	311 > 267	31.1
15-HETE-d8	327 > 226	22.2
LTB4-d4	339 > 197	18.4
PGE2-d4	355 > 275	14.8
PGB2-d4	227 > 179	17.1
8-Iso-PGF2α-d4	357 > 197	13.7
PGE3	349 > 233	13.8

(*continued*)

Table 11.1 (*Continued*)

Compound	MRM transition (*m/z*)	Retention time (min)
PGD3	349 > 189	14.1
PGF3α	351 > 191	13.5
Δ17-6-Keto-PGF1α	367 > 207	10.5
TXB3	367 > 169	12.7
LTB5	333 > 195	17.2
LXA5	349 > 115	14.7
RvE1	349 > 195	12.2
RvE2	333 > 115	16.6
5-HEPE	317 > 115	21.8
8-HEPE	317 > 155	21.2
9-HEPE	317 > 149	21.5
11-HEPE	317 > 167	20.9
12-HEPE	317 > 179	21.2
15-HEPE	317 > 219	20.8
14,15-EpETE	317 > 207	23.1
18-HEPE	317 > 259	20.4
17,18-EpETE	317 > 259	22.5
17,18-Di-HETE	335 > 247	18.1
19-HEPE	317 > 229	19.4
20-HEPE	317 > 287	20.1
4-HDoHE	343 > 101	24.7
7-HDoHE	343 > 141	23.5
8-HDoHE	343 > 189	23.7
10-HDoHE	343 > 153	22.9
11-HDoHE	343 > 121	23.2
13-HDoHE	343 > 193	22.6
14-HDoHE	343 > 205	22.9
16-HDoHE	343 > 233	22.3
17-HDoHE	343 > 245	22.4
16,17-EpDPE	343 > 233	25.1
20-HDoHE	343 > 241	21.9
19,20-EpDPE	343 > 241	24.4
19,20-Di-HDoPE	361 > 229	19.4
21-HDoHE	343 > 255	21.5
22-HDoHE	343 > 269	21.4
4,14-Di-HDoHE	359 > 101	19.1
7,14-Di-HDoHE	359 > 113	18.6
7,17-Di-HDoHE	359 > 199	18.1
10,17-Di-HDoHE(PD1)	359 > 153	18.0
RvD1	375 > 141	15.7
RvD2	375 > 175	15.0
AA	303 > 259	31.3
EPA	301 > 257	28.9
DHA	327 > 283	30.6
DPA *n*-3	329 > 285	32.0
DPA *n*-6	329 > 285	33.4

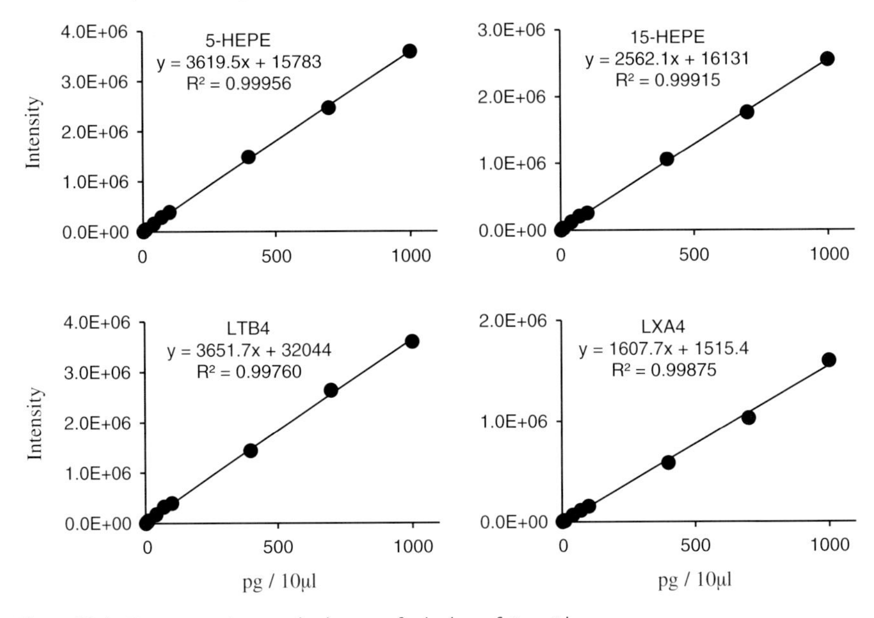

Figure 11.4 Representative standard curves for hydroxy-fatty acids.

Representative chromatograms of mono-HEPEs are depicted in Figure 11.5. Among HEPE isomers, 8-HEPE and 12-HEPE did not resolve well using a C_{18} column (retention time of 21.2 min). However, the choice of structure-specific product ions allowed their differentiation (8-HEPE m/z 317 > 155 and 12-HEPE m/z 317 > 179). Also, the MRM transition used for 18-HEPE (m/z 317 > 215) showed cross-reactivity with 17,18-EpETE, but these hydroxy- and epoxy-fatty acids were resolved well by C_{18} column chromatography (retention times of 20.4 and 22.5 min, respectively). Therefore, the LC-MS/MS system using MRM mode provides structure-specific signal detection and further improves the quantification limits when combined with high-resolution LC separations. New MS instruments such as QTRAP 5500 can conduct hundreds of MRM analyses, and high-resolution LCs such as UPLC can improve quantification by optimal peak separation.

11.4 Mediator Lipidomics in Inflammation and Resolution

In many human diseases, uncontrolled inflammation is suspected as a key component of pathogenesis [14]. Acute inflammation is an indispensable host response to insult or tissue injury. However, excessive or inappropriate inflammatory responses can cause local tissue damage and remodeling, which contribute to a range of chronic diseases. In healthy individuals, acute inflammation is self-limiting and has an active termination program [15]. Therefore, the mechanisms by which acute inflammation is resolved are of interest.

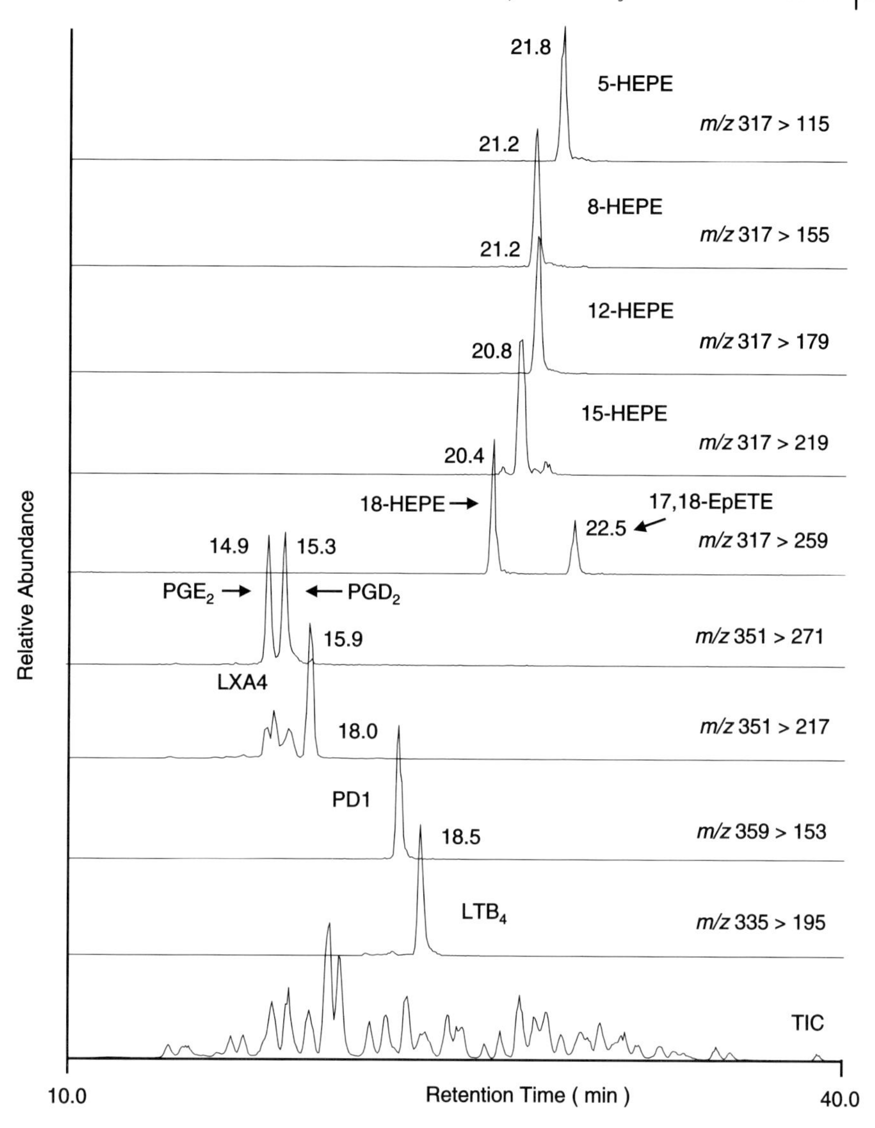

Figure 11.5 Representative MRM chromatograms for hydroxy-fatty acids.

The local inflammatory response is characterized by a sequential release of mediators and the recruitment of different types of leukocytes that become activated at the inflamed site [15, 16]. To better understand the molecular and cellular mechanisms underlying the coordinated processes of inflammation and resolution, we applied LC-ESI-MS/MS-based mediator lipidomics to the self-resolving acute inflammation model, namely, murine zymosan-induced peritonitis [17]. As shown in Figure 11.6, temporal and quantitative differences in the lipid mediator profiles were observed in the course of acute inflammation and resolution. COX pathway

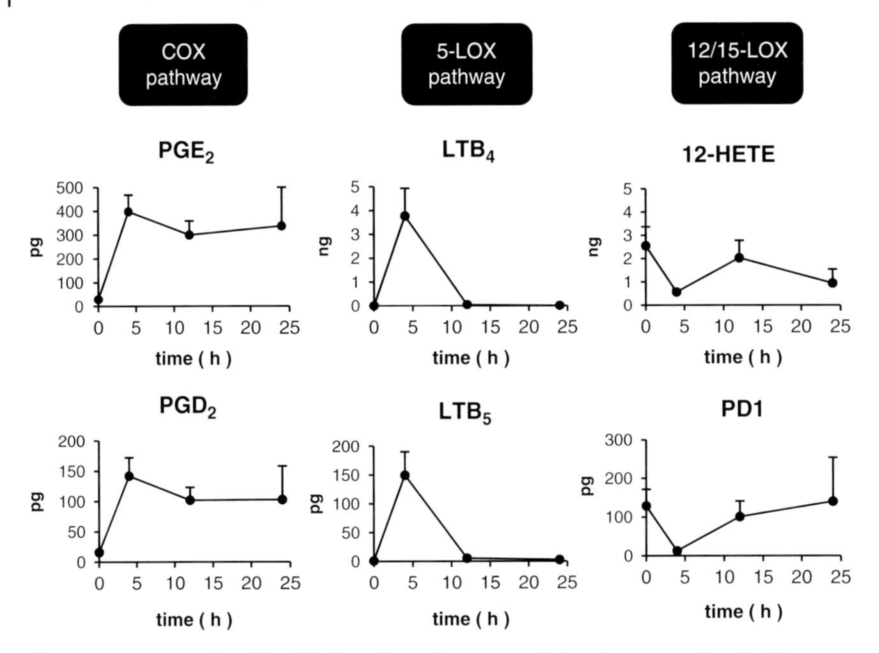

Figure 11.6 Temporal profile of lipid mediators in zymosan-induced mouse peritonitis. Time course of COX products PGE_2 and PGD_2, 5-LOX products LTB_4 and LTB_5, and 12/15-LOX products 12-HETE and PD1 in the peritoneal exudates of zymosan-induced mouse peritonitis.

products such as PGE_2 and PGD_2 appeared upon zymosan stimulation and were present in both the initiation (4 h) and the early resolution (24 h) phases. In contrast, 5-LOX pathway products such as LTB_4 appeared only in the initiation phase. Of interest, the 12/15-LOX pathway products, including DHA-derived mediator protectin D1 (PD1), were present in the lavage of naive mice. Upon zymosan challenge, the concentrations of the 12/15-LOX products decreased in the initiation phase and recovered in the early resolution phase. Especially, the 5-LOX and 12/15-LOX pathway products displayed a reciprocal pattern.

Next we examined the temporal and differential profiles of inflammatory cells (leukocytes in exudates) in the course of acute peritonitis. Acute inflammatory response is characterized by edema formation and neutrophil infiltration followed by monocyte/macrophage accumulation [15]. In zymosan-induced peritonitis, we found that eosinophils are recruited to the inflamed loci during the resolution phase of acute inflammation [17]. *In vivo* depletion of eosinophils by injecting anti-IL-5 monoclonal antibody caused a resolution deficit, and the differential display of lipid mediators using LC-MS/MS revealed that locally activated eosinophils in the early resolution phase produced 12/15-LOX-derived mediators, including DHA-derived anti-inflammatory/proresolving lipid mediator PD1 (Figure 11.7) [13]. Indeed, the predominant population expressing 12/15-LOX in the early resolution phase was eosinophils, and isolated eosinophils produced a significant amount of 12/15-LOX-derived mediators. In the naive peritoneal cavity, resident macrophages

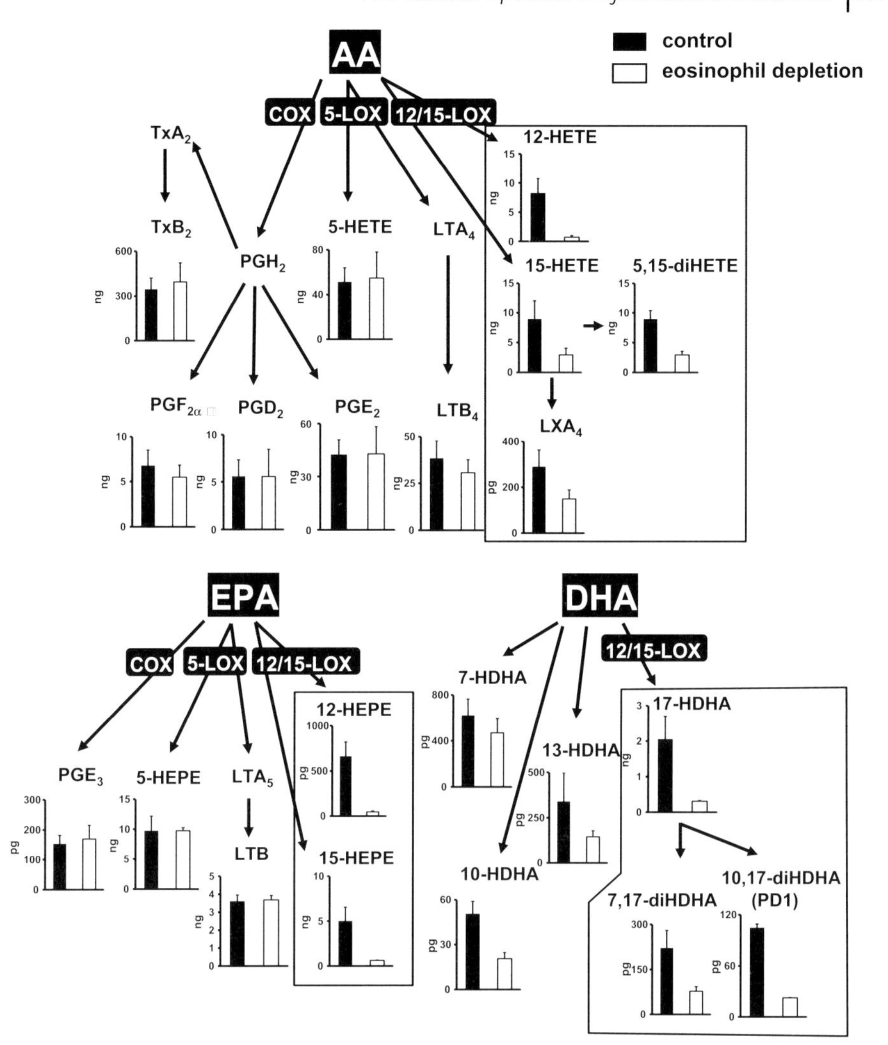

Figure 11.7 Differential display of LC-MS/MS-based mediator lipidomics. Peritoneal lavage cells in the resolution phase from control IgG-treated (black bars) or anti-IL-5 monoclonal antibody-treated mice (white bars) were stimulated *ex vivo* with calcium ionophore. Arachidonic acid (AA)-, eicosapentaenoic acid (EPA)-, and docosahexaenoic acid (DHA)-derived products were quantified by LC-MS/MS. These results indicate that eosinophils in the resolution phase produce 12/15-LOX products, including DHA-derived lipid mediator PD1. 12/15-LOX products are boxed. Reprinted from Ref. [17].

are the predominant cell population expressing relatively high levels of 12/15-LOX, and are likely responsible for the generation of 12/15-LOX products in the lavage of naive mice. The resolution deficit caused by eosinophil depletion was rescued by eosinophil restoration or local administration of PD1, and eosinophils deficient in 12/15-LOX were unable to rescue the resolution phenotype. These results indicate that eosinophils and eosinophil-derived lipid mediators (i.e., 12/15-LOX-derived mediators) have a role in promoting the resolution of acute inflammation [17].

11.5
Conclusion and Future Perspective

The comprehensive lipidomic method described in this chapter has a number of attractive applications. One of them, as shown above, is a coordinated class switching of lipid mediators in the course of acute inflammation and resolution. This technology could potentially identify the metabolic fingerprint of a disease for clinical diagnosis and treatment. Mediator lipidomics concerns the simultaneous and quantitative analysis of bioactive lipid mediators in biological systems. When combined with proteomic, transcriptomic, and genomic profiles (multiomics profiling), it can greatly assist in understanding the role of lipid mediators in certain biological and/or pathological conditions.

References

1 Murakami, M., Taketomi, Y., Miki, Y., Sato, H., Hirabayashi, T., and Yamamoto, K. (2011) Recent progress in phospholipase A2 research: from cells to animals to humans. *Prog. Lipid Res.*, **50**, 152–192.
2 Woodward, D.F., Jones, R.L., and Narumiya, S. (2011) International union of basic and clinical pharmacology. LXXXIII: classification of prostanoid receptors, updating 15 years of progress. *Pharmacol. Rev.*, **63**, 471–538.
3 Smith, W.L., DeWitt, D.L., and Garavito, R.M. (2000) Cyclooxygenases: structural, cellular, and molecular biology. *Annu. Rev. Biochem.*, **69**, 145–182.
4 Murphy, R.C. and Gijon, M.A. (2007) Biosynthesis and metabolism of leukotrienes. *Biochem. J.*, **405**, 379–395.
5 Kuhn, H. and O'Donnell, V.B. (2006) Inflammation and immune regulation by 12/15-lipoxygenases. *Prog. Lipid Res.*, **45**, 334–356.
6 Peters-Golden, M. and Henderson, W.R. (2007) Leukotrienes. *N. Engl. J. Med.*, **357**, 1841–1854.
7 Serhan, C.N. (2005) Lipoxins and aspirin-triggered 15-epi-lipoxins are the first lipid mediators of endogenous anti-inflammation and resolution. *Prostaglandins Leukot. Essent. Fatty Acids.*, **73**, 141–162.
8 Kroetz, D.L. and Zeldin, D.C. (2002) Cytochrome P450 pathways of arachidonic acid metabolism. *Curr. Opin. Lipidol.*, **13**, 273–283.
9 Funk, C.D. (2001) Prostaglandins and leukotrienes: advances in eicosanoid biology. *Science*, **294**, 1871–1875.
10 Simopoulos, A.P. (2002) Omega-3 fatty acids in inflammation and autoimmune diseases. *J. Am. Coll. Nutr.*, **21**, 495–505.
11 Kang, J.X. (2007) Fat-1 transgenic mice: a new model for omega-3 research. *Prostaglandins Leukot. Essent. Fatty Acids.*, **77**, 263–267.
12 Schmitz, G. and Ecker, J. (2008) The opposing effects of *n*-3 and *n*-6 fatty acids. *Prog. Lipid Res.*, **47**, 147–155.
13 Bannenberg, G. and Serhan, C.N. (2010) Specialized pro-resolving lipid mediators in the inflammatory response: an update. *Biochim. Biophys. Acta*, **1801**, 1260–1273.
14 Nathan, C. and Ding, A. (2010) Nonresolving inflammation. *Cell*, **140**, 871–882.
15 Serhan, C.N., Brain, S.D., Buckley, C.D., Gilroy, D.W., Haslett, C., O'Neill, L.A.J.,

Perretti, M., Rossi, A.G., and Wallace, J.L. (2007) Resolution of inflammation: state of the art, definitions and terms. *FASEB J.*, **21**, 325–332.

16 Schwab, J.M., Chiang, N., Arita, M., and Serhan, C.N. (2007) Resolvin E1 and protectin D1 activate inflammation-resolution programmes. *Nature*, **447**, 869–874.

17 Yamada, T., Tani, Y., Nakanishi, H., Taguchi, R., Arita, M., and Arai, H. (2011) Eosinophils promote resolution of acute peritonitis by producing proresolving mediators in mice. *FASEB J.*, **25**, 561–568.

12
Lipidomics for Elucidation of Metabolic Syndrome and Related Lipid Metabolic Disorder

Ryo Taguchi, Kazutaka Ikeda, and Hiroki Nakanishi

12.1
Introduction

Lipidomics is an essential technique to elucidate the recent diseases such as metabolic syndrome in which lipid metabolic disorder is considered as one of the main physiological changes [1–4]. Among them, lipid oxidation is the most prominent event as a result of oxidative stress that is thought to be a major physiological process occurring in the progress of these diseases during aging of the human life [5–7]. Thus, a basic analytical system is very important especially for the analysis of the distribution of polyunsaturated fatty acid (PUFA) containing lipids in several organs where the increase in and the existence of their oxidation and accumulation of their oxidized metabolites such as hydroperoxides, hydrooxides, ketons, epoxides, aldehydes, and carboxylated species are expected.

Sources of PUFA in mammalian tissues are exclusively regulated by external factors, especially food supply. Thus, both qualitative and quantitative states of lipid supply from foods for mammalians are extremely important to the progress of these diseases resulting from lipid metabolic disorder including lipid oxidation.

Recent progress of lipidomics has made it possible to know exact distribution of lipid molecular species containing saturated, monounsaturated, and polyunsaturated fatty acids (SFA, MUFA, and PUFA) even in the states of individual classes in phospholipids (PLs) and triglycerides (TGs) [8–12]. Also, their specified localizations are proved to reflect physiological functions of these molecular species within different individual organs and tissue regions.

Molecular species of PLs in individual organs have been proved to be strictly regulated by localization of phospholipases, acyltransferases, and acyl-CoA concentration specifically regulated by acyl-CoA synthetases, fatty acid elongases, and desaturases. These facts have been revealed recently by the application of global analysis by mass spectrometry (MS). Recent methods in lipidomics by mass spectrometry have been very rapidly progressing, making global and comprehensive analyses of individual lipid molecular species in tissues possible [8–12]. As a result,

Lipidomics, First Edition. Edited by Kim Ekroos.

it was revealed that the big differences can exist in molecular species of PLs in specific localization such as different organs or tissue regions. Moreover, exact substrate specificities of several phosholipases and acyltransferases concerning lipid acyl-chain remodeling have been revealed.

In general, most of the already obtained knowledge of the structure of molecular species of PLs revealed to be true for positional specificities in fatty acids (FAs) of *sn*-1 and *sn*-2 such that the saturated ones are on *sn*-1 and the polyunsaturated ones on *sn*-2. On the other hand, precise analytical data obtained from recent mass analysis techniques have proved the fact that the significant level of unusual opposite structure to normal one specifically exist in special organs or special tissue domains [13]. For instance, significantly higher amounts of molecular species having opposite combination of FAs in *sn*-1 and *sn*-2 have been reported to exist specifically in brain. Furthermore, different position of double bonds such as in oleic acid and *cis*-vaccenic acid for 18:0 has been revealed to be strictly controlled in tissues and cell organelle [14]. A recent new technique in lipidomics such as ozone-induced dissociation makes it possible to elucidate positional isomers of the double bonds in FAs of intact lipid species [14, 15].

In this chapter, we try to introduce recent global analytical methods by mass spectrometry and applied experimental results to several diseases such as metabolic syndrome in which lipid metabolic disorder is one of the major factors. Also, we will introduce recent progress in direct surface analysis for elucidation of diversity of lipid molecular species in different mammalian organs and tissue domains.

12.2
Basic Strategy of Lipidomics for Elucidating Metabolic Changes of Lipids at the Level of their Molecular Species in Metabolic Syndrome and Related Diseases

There are some basic key points to calculate the level of lipid metabolic disorder such as in metabolic syndrome. First, the changes in the ratio of PLs and TGs are very important to elucidate some lipid metabolic changes in obesity-derived diseases [5, 7]. Usually, it will be detected as an increase in TGs against phosphatidylcholine (PC) by the relative ratio in plasma, liver, and adipose tissues. Also, in many cases increased ratio of sphingomyelin (SM) against PC is often observed in obesity and related diseases such as diabetes, arteriosclerosis, and cardial infarction. In many chronic inflammation diseases, such as atherosclerosis, significant increase in lysophospholipids (LPLs) is observed as a result of decrease in PUFA-containing molecular species in PC and phosphatidylethanolamine (PE). This indicated that some specific phospholipase A_2 (PLA_2) is involved in this process [16]. Furthermore, the resulting free FAs and oxidized FAs are also important analytes as they influence the level of lipid mediators relating to both initiation and termination processes of inflammation and their chronic situation. Thus, knowing the level of oxidized lipid metabolites, such as oxidized FAs, PLs, and TGs and their metabolites helps in interpreting the different states of oxidative stress in metabolic syndrome. For instance, membrane PLs can be utilized as a source of arachidonic acid (AA, 20:4), which is the substrate for the formation of lipid mediators, such as leukotriens and prostaglandins. The oxidized metabolites of *n*-3 PUFAs such as eicosapentaenoic acid (EPA, 20:5) and docosahexaenoic acid (DHA, 22:6) are shown to be important antiinflammatory mediators [17, 18]. However, many of their functions remain unclear. Thus, global and comprehensive analysis of PLs by lipidomics is essential to further clarify their functions and the enzymes related to their metabolism.

Recent approaches in lipidomics have been applied to elucidate individual diseases, categorized as lipid metabolic disorders, such as diabetes [19, 20], Alzheimer's disease [21, 22], atherosclerosis [16, 23], heart failure [24], and other lipid-related diseases [25]. However, several different mass analytical approaches should typically be selected to obtain the required basic changes in molecular lipid profiles.

12.3 Analytical Systems by Mass Spectrometry in Lipidomics

12.3.1 LC-MS and LC-MS/MS Analyses for Global Detection of Phospholipids and Triglycerides [26–28]

Current lipidomics technologies are predominantly associated with soft ionization techniques, such as electrospray ionization (ESI). For global analysis of PLs and TGs, we normally select comprehensive liquid chromatography (LC)-MS/MS analysis with data-dependent scanning, using a high separation C18 or C30 reverse-phase (RP) column by LTQ-Orbitrap mass spectrometer (Thermo-Fisher Scientific) with high sensitivity of linear ion trap and high resolution of Fourier transform MS (FTMS) [26]. PL mixtures are typically analyzed by LC-ESIMS/MS using data-dependent scanning.

For example, when using a C18 or C30 RP column, PLs elute in order from the hydrophilic molecules to the hydrophobic molecule. Predominantly, the FA length and the degree of unsaturation influences the elution order in each molecular species. We normally perform the analysis in negative ion mode to detect FAs as fragment ions (i.e., acyl anions) from PLs by MS/MS. Data-dependent scanning is further utilized to automatically identify molecular peaks with high intensity for subsequent product ion scanning. The ion intensities of each molecular ion can be further used for quantitative profiling of molecular species of PLs after proper compensation by standard PLs of same classes. In our laboratory, we have developed a lipid identification and profiling tool called "Lipid Search" to handle all these processes. This tool is described in detail below.

Once LC-MS/MS data are obtained from the same samples, major molecular species in PLs or TGs could be directly identified from single LC-MS data.

12.3.2
Infusion Analysis with Precursor Ion and Neutral Loss Scanning [29–35]

There are two types of structure-related focused analysis by ESIMS/MS. Precursor ion scanning (PIS) and neutral loss scanning (NLS) for head group survey (HGS) and fatty acid survey (FAS), respectively, are typically performed in the positive or negative ion modes. PIS and NLS of the polar head groups of PLs are valuable techniques for detecting most molecular species within the same class of PLs [29–35]. PIS of carbonic anions was also very useful for identifying molecular species of PLs with specified fatty acyl chains as reported by Ekroos *et al.* [29].

12.3.3
Targeted Analysis by Multiple Reaction Monitoring for Oxidized Lipids and Lipid Mediators by LC-MS/MS on Triple-Stage Quadrupole Mass Spectrometers [6, 36–39]

Low-abundance important lipid metabolites such as oxidized lipids are typically difficult to detect by comprehensive and untargeted methods because of their lack of sensitivities in comparison to that of multiple reaction monitoring (MRM) on triple-stage quadrupole mass spectrometers. In this case, we normally select MRM or selected reaction monitoring (SRM) analysis for detecting very small amounts of targeted molecules such as lipid mediators or oxidized lipid metabolites [6, 36–39]. Even in this analysis, the ESI makes it possible to detect more than hundred molecules in a single LC run. In MRM analysis, though the target molecules to be analyzed needs to be defined in advance, even with this method some level of comprehensive approach is possible to expand the detecting target to probable molecular masses theoretically calculated by their structural similarities.

12.4
Lipidomic Data Processing

12.4.1
Strategy of Lipid Search

One of the important processes in global analysis in lipid molecular species is an automated search engine to obtain effective identification results with high reproducibility. Next, we will introduce our automated search engine for lipidomics named “Lipid Search.” The prototype of this search engine was opened to public in 2003.

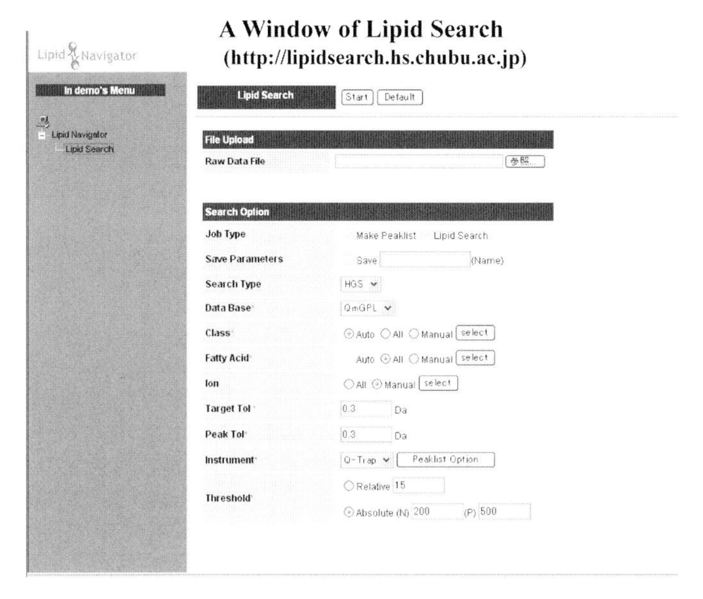

Figure 12.1 Window of Lipid Search, an automated identification tool of lipid molecular species from raw MS data (http://lipidsearch.hs.chubu.ac.jp/) (2010 version).

The "Lipid Search" (http://lipidsearch.hs.chubu.ac.jp) is a search engine for the identification of PL molecular species from MS raw data developed by our group in collaboration with Mitsui Knowledge Industry (http://www.mki.co.jp/eng/) (Figures 12.1 and 12.2). The latest version of "Lipid Search" was opened to the public in November 2009 with subsequent minor changes. This tool can directly handle raw MS data obtained from Q-TOF micro (Waters), LTQ-Orbitrap, 4000 QTRAP (AB Sciex), and IT-TOF (Shimadzu). The identification efficiency of "Lipid Search" highly depends on the quality of MS data. Our database contains more than 200 000 theoretical m/z values of molecular weight-related ions of individual lipid metabolites, and their theoretical fragment ions. Patterns of ion fragments have been theoretically constructed and improved by using experimental data obtained from synthesized standards and many different natural samples.

12.4.2
Application and Identification Results of "Lipid Search"

For demonstration purposes, we analyzed by LC-MS^n lipid mixtures obtained from mouse liver and brain [26]. Data-dependent acquisitions were performed to acquire the fragment ions in MS/MS and MS^3 modes. The raw MS, MS/MS, and MS^n data were automatically analyzed by the latest version of "Lipid Search" with "Lipid Navigator." Finally, the identification results were confirmed manually, using raw MS^n spectra.

By using LTQ-Orbitrap with C30 RP column, even within a single PL class, such as PC, about 100 molecular species were effectively separated and

Examples of analytical results

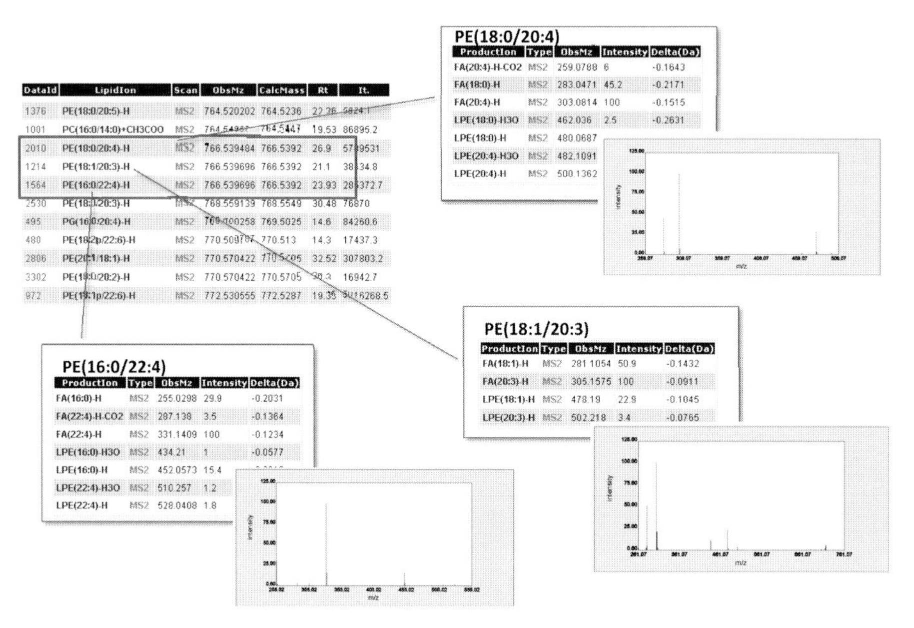

Figure 12.2 Examples of identification results of Lipid Search. Individual identification results were confirmed by consulting their precise qualitative and quantitative data of fragmentation peaks.

accurately identified with a combination of fatty acyl anions on *sn*-1 and *sn*-2 position by data-dependent MS^3 analysis [26]. These quantitative data were automatically merged with qualitative data by the latest version of our search engine "Lipid Search" with "Lipid Navigator." The peak area of each molecular weight-related ion was used for quantitative profiling of molecular species of PLs after compensation for the abundance ratio of monoisotopic ions. The quantitative data obtained from the peak area of each m/z value and identified molecular species from MS/MS data for each m/z vale were merged by coincidence with the accurate m/z value and the elution time on LC. With this method, quantitative and qualitative profiling data for approximately 500 molecular species of PLs were obtained in a single LC run. Furthermore, the odd carbon number of alkenyl-acyl species and the even carbon number of diacyl species were clearly distinguishable using this RPLC-LTQ-Orbitrap system.

The automated search engine "Lipid Search" revealed to be a very useful identification tool for lipidomics. Most of the major peaks of the individual PL molecular species were correctly identified by this search engine within 10 min. Automated identification and quantification are inevitable for quick analysis in lipidomics, otherwise individual peak areas should be manually calculated leading to tedious, laborious processes.

12.5
Analysis of Lipids as Markers of Metabolic Syndrome

12.5.1
Oxidized Phospholipids

PUFAs and their metabolites have a diversity of physiological roles including energy provision, membrane structure, cell signaling, and regulation of gene expression. In addition, lipids containing PUFAs are susceptible to free radical-initiated oxidation and can participate in chain reactions that increase damage to biomolecules. Lipid peroxidation, which often leads to lipid hydroperoxide formation, occurs in response to oxidative stress. Oxidized PLs can exist in the form of many different derivatives and isomers *in vivo*. Recently, a novel family of oxidized PCs was identified as highly specific ligands for scavenger receptor CD36 [40–42]. It was reported that oxidized PCs accumulate *in vivo* and both mediate macrophage foam cell formation and promote platelet hyperreactivity in hyperlipidemia via CD36. Furthermore, oxidized PCs were identified as inducing lung injury and cytokine production by lung macrophages via Toll-like receptor 4-TRIF [43]. Thus, the analysis of oxidized PCs is very important to not only understand these physiological and pathological phenomena but also explore novel bioactive substances for candidate biomarkers of various inflammatory diseases and common diseases. In the next section, we introduce analytical methods for oxidized PCs using LC-MS/MS.

12.5.1.1 Application for Myocardial Ischemia-Reperfusion Model

The mouse heart was subjected to acute oxidative stress induced by ischemia-reperfusion injury. The analysis data are shown in Figure 12.3. The calibration of data was performed using an internal standard (17:0-LPC) and the tissue weight. As indicated in our experiment results, oxidized PCs were already produced in nonischemic myocardium because the heart is always subjected to oxidative stress. As a result, most oxidized PCs were detected in higher amounts in the ischemic myocardium than in the nonischemic myocardium. In brief, in the linoleic acid (LA, 18:2)- or AA (20: 4)-containing oxidized PC series, hydroxide, hydroperoxide, and aldehyde forms were detected at significant levels. On the other hand, DHA (22:6)-containing oxidized PC series consisted mainly of aldehyde forms rather than hydroxide or hydroperoxide forms. The aldehyde forms seemed to be significantly higher in DHA-derived oxidized PCs than for those in LA- or AA-derived oxidized PCs. DHA-containing PLs are abundant in heart tissue. It is believed that DHA plays a critical role as the major target of radical reactions by reactive oxygen species, and their turnover might be involved in oxidative stress within the cell. These results suggest that DHA and their oxidized metabolites provide the protective barrier of the myocardium to oxidative stress [44, 45]. Research on these biological mechanisms and roles is underway.

In the future, it may be possible to discover novel bioactive substances or candidate biomarkers. Furthermore, this technique will be helpful in clarifying the

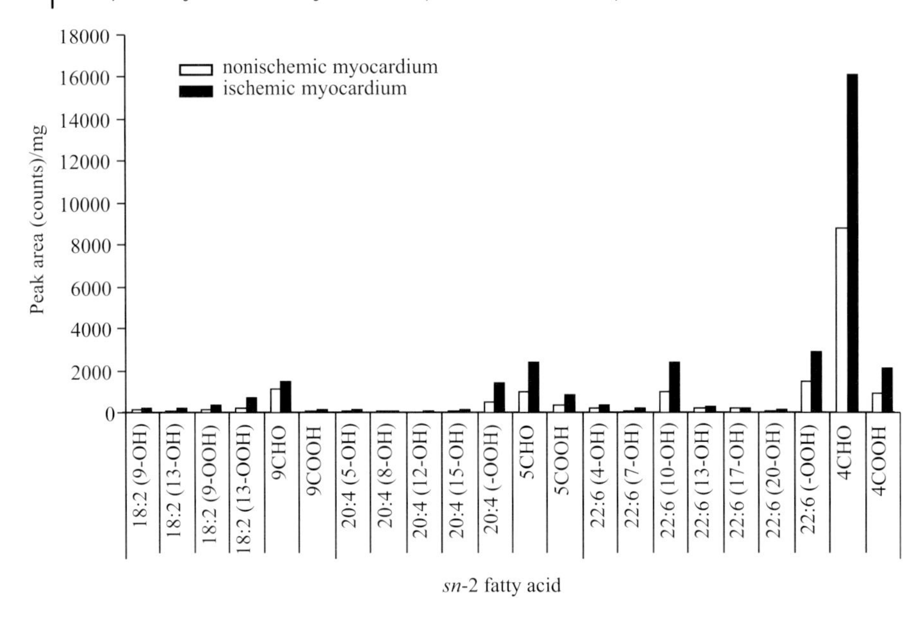

Figure 12.3 Variation analyses of oxidized PCs using myocardial ischemia-reperfusion model mouse [6]. The peak area is shown along the longitudinal axis, which was corrected using an internal standard (17:0-LPC) and the tissue weight. The *sn*-1 fatty acid is palmitic acid (16:0). Most of the oxidized PCs tended to increase in the ischemic myocardium (*closed bars*) versus nonischemic myocardium (*open bars*). A similar result was reproducibly obtained.

physiological and pathological phenomena of various inflammatory diseases and common diseases.

12.5.2 Bioactive Acidic Phospholipids

12.5.2.1 Lysophosphatidic Acid

Lysophosphatidic acid (LPA) plays important roles in many biological processes, such as the function of the nervous system, cancer progression, wound healing, cardiovascular function, and reproduction [46]. As a consequence, LPA is paid attention as a bioactive lipid of the second generation following steroid hormones and eicosanoids. Normally, a saturated or unsaturated fatty acid is esterified at the *sn*-1 or *sn*-2 position of the glycerol backbone; however, *sn*-1 alkyl or alkenyl ether-linked (plasmalogen) LPA species are also present. The biological activities of LPA depend on the carbon chain length and the degree of unsaturation, as well as the position and linkage type of the carbon chain attached to the glycerol backbone. LPA have been shown to play critical roles in multiple cellular processes via G-protein-coupled seven transmembrane receptors (GPCRs). Indeed, six GPCRs for LPA (LPA1–6) have been identified, and recent studies on gene targeting mice have clearly shown that these receptors function individually in pathological and physiological situations [47].

We typically analyze LPA species of tissue extracts using a reverse-phase LC-ESIMS/MS method. LPA ions are selectively detected by MRM and/or SRM mode (e.g., m/z 437.3 for the negatively charged parent ion of 18:0-LPA and m/z 153.0 for the fragment ion of 18:0-LPA) [39, 48].

12.5.2.2 Phosphoinositides

Phosphoinositides (PIPs) are lipids that are present in the cytoplasmic leaflet of a cell's plasma and internal membranes and play pivotal roles in the regulation of a wide variety of cellular processes [49]. The combinatorial phosphorylation of the hydroxyl residues on the inositol ring of PI gives rise to seven PIPs (eight if PI itself is included). In mammals, close to 50 genes encode the phosphoinositide kinases and phosphoinositide phosphatases that regulate PIP metabolism and thus allow cells to respond rapidly and effectively to ever-changing environmental cues [50]. Understanding the distinct and overlapping functions of these PIP-metabolizing enzymes is important for our knowledge of both the normal human physiology and the growing list of human diseases whose etiologies involve these proteins. In recent years, many phenotypes of gene-targeted mice lacking these enzymes have been reported [51–53]. For example, mice that lack INPP4A, a PI (3, 4) P_2 phosphatase, have neurodegeneration in the striatum and suffer from severe involuntary movements, because INPP4A protects neurons from NMDA-type glutamate receptor-mediated excitotoxic cell death. [53].

Special attention in PIP analysis is given to metabolic turnover and degradation. Typically, PIP ions are selectively detected by MRM and/or SRM in the positive ion mode [54, 55].

12.5.3 Oxidative Triglycerides

TGs consist of various molecular species caused by their three fatty acyl chains with a large variety of carbon chain lengths and degrees of unsaturation. High TG level is one of the characteristic traits in metabolic syndrome, along with low high-density lipoprotein cholesterol, high blood pressure, abdominal obesity, and increased fasting glucose levels [56]. Among them, storage of excess TGs in adipose tissue is related to abdominal obesity, and the incidence rate is increasing around the world [57, 58]. For these reasons, global analysis of TGs has become more important in the abnormality of lipid metabolism.

As for analytical methods of TGs, clinical colorimetric kits using lipoprotein lipase, which hydrolyzes fatty acids from TGs, and gas chromatography (GC)-MS are well known [59–61]. However, detailed analysis of TGs at the molecular species level is difficult, because they collectively, but not specifically, analyze fatty acids released from the TGs.

An alternative approach that utilizes ESIMS has received interest for detailed structural analysis of TGs [12, 62]. A recent application of triple quadrupole MS for

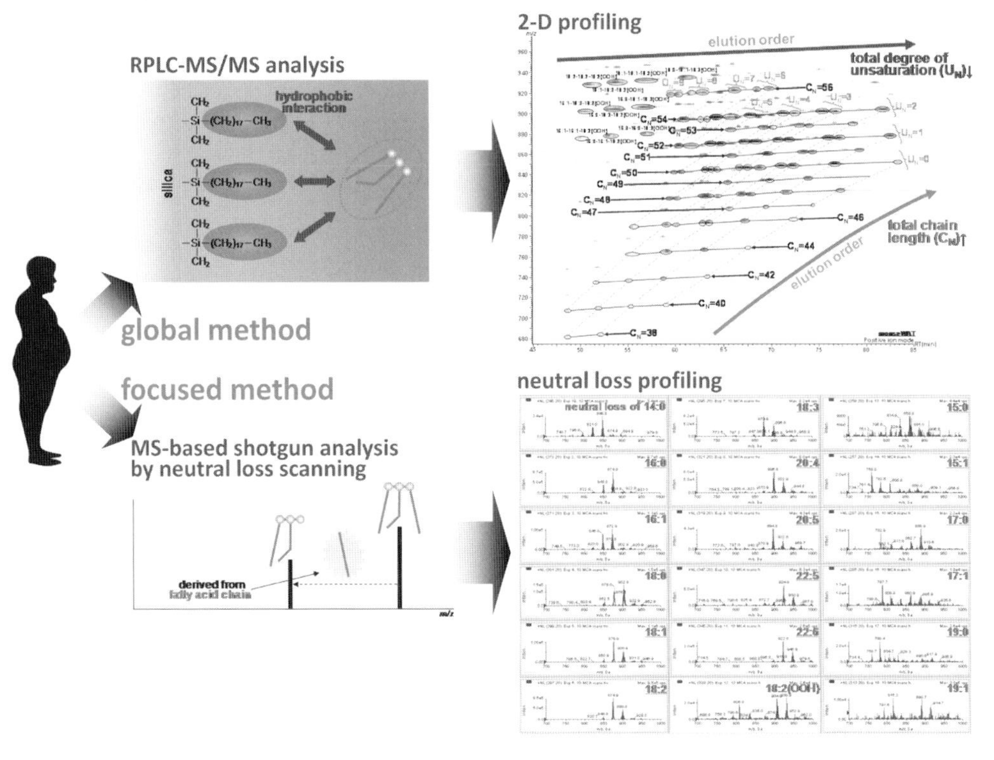

Figure 12.4 TG profiling by lipidomics approaches (adapted from Ref. [5]). In our global method, a combination of RPLC-MS/MS analysis and 2D profiling is used and is effective for detailed profiling of TG molecular species including their structural isomers. Meanwhile, in the focused method, neutral loss scanning is applied for easy distinguishing TG molecular species from complex lipid mixtures.

focused TG analysis involves neutral loss scanning of the individual fatty acyl-chains (Figure 12.4) [30, 63, 64]. The scanning is possible to detect TG molecular species with high sensitivity and directly from complex lipid extracts by neutral loss of the various FAs, which are derived from CID of individual TGs in the collision cell. But it is insufficient to discriminate between the close m/z value TGs such as normal TGs and modified TGs with oxidized acyl carbon chain, and difficult to detect TGs globally and quantitatively.

In the next section, we describe an effective method for global analysis of TG molecular species using RP high-resolution LC coupled with Q-TOF micro (Figure 12.4) [5]. For effective profiling of TG molecular species, sensitive two-dimensional (2D) maps are constructed and individual structures are correctly identified by the elution profile and MS/MS.

12.5.3.1 **Application for Mouse White Adipose Tissue**

For our method application, mouse white adipose tissue (WAT) was designed. WAT in the viscera is thought to serve as the primary energy depot by storing TGs.

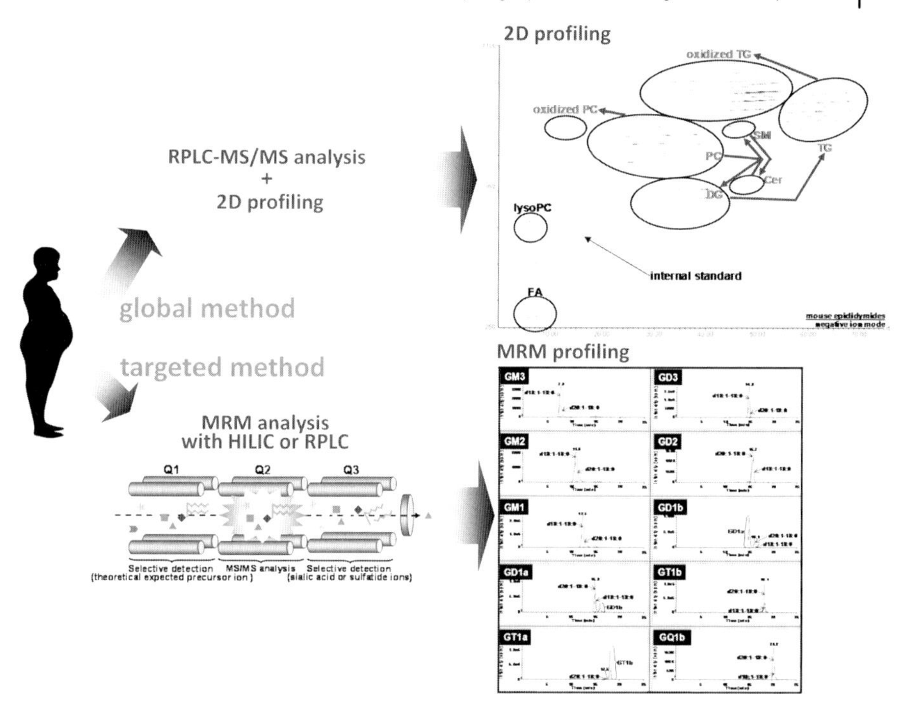

Figure 12.5 Sphingolipid profiling by lipidomics approaches. In our global method, simultaneous searching of sphingolipid metabolic balance is possible by combination of RPLC-MS/MS analysis and 2D profiling. Meanwhile, in the targeted method, HILIC or RPLC-MS with MRM is effective for high-sensitive and quantitative detecting sphingolipids such as gangliosides.

The excess accumulation is associated with obesity, which is linked to the risk of developing diabetes. For this reason, investigation of the distinctive TG distributions might be important in elucidating obesity. We then performed global analysis of TG molecular species in mouse WAT, and TGs including their structural isomers and TGs with oxidized or odd number acyl carbon chain were identified. The sum of them was almost 150 molecular species [5]. Among them, hydroperoxy TGs, which were formed by oxidation of one of the linoleic acyl chains, were detected for the first time in a biological sample [65, 66]. In our preliminary experiment, increase in hydroperoxy TGs was observed in some obese model mice at inflamed sites, and these TGs might be available for indication of obesity and subsequent inflammation. Hydroperoxy TGs in WAT might be sources of precursors for some oxidized FA mediators during this progress. These oxidized TGs were also found in serum in our research [7]. For these reasons, suppression of hydroperoxy TG production might systemically depress oxidative and inflammatory stress and lead to improvement of obesity [67, 68]. Future studies will provide more insights into correlations between oxidized TGs and pathologies of metabolic syndrome.

12.5.4 Sphingolipids

Sphingolipids, a class of lipids containing a backbone of sphingoid bases, are one of the major components of plasma membrane, along with glycerophospholipids, neutral lipids such as TGs, and sterols [69]. Sphingolipids play important roles in cell recognition and signaling [70, 71]. Recent research has revealed that qualitative and quantitative changes in sphingolipids are associated with pathologies of metabolic syndrome [71–73].

Ceramides, a major sphingolipid in mammals, are composed of sphingosine and a fatty acid, and they are synthesized in the endoplasmic reticulum by ceramide synthases [74]. Ceramides have a central function in sphingolipid metabolic pathways and serve as precursors to a large number of lipid-signaling molecules such as sphingomyelin and glycosphingolipids, which are synthesized from ceramides by addition of phosphate and carbohydrate to the terminal primary hydroxyl group [71, 75]. These sphingolipids consist of various molecular species, created by variations in sphingoid base or fatty acyl chain with a large variety of carbon chain lengths and degrees of unsaturation [74]. For these reasons, LC-ESIMS is considered to be effective in extensive detection and identification of these molecular species. Several laboratories have reported targeted analysis of multiple ceramide molecular species by normal-phase (NP) LC-MS/MS [76–78].

Next, we describe our global method by RPLC-MS/MS (applied method of Section 3.1), which is also capable of analyzing not only TGs but also ceramides, SMs, PLs, and diacylglycerols (DGs) at the same time and available for searching sphingolipid metabolic balance (Figure 12.5) [5]. Moreover, we discuss our targeted analysis of sphingolipids by hydrophilic interaction liquid chromatography (HILIC)-MS with MRM for the localization profiling (Figure 12.5).

12.5.4.1 Application for Sphinogolipid Metabolism

Our global method by RPLC-MS/MS is effective in searching sphingolipid metabolic balance, because transfer of phosphocholine from PC to ceramide yields SM and DG [5]. To monitor the sphingolipid metabolism, SM synthase 1 knockout (SMS1-KO) mice were designed [79]. SMS1-KO mice exhibited metabolic abnormality such as severe deficiencies of insulin secretion in the pancreatic islet. Sphingolipid metabolic changes such as increase in SM molecular species and decrease in ceramide and ganglioside molecular species were clearly detected compared to wild-type mice even from only a small amount of the isolated islets. These results suggest that ceramide metabolism is correlated with the pathogenesis of insulin resistance.

Metabolic disorders such as type 2 diabetes have been shown to be linked to membrane microdomain malfunctions due to aberrant expression of gangliosides [80, 81]. Gangliosides are present on cell surfaces and especially localized in glycolipid-enriched membrane microdomains termed lipid rafts to interact with various molecules on plasma membranes and to modulate cell signal transduction [82]. In the adipose tissues of obese/diabetic rodent models such as Zucker *fa/fa* rats and *ob/ob* mice, attenuation of the insulin signaling is considered to be caused by increase in GM3 levels and elimination of insulin receptor from the lipid rafts [83]. For the localization or profiling analysis of gangliosides, our targeted analysis, based on theoretical ganglioside molecular species, by HILIC-MS with MRM will be useful for elucidation of ganglioside metabolism in the metabolic disorder [38].

These analysis techniques by LC-ESIMS are rapidly developing and will improve the detection of molecular sphingolipids with high sensitivity. Future studies will provide more insights into metabolic syndrome processes mediated by sphingolipids.

12.6 Direct Detection of Lipid Molecular Species in Specific Tissue Domains by Disease-Specific Changes

As a new technique, liquid extraction surface analysis (LESA) is a practical method to elucidate lipid molecular species in localized areas on tissue slices from mammalian organs. Several different methods have been applied in surface analysis of lipid molecular species such as MALDI imaging [84, 85], desorption ESI (DESI) [86], and the analysis by laser microdissection (LMD) combination with nano-ESI.

We have evaluated the novel technique called LESA in combination with a NanoMate system (Advion BioSciences) [87]. By LESA, most of lipid molecules were effectively extracted and detected from specific localized position of tissues, even though space resolution is lower than 1 mm in diameter. As a result, mass profiles of PC, PE, and PS in different areas of mouse brain slices were effectively obtained by direct extraction. The profiles in molecular species of glycosphingolipids (gangliosides and sulfatides) and TGs in mouse brain were also effectively obtained by this method. Detection efficiency of LESA was revealed to be very sensitive. The sensitivity is almost compatible to LMD combination with nano-ESI.

This method will be useful for detection of changes in lipid profiles when tissue amounts are limited, such as damaged areas of tissues caused by inflammation or oxidation in metabolic syndrome-related diseases.

12.7 Conclusions

Tissue-specific localization of lipid molecular species and diversity in their distribution are key factors giving rise to the tissue-specific physiological functions. Especially, the diversity in polyunsaturated fatty acid containing lipid metabolites is biologically very important because of their high susceptibility to oxidation under oxidative stress. Precise analysis of the changes in lipid metabolites in specific tissue regions by lipidomics seems to be one of the most important approaches in the study of metabolic syndrome.

References

1 Gross, R.W. and Han, X. (2007) Lipidomics in diabetes and the metabolic syndrome. *Methods Enzymol.*, **433**, 73–90.

2 Graessler, J., Schwudke, D., Schwarz, P.E., Herzog, R., Shevchenko, A., and Bornstein, S.R. (2009) Top-down lipidomics reveals ether lipid deficiency in blood plasma of hypertensive patients. *PLoS One*, **4**, e6261.

3 Mancuso, D.J., Sims, H.F., Yang, K., Kiebish, M.A., Su, X., Jenkins, C.M., Guan, S., Moon, S.H., Pietka, T., Nassir, F., Schappe, T., Moore, K., Han, X., Abumrad, N.A., and Gross, R.W. (2010) Genetic ablation of calcium-independent phospholipase A2gamma prevents obesity and insulin resistance during high fat feeding by mitochondrial uncoupling and increased adipocyte fatty acid oxidation. *J. Biol. Chem.*, **285**, 36495–36510.

4 Meikle, P.J. and Christopher, M.J. (2011) Lipidomics is providing new insight into the metabolic syndrome and its sequelae. *Curr. Opin. Lipidol.*, **22**, 210–215.

5 Ikeda, K., Oike, Y., Shimizu, T., and Taguchi, R. (2009) Global analysis of triacylglycerols including oxidized molecular species by reverse-phase high resolution LC/ESI-QTOF MS/MS. *J. Chromatogr. B.*, **877**, 2639–2647.

6 Nakanishi, H., Iida, Y., Shimizu, T., and Taguchi, R. (2009) Analysis of oxidized phosphatidylcholines as markers for oxidative stress, using multiple reaction monitoring with theoretically expanded data sets with reversed-phase liquid chromatography/tandem mass spectrometry. *J. Chromatogr. B*, **877**, 1366–1374.

7 Ikeda, K., Mutoh, M., Teraoka, N., Nakanishi, H., Wakabayashi, K., and Taguchi, R. (2011) Increase of oxidant-related triglycerides and phosphatidylcholines in serum and small intestinal mucosa during development of intestinal polyp formation in Min mice. *Cancer Sci.*, **102**, 79–87.

8 Brown, H.A. and Murphy, R.C. (2009) Working towards an exegesis for lipids in biology. *Nat. Chem. Biol.*, **5**, 602–606.

9 Han, X. (2009) Lipidomics: developments and applications. *J. Chromatogr. B*, **877**, 2663.

10 Ivanova, P.T., Milne, S.B., Myers, D.S., and Brown, H.A. (2009) Lipidomics: a mass spectrometry based systems level analysis of cellular lipids. *Curr. Opin. Chem. Biol.*, **13**, 526–531.

11 Shevchenko, A. and Simons, K. (2010) Lipidomics: coming to grips with lipid diversity. *Nat. Rev. Mol. Cell Biol.*, **11**, 593–598.

12 Taguchi, R., Nishijima, M., and Shimizu, T. (2007) Basic analytical systems for lipidomics by mass spectrometry in Japan. *Methods Enzymol.*, **432**, 185–211.

13 Nakanishi, H., Iida, Y., Shimizu, T., and Taguchi, R. (2010) Separation and quantification of *sn*-1 and *sn*-2 fatty acid positional isomers in phosphatidylcholine by RPLC-ESIMS/MS. *J. Biochem.*, **147**, 245–256.

14 Blanksby, S.J. and Mitchell, T.W. (2010) Advances in mass spectrometry for lipidomics. *Annu. Rev. Anal. Chem.*, **3**, 433–465.

15 Thomas, M.C., Mitchell, T.W., and Blanksby, S.J. (2006) Ozonolysis of phospholipid double bonds during electrospray ionization: a new tool for structure determination. *J. Am. Chem. Soc.*, **128**, 58–59.

16 Sato, H., Kato, R., Isogai, Y., Saka, G.I., Ohtsuki, M., Taketomi, Y., Yamamoto, K., Tsutsumi, K., Yamada, J., Masuda, S., Ishikawa, Y., Ishii, T., Kobayashi, T., Ikeda, K., Taguchi, R., Hatakeyama, S., Hara, S., Kudo, I., Itabe, H., and Murakami, M. (2008) Analyses of group III secreted phospholipase A2 transgenic mice reveals potential participation of this enzyme in plasma lipoprotein modification, macrophage foam cell formation, and atherosclerosis. *J. Biol. Chem.*, **283**, 33483–33497.

17 Arita, M., Yoshida, M., Hong, S., Tjonahen, E., Glickman, J.N., Petasis, N. A., Blumberg, R.S., and Serhan, C.N. (2005) Resolvin E1, an endogenous lipid mediator derived from omega-3 eicosapentaenoic acid, protects against

2,4,6-trinitrobenzene sulfonic acid-induced colitis. *Proc. Natl. Acad. Sci. U.S. A.*, **102**, 7671–7676.

18 Seki, H., Fukunaga, K., Arita, M., Arai, H., Nakanishi, H., Taguchi, R., Miyasho, T., Takamiya, R., Asano, K., Ishizaka, A., Takeda, J., and Levy, B.D. (2010) The anti-inflammatory and proresolving mediator resolvin E1 protects mice from bacterial pneumonia and acute lung injury. *J. Immunol.*, **184**, 836–843.

19 Talahalli, R., Zarini, S., Sheibani, N., Murphy, R.C., and Gubitosi-Klug, R.A. (2010) Increased synthesis of leukotrienes in the mouse model of diabetic retinopathy. *Invest. Ophthalmol. Vis. Sci*, **51**1699–1708.

20 Han, X., Yang, J., Cheng, H., Yang, K., Abendschein, D.R., and Gross, R.W. (2005) Shotgun lipidomics identifies cardiolipin depletion in diabetic myocardium linking altered substrate utilization with mitochondrial dysfunction. *Biochemistry*, **44**, 16684–16694.

21 Han, X. (2007) Potential mechanisms contributing to sulfatide depletion at the earliest clinically recognizable stage of Alzheimer's disease: a tale of shotgun lipidomics. *J. Neurochem.*, **103**, 171–179.

22 Han, X. (2010) Multi-dimensional mass spectrometry-based shotgun lipidomics and the altered lipids at the mild cognitive impairment stage of Alzheimer's disease. *Biochim. Biophys. Acta.*, **1801**, 774–783.

23 Ekroos, K., Jänis, M., Tarasov, K., Hurme, R., and Laaksonen, R. (2010) Lipidomics: a tool for studies of atherosclerosis. *Curr. Atheroscler. Rep.*, **12**, 273–281.

24 Zachman, D.K., Chicco, A.J., McCune, S. A., Murphy, R.C., Moore, R.L., and Sparagna, G.C. (2010) The role of calcium-independent phospholipase A2 in cardiolipin remodeling in the spontaneously hypertensive heart failure rat heart. *J. Lipid Res.*, **51**, 525–534.

25 Gross, R.W. and Han, X. (2009) Shotgun lipidomics of neutral lipids as an enabling technology for elucidation of lipid-related diseases. *Am. J. Physiol. Endocrinol. Metab.*, **297**, E297–303.

26 Taguchi, R. and Ishikawa, M. (2010) Precise and global identification of phospholipid molecular species by an Orbitrap mass spectrometer and automated search engine Lipid Search. *J. Chromatogr. A.*, **1217**, 4229–4239.

27 Astarita, G., Ahmed, F., and Piomelli, D. (2009) Lipidomic analysis of biological samples by liquid chromatography coupled to mass spectrometry. *Methods Mol. Biol.*, **579**, 201–219.

28 Houjou, T., Yamatani, K., Imagawa, M., Shimizu, T., and Taguchi, R. (2005) A shotgun tandem mass spectrometric analysis of phospholipids with normal-phase and/or reverse-phase liquid chromatography/electrospray ionization mass spectrometry. *Rapid Commun. Mass Spectrom.*, **19**, 654–666.

29 Ekroos, K., Chernushevich, I.V., Simons, K., and Shevchenko, A. (2002) Quantitative profiling of phospholipids by multiple precursor ion scanning on a hybrid quadrupole time-of-flight mass spectrometer. *Anal. Chem.*, **74**, 941–949.

30 Han, X. and Gross, R.W. (2003) Global analyses of cellular lipidomes directly from crude extracts of biological samples by ESI mass spectrometry: a bridge to lipidomics. *J. Lipid Res.*, **44**, 1071–1079.

31 Han, X. and Gross, R.W. (2005) Shotgun lipidomics: electrospray ionization mass spectrometric analysis and quantitation of cellular lipidomes directly from crude extracts of biological samples. *Mass Spectrom. Rev.*, **24**, 367–412.

32 Taguchi, R., Houjou, T., Nakanishi, H., Yamazaki, T., Ishida, M., Imagawa, M., and Shimizu, T. (2005) Focused lipidomics by tandem mass spectrometry. *J. Chromatogr. B.*, **823**, 26–36.

33 Han, X., Yang, J., Cheng, H., Ye, H., and Gross, R.W. (2004) Toward fingerprinting cellular lipidomes directly from biological samples by two-dimensional electrospray ionization mass spectrometry. *Anal. Biochem.*, **330**, 317–331.

34 Ejsing, C.S., Sampaio, J.L., Surendranath, V., Duchoslav, E., Ekroos, K., Klemm, R. W., Simons, K., and Shevchenko, A. (2009) Global analysis of the yeast lipidome by quantitative shotgun mass spectrometry. *Proc. Natl. Acad. Sci. USA*, **106**, 2136–2141.

35 Mitchell, T.W. (2009) Tracking the glycerophospholipid distribution of docosahexaenoic acid by shotgun lipidomics. *Methods Mol. Biol.*, **579**, 19–31.

36 Ikeda, K., Shimizu, T., and Taguchi, R. (2008) Targeted analysis of ganglioside and sulfatide molecular species by LC/ESI-MS/MS with theoretically expanded multiple reaction monitoring. *J. Lipid Res.*, **49**, 2678–2689.

37 Dong, Y., Ikeda, K., Hamamura, K., Zhang, Q., Kondo, Y., Matsumoto, Y., Ohmi, Y., Yamauchi, Y., Furukawa, K., Taguchi, R., and Furukawa, K. (2010) GM1/GD1b/GA1 synthase expression results in the reduced cancer phenotypes with modulation of composition and raft-localization of gangliosides in a melanoma cell line. *Cancer Sci.*, **101**, 2039–2047.

38 Ikeda, K. and Taguchi, R. (2010) Highly sensitive localization analysis of gangliosides and sulfatides including structural isomers in mouse cerebellum sections by combination of laser microdissection and hydrophilic interaction liquid chromatography/electrospray ionization mass spectrometry with theoretically expanded multiple reaction monitoring. *Rapid Commun. Mass Spectrom.*, **24**, 2957–2965.

39 Nakanishi, H., Ogiso, H., and Taguchi, R. (2009) Quantitative analysis of phospholipids by LC-MS for Lipidomics. *Methods Mol. Biol.*, **579**, 287–313.

40 Kar, N.S., Ashraf, M.Z., Valiyaveettil, M., and Podrez, E.A. (2008) Mapping and characterization of the binding site for specific oxidized phospholipids and oxidized low density lipoprotein of scavenger receptor CD36. *J. Biol. Chem.*, **283**, 8765–8771.

41 Ashraf, M.Z., Kar, N.S., Chen, X., Choi, J., Salomon, R.G., Febbraio, M., and Podrez, E.A. (2008) Specific oxidized phospholipids inhibit scavenger receptor bi-mediated selective uptake of cholesteryl esters. *J. Biol. Chem.*, **283**, 10408–10414.

42 Hazen, S.L. (2008) Oxidized phospholipids as endogenous pattern recognition ligands in innate immunity. *J. Biol. Chem.*, **283**, 15527–15531.

43 Imai, Y., Kuba, K., Neely, G.G., Yaghubian-Malhami, R., Perkmann, T., van Loo, G., Ermolaeva, M., Veldhuizen, R., Leung, Y. H., Wang, H., Liu, H., Sun, Y., Pasparakis, M., Kopf, M., Mech, C., Bavari, S., Peiris, J.S., Slutsky, A.S., Akira, S., Hultqvist, M., Holmdahl, R., Nicholls, J., Jiang, C., Binder, C.J., and Penninger, J.M. (2008) Identification of oxidative stress and Toll-like receptor 4 signaling as a key pathway of acute lung injury. *Cell*, **133**235–249.

44 Yin, H., Musiek, E.S., Gao, L., Porter, N. A., and Morrow, J.D. (2005) Regiochemistry of neuroprostanes generated from the peroxidation of docosahexaenoic acid *in vitro* and *in vivo*. *J. Biol. Chem.*, **280**, 26600–26611.

45 Bazan, N.G. (2007) Omega-3 fatty acids, pro-inflammatory signaling and neuroprotection. *Curr. Opin. Clin. Nutr. Metab. Care*, **10**, 136–141.

46 Ishii, S., Noguchi, K., and Yanagida, K. (2009) Non-Edg family lysophosphatidic acid (LPA) receptors. *Prostaglandins Other Lipid Mediat.*, **89**, 57–65.

47 Lee, H.J., Wall, B., and Chen, S. (2008) G-protein-coupled receptors and melanoma. *Pigment Cell Melanoma Res.*, **21**, 415–428.

48 Inoue, A., Arima, N., Ishiguro, J., Prestwich, G.D., Arai, H., and Aoki, J. (2011) LPA-producing enzyme PA-PLA$_1\alpha$ regulates hair follicle development by modulating EGFR signalling. *EMBO J.*, **30**, 4248–4260.

49 Sasaki, T., Sasaki, J., Sakai, T., Takasuga, S., and Suzuki, A. (2007) The physiology of phosphoinositides. *Biol. Pharm. Bull.*, **9**, 1599–1604.

50 Sasaki, T., Takasuga, S., Sasaki, J., Kofuji, S., Eguchi, S., Yamazaki, M., and Suzuki, A. (2009) Mammalian phosphoinositide kinases and phosphatases. *Prog. Lipid Res.*, **48**, 307–343.

51 Clement, S., Krause, U., Desmedt, F., Tanti, J.F., Behrends, J., Pesesse, X., Sasaki, T., Penninger, J., Doherty, M., Malaisse, W., Dumont, J.E., Le Marchand-Brustel, Y., Erneux, C., Hue, L., and Schurmans, S. (2001) The lipid phosphatase SHIP2 controls insulin sensitivity. *Nature*, **409**, 92–97.

52 Suzuki, A., Nakano, T., Mak, T.W., and Sasaki, T. (2008) Portrait of PTEN:

messages from mutant mice. *Cancer Sci.*, **99**, 209–213.

53 Sasaki, J., Kofuji, S., Itoh, R., Momiyama, T., Takayama, K., Murakami, H., Chida, S., Tsuya, Y., Takasuga, S., Eguchi, S., Asanuma, K., Horie, Y., Miura, K., Davies, E.M., Mitchell, C., Yamazaki, M., Hirai, H., Takenawa, T., Suzuki, A., and Sasaki, T. (2010) The PtdIns(3,4)P(2) phosphatase INPP4A is a suppressor of excitotoxic neuronal death. *Nature*, **465**497–501.

54 Ogiso, H. and Taguchi, R. (2008) Reversed-phase LC/MS method for polyphosphoinositide analyses: changes in molecular species levels during epidermal growth factor activation in A431 cells. *Anal. Chem.*, **80**, 9226–9232.

55 Clark, J., Anderson, K.E., Juvin, V., Smith, T.S., Karpe, F., Wakelam, M.J., Stephens, L.R., and Hawkins, P.T. (2011) Quantification of PtdInsP3 molecular species in cells and tissues by mass spectrometry. *Nat. Methods*, **8**, 267–272.

56 Rosen, E.D. and Spiegelman, B.M. (2006) Adipocytes as regulators of energy balance and glucose homeostasis. *Nature*, **444**, 847–853.

57 Unger, R.H. (2003) Lipid overload and overflow: metabolic trauma and the metabolic syndrome. *Trends Endocrinol. Metab.*, **14**, 398–403.

58 Popkin, B.M. (2001) The nutrition transition and obesity in the developing world. *J. Nutr.*, **131**, 871S–873S.

59 Schwartz, D.M. and Wolins, N.E. (2007) A simple and rapid method to assay triacylglycerol in cells and tissues. *J. Lipid Res.*, **48**, 2514–2520.

60 Snyder, F. and Stephens, N. (1959) A simplified spectrophotometric determination of ester groups in lipids. *Biochim. Biophys. Acta.*, **34**, 244–245.

61 Pouteau, E., Beysen, C., Saad, N., and Turner, S. (2009) Dynamics of adipose tissue development by 2H2O labeling. *Methods Mol. Biol.*, **579**, 337–358.

62 Han, X. and Gross, R.W. (2001) Quantitative analysis and molecular species fingerprinting of triacylglyceride molecular species directly from lipid extracts of biological samples by electrospray ionization tandem mass spectrometry. *Anal. Biochem.*, **295**, 88–100.

63 Ikeda, K., Kubo, A., Akahoshi, N., Yamada, H., Miura, N., Hishiki, T., Nagahata, Y., Matsuura, T., Suematsu, M., Taguchi, R., and Ishii, I. (2011) Triacylglycerol/phospholipid molecular species profiling of fatty livers and regenerated non-fatty livers in cystathionine beta-synthase-deficient mice: an animal model for homocysteinemia/homocystinuria. *Anal. Bioanal. Chem.*, **400**, 1853–1863.

64 Murphy, R.C., James, P.F., McAnoy, A.M., Krank, J., Duchoslav, E., and Barkley, R.M. (2007) Detection of the abundance of diacylglycerol and triacylglycerol molecular species in cells using neutral loss mass spectrometry. *Anal. Biochem.*, **366**, 59–70.

65 Giuffrida, F., Destaillats, F., Skibsted, L. H., and Dionisi, F. (2004) Structural analysis of hydroperoxy- and epoxy-triacylglycerols by liquid chromatography mass spectrometry. *Chem. Phys. Lipids.*, **131**, 41–49.

66 Nagare, T., Sakaue, H., Matsumoto, M., Cao, Y., Inagaki, K., Sakai, M., Takashima, Y., Nakamura, K., Mori, T., Okada, Y., Matsuki, Y., Watanabe, E., Ikeda, K., Taguchi, R., Kamimura, N., Ohta, S., Hiramatsu, R., and Kasuga, M. (2011) Overexpression of KLF15 in adipocytes of mice results in down-regulation of SCD1 expression in adipocytes and consequent enhancement of glucose-induced insulin secretion. *J. Biol. Chem.*, **286** (43), 37458–37469.

67 Hotamisligil, G.S. (2006) Inflammation and metabolic disorders. *Nature*, **444**, 860–867.

68 Lumeng, C.N., Bodzin, J.L., and Saltiel, A. R. (2007) Obesity induces a phenotypic switch in adipose tissue macrophage polarization. *J. Clin. Invest.*, **117**, 175–184.

69 Lahiri, S. and Futerman, A.H. (2007) The metabolism and function of sphingolipids and glycosphingolipids. *Cell Mol. Life Sci.*, **64**, 2270–2284.

70 Hannun, Y.A. and Obeid, L.M. (2008) Principles of bioactive lipid signalling: lessons from sphingolipids. *Nat. Rev. Mol. Cell Biol.*, **9**, 139–150.

71 Cowart, L.A. (2009) Sphingolipids: players in the pathology of metabolic disease. *Trends Endocrinol. Metab.*, **20**, 34–42.

72 Summers, S.A. and Nelson, D.H. (2005) A role for sphingolipids in producing the common features of type 2 diabetes, metabolic syndrome X, and Cushing's syndrome. *Diabetes*, **54**, 591–602.

73 Holland, W.L., Brozinick, J.T., Wang, L.P., Hawkins, E.D., Sargent, K.M., Liu, Y., Narra, K., Hoehn, K.L., Knotts, T.A., Siesky, A., Nelson, D.H., Karathanasis, S. K., Fontenot, G.K., Birnbaum, M.J., and Summers, S.A. (2007) Inhibition of ceramide synthesis ameliorates glucocorticoid-, saturated-fat-, and obesity-induced insulin resistance. *Cell Metab.*, 5167–179.

74 Menaldino, D.S., Bushnev, A., Sun, A., Liotta, D.C., Symolon, H., Desai, K., Dillehay, D.L., Peng, Q., Wang, E., Allegood, J., Trotman-Pruett, S., Sullards, M.C., and Merrill, A.H., Jr. (2003) Sphingoid bases and *de novo* ceramide synthesis: enzymes involved, pharmacology and mechanisms of action. *Pharmacol Res*, **47**, 373–381.

75 Hannun, Y.A. and Obeid, L.M. (2002) The ceramide-centric universe of lipid-mediated cell regulation: stress encounters of the lipid kind. *J. Biol. Chem.*, **277**, 25847–25850.

76 Scherer, M., Leuthäuser-Jaschinski, K., Ecker, J., Schmitz, G., and Liebisch, G. (2010) A rapid and quantitative LC-MS/MS method to profile sphingolipids. *J. Lipid Res.*, **51**, 2001–2011.

77 Haynes, C.A., Allegood, J.C., Park, H., and Sullards, M.C. (2009) Sphingolipidomics: methods for the comprehensive analysis of sphingolipids. *J. Chromatogr. B*, **877**, 2696–2708.

78 Masukawa, Y., Narita, H., Sato, H., Naoe, A., Kondo, N., Sugai, Y., Oba, T., Homma, R., Ishikawa, J., Takagi, Y., and Kitahara, T. (2009) Comprehensive quantification of ceramide species in human stratum corneum. *J. Lipid Res.*, **50**, 1708–1719.

79 Yano, M., Watanabe, K., Yamamoto, T., Ikeda, K., Senokuchi, T., Lu, M., Kadomatsu, T., Tsukano, H., Ikawa, M., Okabe, M., Yamaoka, S., Okazaki, T., Umehara, H., Gotoh, T., Song, W.J., Node, K., Taguchi, R., Yamagata, K., and Oike, Y. (2011) Mitochondrial dysfunction and increased reactive oxygen species impair insulin secretion in sphingomyelin synthase 1-null mice. *J. Biol. Chem.*, **286**, 3992–4002.

80 Inokuchi, J. (2010) Membrane microdomains and insulin resistance. *FEBS Lett.*, **584**, 1864–1871.

81 Kabayama, K., Sato, T., Saito, K., Loberto, N., Prinetti, A., Sonnino, S., Kinjo, M., Igarashi, Y., and Inokuchi, J. (2007) Dissociation of the insulin receptor and caveolin-1 complex by ganglioside GM3 in the state of insulin resistance. *Proc. Natl. Acad. Sci. USA*, **104**, 13678–13683.

82 Regina, TodeschiniA. and Hakomori, S.I. (2008) Functional role of glycosphingolipids and gangliosides in control of cell adhesion, motility, and growth, through glycosynaptic microdomains. *Biochim. Biophys. Acta.*, **1780**, 421–433.

83 Tagami, S., Inokuchi, J.J., Kabayama, K., Yoshimura, H., Kitamura, F., Uemura, S., Ogawa, C., Ishii, A., Saito, M., Ohtsuka, Y., Sakaue, S., and Igarashi, Y. (2002) Ganglioside GM3 participates in the pathological conditions of insulin resistance. *J. Biol. Chem.*, **277**, 3085–3092.

84 Puolitaival, S.M., Burnum, K.E., Cornett, D.S., and Caprioli, R.M. (2008) Solvent-free matrix dry-coating for MALDI imaging of phospholipids. *J. Am. Soc. Mass Spectrom.*, **19**, 882–886.

85 Sugiura, Y., Konishi, Y., Zaima, N., Kajihara, S., Nakanishi, H., Taguchi, R., and Setou, M. (2009) Visualization of the cell-selective distribution of PUFA-containing phosphatidylcholines in mouse brain by imaging mass spectrometry. *J. Lipid Res.*, **50**, 1776–1788.

86 Manicke, N.E., Wiseman, J.M., Ifa, D.R., and Cooks, R.G. (2008) Desorption electrospray ionization (DESI) mass spectrometry and tandem mass spectrometry (MS/MS) of phospholipids and sphingolipids: ionization, adduct formation, and fragmentation. *J. Am. Soc. Mass Spectrom.*, **19**, 531–543.

87 Eikel, D. and Henion, J. (2011) Liquid extraction surface analysis (LESA) of food surfaces employing chip-based nano-electrospray mass spectrometry. *Rapid. Commun. Mass Spectrom.*, **25**, 2345–2354.

13
Lipidomics in Atherosclerotic Vascular Disease

Minna T. Jänis and Reijo Laaksonen

13.1
Introduction

Dyslipidemia and atherosclerotic vascular disease are thought to originate from the imbalance of lipids in the affected organism. Elevated serum total cholesterol and low-density lipoprotein cholesterol (LDL-C) and decreased levels of serum high-density lipoprotein cholesterol (HDL-C) have been established as risk factors for atherosclerosis, which is a multifactorial, systemic disease of the large and midsize arteries. The retention of LDL particles and accumulation of calcium, cell waste products, and other substances on subendothelial space of an arterial wall is a culmination point in atherogenesis and plaque formation [1]. The foamy core region of an atherosclerotic plaque consists of extracellular lipid droplets and lipid-laden macrophages called foam cells. The core region is surrounded by a fibrous cap consisting of smooth muscle cells, collagen, and calcium that increases stiffness of the arterial wall. Immune cells including T cells, monocytes, and mast cells infiltrate the lesion and produce inflammatory cytokines and proteolytic enzymes further accelerating the inflammation [2, 3]. This leads to weakening of the fibrous cap and finally to a slow, progressive occlusion of the vessel lumen causing acute vascular events such as strokes, heart attacks, or peripheral vascular complications (Figure 13.1). The erosion and rupture of a vulnerable plaque concurrent with thrombosis and subsequent ischemic injury convert a stable disease to a life-threatening condition [4].

Different LDL-C lowering treatments have been successfully used in clinical trials to treat and prevent coronary artery disease (CAD) and other clinical atherosclerotic manifestations. However, a number of atherosclerosis patients have LDL-C levels within the recommended range suggesting the need for additional diagnostic measures of the residual risk. Recently, treatment of CAD has been extended and targeted to other risk factors such as low serum HDL-C level [6]. In the circulation, lipoprotein particles undergo multiple modifications such as hydrolysis and oxidation resulting in a very heterogeneous group of particles displaying different physiological functions. The constant modification of lipoproteins is likely to affect the atherosclerosis development. Thus, retracing the steps in

Lipidomics, First Edition. Edited by Kim Ekroos.

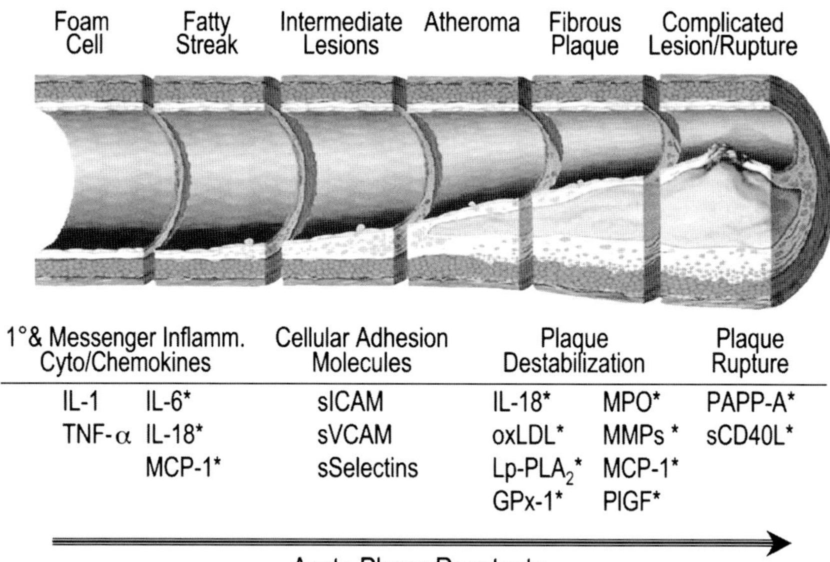

Figure 13.1 Development of an atherosclerotic plaque from foam cells to plaque rupture. Markers of inflammation and plaque instability: IL, interleukin; TNF-α, tumor necrosis factor-α; MCP-1, monocyte chemoattractant protein-1; sICAM, soluble intercellular adhesion molecule-1; sVCAM, soluble vascular cell adhesion molecule; oxLDL, oxidized low-density lipoprotein; Lp-PLA2, lipoprotein-associated phospholipase A2; GPx-1, glutathione peroxidase; MPO, myeloperoxidase; MMPs, matrix metalloproteinases; PlGF, placental growth factor; PAPP-A, pregnancy-associated plasma protein-A; sCD40L, soluble CD40 ligand; CRP, C-reactive protein; sPLA2, secretory type II phospholipase A2; SAA, serum amyloid A; WBCC, white blood cell count. *Biomarkers, which are covered in the review of Koenig and Khuseyinova [5]. Reproduced with the permission of American Heart Association.

atherogenesis and plaque development necessitates the structural and functional understanding of the lipoproteins carrying the putative plaque building blocks, lipids, which are the key components in atherosclerosis.

Molecular lipids play a pivotal role in atherogenesis as they are abundantly present in vascular plaques. However, little is known about the retention of lipid species rather than lipid classes such as LDL-C in atherosclerotic lesions during the disease development. Recently, Didangelos *et al.* reviewed the use of proteomics and lipidomics applications in vascular research depositing special emphasis on extracellular matrix remodeling and lipoprotein interaction with vascular proteoglycans [7]. As different lipid species are known to exhibit various bioactive properties affecting the course of the disease via specific biologic mechanisms such as apoptosis, it is crucial to decipher the actual atherosclerotic plaque composition at various stages of its development at the level of individual lipid species. The lipid structure is an essential determinant of the biological effect, and the biological processes leading to plaque development and rupture may be fine-tuned at molecular lipid level.

Therefore, molecular lipidomics is an essential advantage supporting both basic and pharmaceutical drug research leading potentially to new innovations in both diagnostics and drug development.

The use of metabolomics in cardiac research has been recently reviewed [8]. Metabolomics measures all small molecules, metabolites, in an organism, while transcriptomics and proteomics measure mRNA and proteins, respectively. However, metabolomics applications are still predominantly qualitative while detailed quantitative information about factors affecting atherogenesis and plaque development would be needed. Advanced methods in lipidomics enable the simultaneous high-throughput identification and quantification of hundreds of molecular lipid species in a number of lipid classes with various structural and functional roles at a level of detail not achievable with classical analytical approaches [9, 10]. Lipidomic studies quantify the precise lipid constituents of lipoproteins and tissue lipidomes, identify lipid cellular distribution, and describe their interactions and dynamics in animal models and clinical specimen [11]. Still, the challenge remains in maintaining the quality also in a high-throughput setting. We have previously described a high-throughput molecular lipidomic application based on robotic-assisted sample preparation and lipid extraction and multiple lipidomic platforms integrated with a sophisticated bioinformatics system [12]. We anticipate that this tool-kit will offer a number of opportunities not only for understanding the cellular processes in health and disease but also for enabling the biomarker screens for disease progression and identification of drug targets in future studies related both to atherosclerosis and to other diseases. The quantification at the level of lipid class or brutto lipid species is simply not enough for deciphering the culprits in the involved pathological cascades but high-precision quantification of molecular lipids is a necessity. The available technology allows high-throughput screening of large-scale biobanked samples and new cardiovascular sample collections, which will open a new avenue for lipid research in the field of atherosclerosis. It is expected that such screenings will add significantly to current knowledge on cell biology behind atherosclerosis and plaque development. Improved understanding about the mechanisms involved in atherogenesis has great potential to lead to development of novel biomarkers and discovery of new drug targets and therapies. In this chapter, we explore the role of molecular lipids in the development of atherosclerotic vascular disease and in the screen for novel biomarkers.

13.2 Lipids and Atherosclerotic Vascular Disease

Atherosclerotic vascular diseases have been studied for decades, but only in recent years a clear progress has been reached with the help of epidemiological studies of the disease, research in animal models, and modern analytical approaches. The origin and development of the disease have not been fully clarified, but an injury to the endothelium and the retention of atherogenic lipoprotein particles into subendothelial space are known to play a critical role in atherogenesis [1, 13]. The disease begins typically in later childhood and its progression is normally slow and silent until the rupture of a vulnerable plaque causing potentially fatal vascular events. Although early identification of individuals at risk is challenging, there is a quest for improved diagnostic tools. The atherosclerosis-related vascular complications including ischemic heart disease and cerebrovascular diseases are the commonest causes of death globally and cause great economic burden on health care system. The lengthening of life expectancy will further increase the incidence of atherosclerotic diseases. To that end, sensitive biomarkers with good specificity should be developed.

13.2.1 Lipoproteins

Lipoproteins are water-soluble particles consisting of a hydrophobic core of triacylglycerols and cholesteryl esters, and a hydrophilic surface of a monolayer of phospholipids, sphingolipids, free cholesterol, and apolipoproteins embedded in the lipid membrane. Lipoproteins are required to transport lipids and other hydrophobic molecules from one tissue site to another. Numerous proteins and enzymes continuously modify the lipoprotein structure and lipid composition affecting the lipoprotein properties and functions. This is described in more detail in Chapter 9 of this book.

Several lipoprotein isolation protocols are available to study the lipid composition of these particles. Different precipitation techniques, sequential flotation ultracentrifugation, size-exclusion gel chromatography, and isopycnic density gradient ultracentrifugation are most widely used. However, different isolation techniques might affect the lipid composition of lipoprotein particles that is important to take into account when selecting an isolation technique. In recent years, new techniques have arisen and several commercial applications are available for studying the distribution and composition of lipoprotein subfractions. The advantages of these new techniques, which often are based on gel-electrophoresis, are that they enable a rapid and relatively easy assessment of a large number of samples especially in clinical laboratories and, for example, the identification of highly atherogenic small LDL particles [14] or oxidized LDL (oxLDL). The oxLDL has been shown to contribute to atherosclerosis development, which has prompted attempts to use circulating oxLDL as a disease biomarker. Santos *et al.* examined the relationship of human immunoglobulin G (IgG) anti-oxLDL antibodies with cardiovascular disease risk markers in stable subjects and in patients with acute coronary syndrome (ACS) [15]. They observed that acute inflammatory and metabolic conditions decrease titers of human antibodies of IgG class against oxLDL. They also suggested that circulating anti-oxLDL antibodies could be associated with a protective role in atherosclerosis.

On the other hand, ultracentrifugal separation of lipoproteins on density gradients based either on salt [16] or on sucrose [17] is most popular as it enables further investigations of the lipoproteins and their use, for example, in different *in vitro* and *in vivo* functional studies. Moreover, an advantage of the sucrose-based density gradient ultracentrifugation is that the isolated fractions are compatible with direct analyses with most electrophoretic, chromatographic, and mass spectrometric applications without further sample preparation procedures. Recently, a novel lipidomics technology was applied to investigate the proatherogenic role of apolipoprotein (apo)CIII [18]. The authors demonstrated that diabetic ApoCIII-enriched LDL particles displayed changes in lipid composition leading to increased susceptibility to the proatherosclerotic sphingomyelinase. Indeed, serum fractionation for lipoprotein isolation is necessary for analysis of compositional changes of lipoproteins during disease progression or in different disease states.

13.2.2 Atherosclerotic Plaque

Despite improved laboratory and imaging technologies, risk estimation of cardiovascular disease (CVD) remains challenging. LDL-C and HDL-C measurements have traditionally been used in CVD risk screening. In type 2 diabetics, it has recently been demonstrated that plasma apoB, but not LDL-C, levels were associated with coronary artery calcification (CAC) scores [19]. Thus, the authors suggested that apoB levels might be particularly useful in assessing the atherosclerotic burden and cardiovascular risk in type 2 diabetic subjects. However, in a large Framingham population-based cohort, it was reported that the overall performance of apoB:apoA-I ratio for prediction of CVD was comparable with that of traditional lipid ratios with no incremental utility over total cholesterol:HDL-C [20].

In the PIVUS study (Prospective Investigation of the Vasculature in Uppsala Seniors), carotid plaques were characterized by ultrasound for size and echogenicity, which are both predictors of cardiovascular events [21]. Analysis of carotid plaque size and echogenicity was performed in 1016 subjects aged 70 years and compared with traditional risk factors and inflammation markers. Low HDL, increased body mass index (BMI) and decreased glutathione levels were associated with the echolucency of carotid plaques, suggesting the role of metabolic factors in plaque composition. On the other hand, markers of inflammation were related to plaque size alone, implying inflammation to be predominantly associated with the degree of atherosclerosis. These data suggest that plaque size and echogenicity are influenced by different risk factors.

Previously, Fagerberg *et al.* studied the heterogeneous structure of carotid atherosclerotic plaques in symptomatic carotid plaques [22]. They aimed to relate blood flow variations with differences in plaque morphology and composition between sites located up- and downstream of the maximum stenosis. The authors observed that the location of maximum stenosis relative to the carotid bifurcation varied considerably between plaques. Furthermore, they reported that compared to the downstream side, the upstream side of the stenosis had higher incidence of severe lesions with cap rupture and intraplaque hemorrhage, more macrophages, less smooth muscle cells, and more collagen. The major implication of these findings to plaque studies is that the intraplaque location may be an important confounding factor and that different plaque sections should be studied separately when assessing the lipid building blocks of plaque specimens.

Stegemann *et al.* recently demonstrated the power of lipidomics in plaque analysis for unraveling the lipid heterogeneity within atherosclerotic plaques [23]. They were able to show that linoleic acid containing cholesteryl esters (CE 18:2) and certain sphingomyelin (SM) species were enriched in plaques. Besides CE and SM species, phosphatidylcholines (PC) and lysophosphatidylcholines (LPC) accounted for the major differences between control and diseased arteries. In addition, the concentrations of the CE 18:0 and the CE 20:3 were shown to differ significantly between the vulnerable and the stable plaque areas. While the CE 18:0 was enriched in unstable plaque areas, the CE 20:3 displayed the opposite pattern and was significantly less abundant in unstable versus stable plaque regions. Indeed, it seems necessary to take the intraplaque location of different lipids into account while analyzing the results of plaque lipidomic screenings. In the best case scenario, plaque lipids related either to vulnerability or to disease development will associate with the corresponding plasma lipid levels, which naturally would increase the clinical utility of such lipid markers.

Plaque lipidomics can also be used to study cellular mechanisms and illuminate metabolic pathways related to atherosclerosis, thereby facilitating drug target identification. Current efforts with new drug targets such as phospholipase A_2 (PLA_2) or inflammatory mediators such as eicosanoids are likely to benefit from lipidomic analyses.

13.2.3 Molecular Lipids

13.2.3.1 Eicosanoids

Nowadays, atherosclerosis is generally considered as an inflammatory disease [24]. Eicosanoids, the metabolites of arachidonic acid, are known to display signaling properties and to be involved in inflammatory processes [25]. Since inflammation plays a crucial role in atherogenesis, the physiological roles of arachidonic acid (AA) and its metabolites have been studied intensively. Leukotrienes (LT) and prostaglandins (PGs) are produced from AA via the lipoxygenase (LOX) and cyclooxygenase (COX) pathways, respectively [26].

The 5-lipoxygenase (5-LO) plays an important role in atherosclerosis progression [27] and is the rate-limiting enzyme in leukotriene biosynthesis. Leukotrienes are potent proinflammatory lipid mediators, which have been shown to play a role in several pathophysiological conditions including asthma and atherosclerosis [28]. 5-LO generates LTA4 as an intermediate product, which is subsequently transformed into LTB4, LTC4, and other cysteinyl leukotrienes. Recently, LTB4 has been shown to induce reactive oxygen species production and chemotaxis of smooth muscle cells through integrin transactivation [29]. Besides 5-LO, the 8-, 12-, and 15-lipoxygenases (8-LO, 12-LO, and 15-LO, respectively) have been linked to inflammatory changes and atherogenesis in both mouse and human. Mouse and human express different lipoxygenase homologues. For instance, the 15S-lipoxygenase-2 (15-LO-2) is a human homologue of mouse phorbol ester-inducible 8S-lipoxygenase (8-LO) and 12/15-lipoxygenase is the mouse orthologue of human 15-lipoxygenase-1 (15-LO-1) [30, 31]. Also, the expression of lipoxygenases appears to be differently regulated in mouse and human [32].

Diverse physiological and pathological stimuli liberate arachidonic acid from membrane phospholipids. In addition to leukotrienes, AA can be converted to prostanoids including PGs and thromboxanes (TX) via the COX pathway. In COX pathway, the arachidonic acid is first oxygenated to form PGG2 and subsequently PGH2 by either constitutive COX-1 or inducible COX-2. The PGH2 is an unstable intermediate product that is, in turn, metabolized to PGD_2, PGE_2, $PGF_{2\alpha}$, PGI_2, and TxA_2 by cell-specific PG isomerases and synthases [33].

There are numerous potential targets including 5-LO activating protein (FLAP) that could be useful in investigation of eicosanoid metabolism and development of atherosclerosis. However, the challenge might be to find a target that would block a specific eicosanoid pathway rather than multiple pathways, especially as different eicosanoid species, such as PGD2 and PGE2, are known to display counteracting properties in progression of asthma and atherosclerosis [34–37].

13.2.3.2 **Sphingolipids and Cholesterol**

Sphingolipids represent a heterogeneous class of biomolecules that can be defined by at least five different long-chain base moieties in mammalian cells, more than 20 species of amide-linked fatty acids, and around 500 different polar head group structures [38, 39]. Due to the molecular complexity, much of the biological information still remains unknown. The involvement of the bioactive lipids such as sphingosine, sphingosine-1-phosphate, ceramide, and ceramide-1-phosphate in cell fate determination has been studied by Hannun and coworker [40]. Recently, it has been shown that sphingosine and ceramide promote apoptosis and inhibit cellular growth whereas, for example, sphingosine-1-phosphate and ceramide-1-phosphate play a role in cell survival, proliferation, differentiation, and migration [41].

In patients with familial hypercholesterolemia (FH), elevated plasma glycosphingolipids such as glucosylceramides (GlcCer) and lactosylceramides (LacCer) have been reported and the elevated glycosphingolipid levels have been shown to correlate with elevated plasma cholesterol and LDL levels [42]. In addition, it has been demonstrated that glycosphingolipids accumulate in atherosclerotic lesions in man and apoE knockout mouse [43, 44]. Indeed, numerous diseases including diabetes, atherosclerosis, and hypertension have been linked to dysfunctional sphingolipid metabolism underscoring the importance of lipid biochemistry for better understanding of the molecular basis of disease [45].

Sphingolipids and glycosphingolipids have a tendency to self-aggregate with cholesterol into membrane microdomains termed lipid rafts [46]. One of the linkages between GSL and cholesterol has been suggested to be the sterol regulatory element-binding protein (SREBP) 1 or 2, which activate the genes of sterol biosynthesis pathway [47]. Pharmacological inhibition of glucosylceramide synthase has been demonstrated to induce SREBP-regulated gene expression and cholesterol synthesis in HepG2 cells. Interestingly, the authors reported that despite activation of SREBP target genes and cholesterol biosynthesis, the cellular cholesterol content did not increase.

Indeed, GSL such as LacCer and GlcCer, but not gangliosides, have been shown to play a role in cholesterol metabolism by suppressing macrophage apoE production, which affects the reverse cholesterol transport [48, 49]. In addition, LacCer has been shown to inhibit cholesterol efflux from peripheral cells through ATP-binding cassette transporter A1 [50], to induce monocyte adhesion to endothelial cells [51], and to stimulate vascular smooth muscle cell proliferation [52], all these properties playing a role in atherogenesis. Ganglioside GM3 has been shown to accelerate the LDL uptake by macrophages resulting in foam cell formation [53] and to stimulate platelet adhesion to sites of lesion formation on arterial wall [54]. Sulfatides have been suggested to have anticoagulant activity and play a physiological role in inflammation upon vascular injury [55, 56]. Recently, it has been shown that inhibition of glycosphingolipid synthesis by targeting the glucosylceramide synthase reduces plasma cholesterol levels and inhibits atherogenesis in ApoE3 Leiden and LDL receptor knockout mice [57]. In apoE KO mice, however, reduced plasma glycosphingolipid levels were not shown to affect the lesion development [58].

13.2.3.3 **Phospholipids**

Hydrolysis, aggregation, and oxidative and enzymatic modifications of LDL particles lead to release of inflammatory phospholipids [59] and activation of endothelial cells expressing several leukocyte adhesion molecules [60, 61]. This results in migration of monocytes and other immune cells into the subendothelial space where monocytes eventually differentiate into macrophages. Oxidized phospholipids have been shown to mediate the uptake of modified LDL particles through macrophage scavenger receptor CD36 resulting in accumulation of excess intracellular cholesterol and foam cell formation [62]. In addition, oxidized phospholipids have been shown to affect the atherogenicity of lipoprotein (a) [63]. Recently, gut-flora-dependent metabolism of dietary phosphatidylcholine (PC) has been shown to be linked to atherosclerosis development [64]. Also, in apoE knockout mice, impaired hepatic PC synthesis has been shown to result in reduced atherosclerosis and diminished cardiac accumulation of triacylglycerols [65].

Previously, it has been suggested that generation of lysophosphatidylcholine (LPC), either from LDL or from cellular membrane phospholipids, could play a role both in general inflammatory processes and in promoting atherogenesis [66]. It has been demonstrated that LPC, which is generated by a phospholipase A2 activity during LDL oxidation and is the major phospholipid component of oxidized LDL, is a potent chemotactic factor for monocytes recruiting them into the arterial wall during the early stages of atherogenesis [67].

These data indicate that molecular lipids across numerous lipid classes affect the course of disease development. Therefore, for revealing the role of lipids in atherogenesis and plaque development, it is crucial to identify the lipid molecular structures at the level of fatty acid composition and double bond position.

13.2.4 Animal Models of Atherosclerotic Research

To study atherosclerotic plaque formation and treatment outcome, a model with good translational properties should be used as the access to human atherosclerotic plaque material is limited particularly for interventional studies. Lipidomics offers tools for disease model selection and translational medicine as it allows a detailed comparison of molecular lipids in plaques obtained from available animal models. We have compared the lipid profiles of the human mammary artery and rat aorta [68]. The data indicated that there was a substantial deviation already at the lipid class level between human and rat tissues. Moreover, a detailed analysis of molecular lipid species indicated that the number of recorded molecular lipids and their molecular percentage distribution was quite different between human and rat (Figure 13.2). Thus,

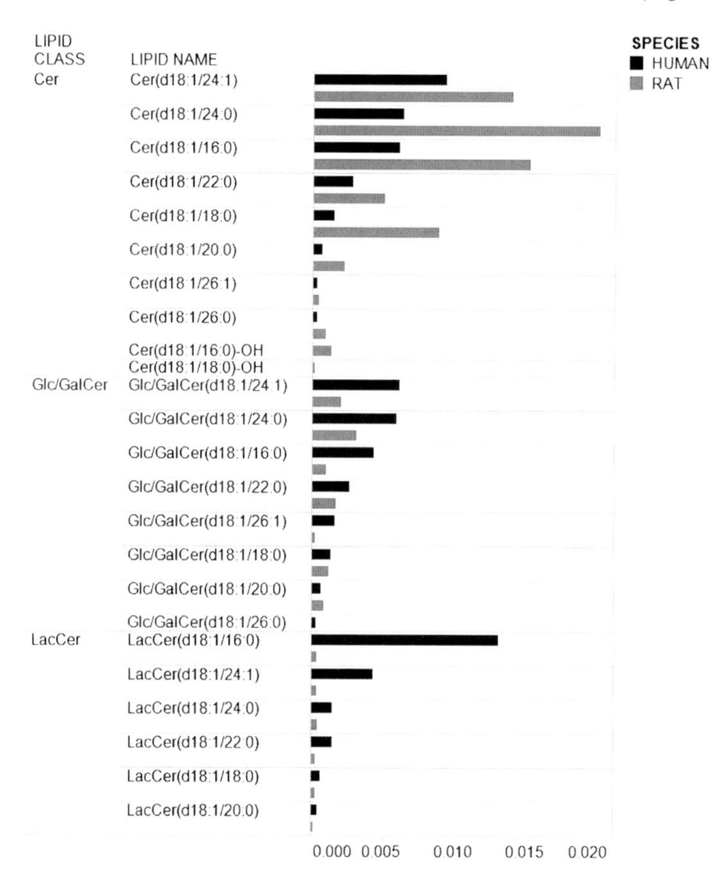

Figure 13.2 Lipidomics analysis conducted on human left internal mammary artery (LIMA) and rat aorta samples show distinct lipid species-specific composition. Cer, ceramide; Glc/GalCer, glucosyl/galactosylceramide; LacCer, lactosylceramide.

lipidomic profiling could be used for demonstration that rat arteries are not a preferential material to study human atherosclerosis due to differently expressed lipids.

The apoE knockout (KO) mouse has been widely used in atherosclerosis studies. The apoE KO mouse spontaneously develops hypercholesterolemia and atherosclerotic lesions even on regular chow diet [69, 70]. The lesion development has been shown to be similar between apoE KO mice and humans [71]. However, the pathology of a lesion in a tiny 0.3 mm mouse vessel could well be different from the pathology of lesions in larger human coronary arteries. The most significant difference between human and mouse lesions seems to be the absence in animal models of fibrin formation either within the lesion or within the lumen [72]. Thus, it is still somewhat unclear how well the data obtained from the ApoE-KO model translate to humans.

Although the prevalence of genetically modified animals such as mice and rats has contributed greatly to research in different disease fields, not all human diseases can properly be modeled in mice or rats. In atherosclerosis research, mice and rats can be considered to be inappropriate because of marked differences in lipoprotein metabolism and atherosclerosis pathophysiology compared to humans. In rabbits, however, lipoprotein metabolism and atherosclerotic lesions closely resemble those in humans. The use of the Watanabe heritable hyperlipidemic (WHHL) rabbit, which develops hypercholesterolemia and atherosclerosis spontaneously due to genetic and functional deficiencies of the LDL receptor has recently been reviewed by Kobayashi *et al.* [73]. The use of lipidomics application for plaque analysis could reveal valuable information about differences in plaque formation and disease progression between humans and different animal models including also bigger animals such as pigs and nonhuman primates.

13.3
Diagnostics and Treatment

13.3.1
Diagnostic Biomarkers of Atherosclerosis

Recently, metabolomic screenings have been used in order to identify novel CVD biomarkers. Zhang *et al.* utilized an ultrafast liquid chromatography coupled with IT-TOF mass spectrometry (UFLC/MS-IT-TOF) to study plasma and urine metabolic profiles of atherosclerotic rats [74]. Their observations suggest that abnormal metabolism of phenylalanine, tryptophan, bile acids, and amino acids might be related to atherosclerosis development. In addition, Zha *et al.* indentified a metabolic fingerprint of 21 compounds in hamsters that could be a potential marker for the development of atherosclerosis [75]. Recently, dietary phosphatidylcholine and its metabolites produced in gut flora have been shown to be potential biomarkers of atherosclerosis [64]. Moreover, it has been suggested that by using desorption electrospray ionization mass spectrometry (DESI-MS) for profiling and imaging of arterial plaques lipid distributions may correlate with the stability or vulnerability of particular region of the plaque [76]. Recently, Meikle *et al.* demonstrated that patients with stable and unstable CAD can be distinguished on the basis of their plasma molecular lipid profile [77]. The concentrations of certain alkyl- and alkenyl-linked phospholipids were found to be markedly lower in samples from acute coronary syndrome (ACS) patients than those with stable angina. As alkyl- and alkenyl-linked phospholipids are known to be prone to oxidation [78], it is possible that patients with vulnerable plaque display higher oxidative stress in vascular system.

Also, according to our unpublished data, it appears that molecular lipids may have greater diagnostic and prognostic value than LDL-C. Moreover, lipidomics in atherosclerotic plaques will provide another window for understanding plaque structure, development, and stability. Similar analyses of material obtained from different species will help in translation of atherosclerotic studies between species. We believe that detailed lipid information will substantially improve CVD risk stratification both on individual and on population level. Detailed lipidomic information will also be useful in tailoring right treatments and in search for new drug targets.

13.3.2
Lipidomics in Efficacy and Safety Measurements

To date, lipidomics has not been exploited for the benefit of clinical trials. As new and more precise CVD risk lipid biomarkers are emerging, biochemical monitoring of drug efficacy by traditional LDL-C and HDL-C assays will also be enhanced. Interestingly, lipidomics is able to distinguish the efficacy profiles of various members of a drug class such as statins. In fact, first attempts to use lipidomic measurements to compare lipid-lowering profiles between different statins have been performed [79, 80]. Recently, patient baseline lipid profile has been shown to correlate with the treatment response. Kaddurah-Daouk *et al.* investigated the lipid profiles in plasma samples obtained from patients before and after treatment with simvastatin [81]. They were able to identify distinct lipid signatures for treatment efficacy. Cholesteryl ester, phospholipid, and diacylglycerol levels were shown to predict statin-induced changes in LDL-C, while baseline plasmalogen concentration, in particular, was shown to correlate with CRP response. In addition, lipidomics application has recently been applied for identification of a lipid biomarker for fatty acid-binding protein 4 (FABP4) inhibition [82]. Suhre *et al.* identified several lipid ratios of different sphingomyelin and phosphatidylcholine species that were significantly and dose-dependently affected by FABP4 inhibiting drug treatment. In apoE knockout mice, total or macrophage-specific FABP4 knockout has been shown to protect against atherogenesis [83]. However, these attempts have so far been more hypothesis generating than informative as the role of most of the molecular lipids detected in these studies has yet to be established.

Indeed, our vision is that lipidomic analyses will play a significant role in evaluating novel HDL modulating compounds such as cholesteryl ester transfer protein (CETP) inhibitors. Furthermore, we suspect that current and new LDL-C lowering treatments and their combinations will be carefully reevaluated as it is very likely that these treatments trigger multiple changes in the lipoprotein lipidome despite similar LDL-C lowering effects.

Lipidomics has also proven to be an efficient way of detecting adverse drug actions. So far, drug-induced liver and muscle toxicities have been demonstrated in plasma lipidomic profiles. The PC to phosphatidylethanolamine (PE) ratio has been observed to reflect hepatocyte membrane fluidity changes allowing a sensitive measure of liver toxicity [84]. Also, taking advantage of the molecular lipid information, the authors found new insight into the mechanisms behind the membrane fluidity alterations. We have reported earlier that a distinct plasma lipid pattern can be used to monitor statin-induced gene expression changes in human skeletal muscle and, thus, to detect statin-induced muscle toxicity [79]. Our present data suggest that this testing can be based on detection of single or couple of molecular lipids and that the lipid biomarkers are more sensitive than current markers including liver enzymes and creatine kinase (CK).

13.4 Conclusions

Novel lipidomics applications add a new dimension of detail and understanding to atherosclerosis studies. The pathogenicity of atherosclerosis is centered on lipid metabolic pathways. At present, lipidomics attempts to characterize structurally defined molecular lipids from a single analysis [85] and localize lipids in tissues such as aortic wall [86] and arterial plaques [23, 76]. Development of lipidomics applications and evolution of new lipid labeling techniques will not only improve lipid metabolism characterization but also identify the deleterious molecular lipids involved in the dysfunction, where this occurs inside the organism and at what rate. Thus, lipidomics is a prerequisite for advancing the biology and mechanisms of atherogenesis and promoting the discovery of cardiovascular disease biomarkers.

Lipidomics in combination with the appropriate clinical samples is used today to address the many unmet needs of disease diagnostics. Lipid biomarkers may serve as a readout of experimental or approved therapies, while the bioactive role of certain lipid molecules has the potential of identifying novel drug targets. Lipids are also excellent candidates for companion diagnostics in drug research that is moving increasingly toward the specialized therapeutics model. Lipidomics enables patient differentiation based on their molecular lipid profiles and facilitates personalized medicine. Such information can be used for ensuring that the right individual receives the right drug at the right time and dose. Lipidomics has also been used to study translational medicine and thereby is helpful in assessing the utility of various experimental animal models.

Quantification of molecular lipids by novel lipidomics applications will provide an important layer of phenotype information that will help us better understand numerous gene–gene, gene–environment, and gene–protein interactions that are involved in the development of atherosclerosis. Therefore, lipidomics will be one of the cornerstones in the next generation of mechanistic studies of atherosclerosis.

References

1 Williams, K.J. and Tabas, I. (1995) The response-to-retention hypothesis of early atherogenesis. *Arterioscler. Thromb. Vasc. Biol.*, **15**, 551–561.

2 Frostegard, J., Ulfgren, A.K., Nyberg, P., Hedin, U., Swedenborg, J., Andersson, U., and Hansson, G.K. (1999) Cytokine expression in advanced human atherosclerotic plaques: dominance of pro-inflammatory (Th1) and macrophage-stimulating cytokines. *Atherosclerosis*, **145**, 33–43.

3 Kovanen, P.T., Kaartinen, M., and Paavonen, T. (1995) Infiltrates of activated mast cells at the site of coronary atheromatous erosion or rupture in myocardial infarction. *Circulation*, **92**, 1084–1088.

4 Falk, E., Shah, P.K., and Fuster, V. (1995) Coronary plaque disruption. *Circulation*, **92**, 657–671.

5 Koenig, W. and Khuseyinova, N. (2007) Biomarkers of atherosclerotic plaque instability and rupture. *Arterioscler. Thromb. Vasc. Biol.*, **27**, 15–26.

6 Linsel-Nitschke, P. and Tall, A.R. (2005) HDL as a target in the treatment of atherosclerotic cardiovascular disease. *Nat. Rev. Drug Discov.*, **4**, 193–205.

7 Didangelos, A., Stegemann, C., and Mayr, M. (2012) The -omics era: proteomics and lipidomics in vascular research. *Atherosclerosis*, **221** (1), 12–17.

8 Griffin, J.L., Atherton, H., Shockcor, J., and Atzori, L. (2011) Metabolomics as a tool for cardiac research. *Nat. Rev. Cardiol.*, **8**, 630–643.

9 Ejsing, C.S., Sampaio, J.L., Surendranath, V., Duchoslav, E., Ekroos, K., Klemm, R.W., Simons, K., and Shevchenko, A. (2009) Global analysis of the yeast lipidome by quantitative shotgun mass spectrometry. *Proc. Natl. Acad. Sci. USA*, **106**, 2136–2141.

10 Stahlman, M., Ejsing, C.S., Tarasov, K., Perman, J., Boren, J., and Ekroos, K. (2009) High-throughput shotgun lipidomics by quadrupole time-of-flight mass spectrometry. *J. Chromatogr. B Analyt. Technol. Biomed. Life Sci.*, **877** (26), 2664–2672.

11 Han, X. and Gross, R.W. (2003) Global analyses of cellular lipidomes directly from crude extracts of biological samples by ESI mass spectrometry: a bridge to lipidomics. *J. Lipid Res.*, **44**, 1071–1079.

12 Jung, H.R., Sylvanne, T., Koistinen, K.M., Tarasov, K., Kauhanen, D., and Ekroos, K. (2011) High throughput quantitative molecular lipidomics. *Biochim. Biophys. Acta*, **1811** (11), 925–934.

13 Ross, R. (1993) The pathogenesis of atherosclerosis: a perspective for the 1990s. *Nature*, **362**, 801–809.

14 Moon, J.Y., Kwon, H.M., Kwon, S.W., Yoon, S.J., Kim, J.S., Lee, S.J., Park, J.K., Rhee, J.H., Yoon, Y.W., Hong, B.K., Rim, S.J., and Kim, H.S. (2007) Lipoprotein(a) and LDL particle size are related to the severity of coronary artery disease. *Cardiology*, **108**, 282–289.

15 Santos, A.O., Fonseca, F.A., Fischer, S.M., Monteiro, C.M., Brandao, S.A., Povoa, R. M., Bombig, M.T., Carvalho, A.C., Monteiro, A.M., Ramos, E., Gidlund, M., Figueiredo Neto, A.M., and Izar, M.C. (2009) High circulating autoantibodies against human oxidized low-density lipoprotein are related to stable and lower titers to unstable clinical situation. *Clin. Chim. Acta*, **406**, 113–118.

16 Havel, R.J., Eder, H.A., and Bragdon, J.H. (1955) The distribution and chemical composition of ultracentrifugally separated lipoproteins in human serum. *J. Clin. Invest.*, **34**, 1345–1353.

17 Stahlman, M., Davidsson, P., Kanmert, I., Rosengren, B., Boren, J., Fagerberg, B., and Camejo, G. (2008) Proteomics and lipids of lipoproteins isolated at low salt concentrations in D2O/sucrose or in KBr. *J. Lipid Res.*, **49**, 481–490.

18 Hiukka, A., Stahlman, M., Pettersson, C., Levin, M., Adiels, M., Teneberg, S., Leinonen, E.S., Hulten, L.M., Wiklund, O., Oresic, M., Olofsson, S.O., Taskinen, M.R., Ekroos, K., and Boren, J. (2009) ApoCIII-enriched LDL in type 2 diabetes displays altered lipid composition, increased susceptibility for sphingomyelinase, and increased binding to biglycan. *Diabetes*, **58**, 2018–2026.

19 Martin, S.S., Qasim, A.N., Mehta, N.N., Wolfe, M., Terembula, K., Schwartz, S., Iqbal, N., Schutta, M., Bagheri, R., and Reilly, M.P. (2009) Apolipoprotein B but not LDL cholesterol is associated with coronary artery calcification in type 2 diabetic whites. *Diabetes*, **58**, 1887–1892.

20 Ingelsson, E., Schaefer, E.J., Contois, J.H., McNamara, J.R., Sullivan, L., Keyes, M.J., Pencina, M.J., Schoonmaker, C., Wilson, P.W., D'Agostino, R.B., and Vasan, R.S. (2007) Clinical utility of different lipid measures for prediction of coronary heart disease in men and women. *JAMA*, **298**, 776–785.

21 Andersson, J., Sundstrom, J., Kurland, L., Gustavsson, T., Hulthe, J., Elmgren, A., Zilmer, K., Zilmer, M., and Lind, L. (2009) The carotid artery plaque size and echogenicity are related to different cardiovascular risk factors in the elderly: the Prospective Investigation of the Vasculature in Uppsala Seniors (PIVUS) study. *Lipids*, **44**, 397–403.

22 Fagerberg, B., Ryndel, M., Kjelldahl, J., Akyurek, L.M., Rosengren, L., Karlstrom, L., Bergstrom, G., and Olson, F.J. (2009) Differences in lesion severity and cellular composition between *in vivo* assessed upstream and downstream sides of human symptomatic carotid atherosclerotic plaques. *J. Vasc. Res.*, **47**, 221–230.

23 Stegemann, C., Drozdov, I., Shalhoub, J., Humphries, J., Ladroue, C., Didangelos, A., Baumert, M., Allen, M., Davies, A.H., Monaco, C., Smith, A., Xu, Q., and Mayr, M. (2011) Comparative lipidomics profiling of human atherosclerotic plaques. *Circ. Cardiovasc. Genet.*, **4**, 232–242.

24 Hansson, G.K. (2005) Inflammation, atherosclerosis, and coronary artery disease. *N. Engl. J. Med.*, **352**, 1685–1695.

25 Back, M. (2009) Leukotriene signaling in atherosclerosis and ischemia. *Cardiovasc. Drugs Ther.*, **23**, 41–48.

26 Murakami, M. (2011) Lipid mediators in life science. *Exp. Anim.*, **60**, 7–20.

27 Mehrabian, M., Allayee, H., Wong, J., Shi, W., Wang, X.P., Shaposhnik, Z., Funk, C.D., and Lusis, A.J. (2002) Identification of 5-lipoxygenase as a major gene contributing to atherosclerosis susceptibility in mice. *Circ. Res.*, **91**, 120–126.

28 Goetzl, E.J., An, S., and Smith, W.L. (1995) Specificity of expression and effects of eicosanoid mediators in normal physiology and human diseases. *FASEB J.*, **9**, 1051–1058.

29 Moraes, J., Assreuy, J., Canetti, C., and Barja-Fidalgo, C. (2010) Leukotriene B4 mediates vascular smooth muscle cell migration through $\alpha v\beta 3$ integrin transactivation. *Atherosclerosis*, **212**, 406–413.

30 Jisaka, M., Kim, R.B., Boeglin, W.E., and Brash, A.R. (2000) Identification of amino acid determinants of the positional specificity of mouse 8S-lipoxygenase and human 15S-lipoxygenase-2. *J. Biol. Chem.*, **275**, 1287–1293.

31 Conrad, D.J. (1999) The arachidonate 12/15 lipoxygenases. a review of tissue expression and biologic function. *Clin. Rev. Allergy Immunol.*, **17**, 71–89.

32 Jawien, J. and Korbut, R. (2010) The current view on the role of leukotrienes in atherogenesis. *J. Physiol. Pharmacol.*, **61**, 647–650.

33 Fitzpatrick, F.A. and Soberman, R. (2001) Regulated formation of eicosanoids. *J. Clin. Invest.*, **107**, 1347–1351.

34 Wang, M., Zukas, A.M., Hui, Y., Ricciotti, E., Pure, E., and FitzGerald, G.A. (2006) Deletion of microsomal prostaglandin E synthase-1 augments prostacyclin and retards atherogenesis. *Proc. Natl. Acad. Sci. USA*, **103**, 14507–14512.

35 Ragolia, L., Palaia, T., Hall, C.E., Maesaka, J.K., Eguchi, N., and Urade, Y. (2005) Accelerated glucose intolerance, nephropathy, and atherosclerosis in prostaglandin D2 synthase knock-out mice. *J. Biol. Chem.*, **280**, 29946–29955.

36 Kunikata, T., Yamane, H., Segi, E., Matsuoka, T., Sugimoto, Y., Tanaka, S., Tanaka, H., Nagai, H., Ichikawa, A., and Narumiya, S. (2005) Suppression of allergic inflammation by the prostaglandin E receptor subtype EP3. *Nat. Immunol.*, **6**, 524–531.

37 Matsuoka, T., Hirata, M., Tanaka, H., Takahashi, Y., Murata, T., Kabashima, K., Sugimoto, Y., Kobayashi, T., Ushikubi, F.,

Aze, Y., Eguchi, N., Urade, Y., Yoshida, N., Kimura, K., Mizoguchi, A., Honda, Y., Nagai, H., and Narumiya, S. (2000) Prostaglandin D2 as a mediator of allergic asthma. *Science*, **287**, 2013–2017.

38 Futerman, A.H. and Hannun, Y.A. (2004) The complex life of simple sphingolipids. *EMBO Rep.*, **5**, 777–782.

39 Zheng, W., Kollmeyer, J., Symolon, H., Momin, A., Munter, E., Wang, E., Kelly, S., Allegood, J.C., Liu, Y., Peng, Q., Ramaraju, H., Sullards, M.C., Cabot, M., and Merrill, A.H., Jr. (2006) Ceramides and other bioactive sphingolipid backbones in health and disease: lipidomic analysis, metabolism and roles in membrane structure, dynamics, signaling and autophagy. *Biochim. Biophys. Acta*, **1758**, 1864–1884.

40 Hannun, Y.A. and Obeid, L.M. (2008) Principles of bioactive lipid signalling: lessons from sphingolipids. *Nat. Rev. Mol. Cell Biol.*, **9**, 139–150.

41 Fyrst, H. and Saba, J.D. (2010) An update on sphingosine-1-phosphate and other sphingolipid mediators. *Nat. Chem. Biol.*, **6**, 489–497.

42 Dawson, G., Kruski, A.W., and Scanu, A. M. (1976) Distribution of glycosphingolipids in the serum lipoproteins of normal human subjects and patients with hypo- and hyperlipidemias. *J. Lipid Res.*, **17**, 125–131.

43 Mukhin, D.N., Chao, F.F., and Kruth, H. S. (1995) Glycosphingolipid accumulation in the aortic wall is another feature of human atherosclerosis. *Arterioscler. Thromb. Vasc. Biol.*, **15**, 1607–1615.

44 Garner, B., Priestman, D.A., Stocker, R., Harvey, D.J., Butters, T.D., and Platt, F.M. (2002) Increased glycosphingolipid levels in serum and aortae of apolipoprotein E gene knockout mice. *J. Lipid Res.*, **43**, 205–214.

45 Alewijnse, A.E. and Peters, S.L. (2008) Sphingolipid signalling in the cardiovascular system: good, bad or both? *Eur. J. Pharmacol.*, **585**, 292–302.

46 Lahiri, S. and Futerman, A.H. (2007) The metabolism and function of sphingolipids and glycosphingolipids. *Cell Mol. Life Sci.*, **64**, 2270–2284.

47 Bijl, N., Scheij, S., Houten, S., Boot, R.G., Groen, A.K., and Aerts, J.M. (2008) The glucosylceramide synthase inhibitor *N*-(5-adamantane-1-yl-methoxy-pentyl)-deoxynojirimycin induces sterol regulatory element-binding protein-regulated gene expression and cholesterol synthesis in HepG2 cells. *J. Pharmacol. Exp. Ther.*, **326**, 849–855.

48 Garner, B., Mellor, H.R., Butters, T.D., Dwek, R.A., and Platt, F.M. (2002) Modulation of THP-1 macrophage and cholesterol-loaded foam cell apolipoprotein E levels by glycosphingolipids. *Biochem. Biophys. Res. Commun.*, **290**, 1361–1367.

49 Bielicki, J.K., McCall, M.R., and Forte, T. M. (1999) Apolipoprotein A-I promotes cholesterol release and apolipoprotein E recruitment from THP-1 macrophage-like foam cells. *J. Lipid Res.*, **40**, 85–92.

50 Glaros, E.N., Kim, W.S., Quinn, C.M., Wong, J., Gelissen, I., Jessup, W., and Garner, B. (2005) Glycosphingolipid accumulation inhibits cholesterol efflux via the ABCA1/apolipoprotein A-I pathway: 1-phenyl-2-decanoylamino-3-morpholino-1-propanol is a novel cholesterol efflux accelerator. *J. Biol. Chem.*, **280**, 24515–24523.

51 Gong, N., Wei, H., Chowdhury, S.H., and Chatterjee, S. (2004) Lactosylceramide recruits PKCalpha/epsilon and phospholipase A2 to stimulate PECAM-1 expression in human monocytes and adhesion to endothelial cells. *Proc. Natl. Acad. Sci. USA*, **101**, 6490–6495.

52 Bhunia, A.K., Han, H., Snowden, A., and Chatterjee, S. (1997) Redox-regulated signaling by lactosylceramide in the proliferation of human aortic smooth muscle cells. *J. Biol. Chem.*, **272**, 15642–15649.

53 Prokazova, N.V., Mikhailenko, I.A., and Bergelson, L.D. (1991) Ganglioside GM3 stimulates the uptake and processing of low density lipoproteins by macrophages. *Biochem. Biophys. Res. Commun.*, **177**, 582–587.

54 Wen, F.Q., Jabbar, A.A., Patel, D.A., Kazarian, T., and Valentino, L.A. (1999) Atherosclerotic aortic gangliosides enhance integrin-mediated platelet

adhesion to collagen. *Arterioscler. Thromb. Vasc. Biol.*, **19**, 519–524.
55 Hara, A. and Taketomi, T. (1993) Sulphatide as a major glycosphingolipid in WHHL rabbit serum lipoproteins and its anticoagulant activity. *Indian J. Biochem. Biophys.*, **30**, 353–357.
56 Inoue, T., Taguchi, I., Abe, S., Li, G., Hu, R., Nakajima, T., Hara, A., Aoyama, T., Kannagi, R., Kyogashima, M., and Node, K. (2010) Sulfatides are associated with neointimal thickening after vascular injury. *Atherosclerosis*, **211**, 291–296.
57 Bietrix, F., Lombardo, E., van Roomen, C.P., Ottenhoff, R., Vos, M., Rensen, P.C., Verhoeven, A.J., Aerts, J.M., and Groen, A.K. (2010) Inhibition of glycosphingolipid synthesis induces a profound reduction of plasma cholesterol and inhibits atherosclerosis development in APOE∗3 Leiden and low-density lipoprotein receptor−/− mice. *Arterioscler. Thromb. Vasc. Biol.*, **30**, 931–937.
58 Glaros, E.N., Kim, W.S., Rye, K.A., Shayman, J.A., and Garner, B. (2008) Reduction of plasma glycosphingolipid levels has no impact on atherosclerosis in apolipoprotein E-null mice. *J. Lipid Res.*, **49**, 1677–1681.
59 Leitinger, N. (2003) Oxidized phospholipids as modulators of inflammation in atherosclerosis. *Curr. Opin. Lipidol.*, **14**, 421–430.
60 Cybulsky, M.I. and Gimbrone, M.A., Jr. (1991) Endothelial expression of a mononuclear leukocyte adhesion molecule during atherogenesis. *Science*, **251**, 788–791.
61 Nakashima, Y., Raines, E.W., Plump, A.S., Breslow, J.L., and Ross, R. (1998) Upregulation of VCAM-1 and ICAM-1 at atherosclerosis-prone sites on the endothelium in the ApoE-deficient mouse. *Arterioscler. Thromb. Vasc. Biol.*, **18**, 842–851.
62 Podrez, E.A., Poliakov, E., Shen, Z., Zhang, R., Deng, Y., Sun, M., Finton, P.J., Shan, L., Febbraio, M., Hajjar, D.P., Silverstein, R.L., Hoff, H.F., Salomon, R.G., and Hazen, S.L. (2002) A novel family of atherogenic oxidized phospholipids promotes macrophage foam cell formation via the scavenger receptor CD36 and is enriched in atherosclerotic lesions. *J. Biol. Chem.*, **277**, 38517–38523.
63 Tsimikas, S. and Witztum, J.L. (2008) The role of oxidized phospholipids in mediating lipoprotein(a) atherogenicity. *Curr. Opin. Lipidol.*, **19**, 369–377.
64 Wang, Z., Klipfell, E., Bennett, B.J., Koeth, R., Levison, B.S., Dugar, B., Feldstein, A. E., Britt, E.B., Fu, X., Chung, Y.M., Wu, Y., Schauer, P., Smith, J.D., Allayee, H., Tang, W.H., DiDonato, J.A., Lusis, A.J., and Hazen, S.L. (2011) Gut flora metabolism of phosphatidylcholine promotes cardiovascular disease. *Nature*, **472**57–63.
65 Cole, L.K., Dolinsky, V.W., Dyck, J.R., and Vance, D.E. (2011) Impaired phosphatidylcholine biosynthesis reduces atherosclerosis and prevents lipotoxic cardiac dysfunction in ApoE−/− mice. *Circ. Res.*, **108**, 686–694.
66 Matsumoto, T., Kobayashi, T., and Kamata, K. (2007) Role of lysophosphatidylcholine (LPC) in atherosclerosis. *Curr. Med. Chem.*, **14**, 3209–3220.
67 Quinn, M.T., Parthasarathy, S., and Steinberg, D. (1988) Lysophosphatidylcholine: a chemotactic factor for human monocytes and its potential role in atherogenesis. *Proc. Natl. Acad. Sci. USA*, **85**, 2805–2809.
68 Ekroos, K., Janis, M., Tarasov, K., Hurme, R., and Laaksonen, R. (2010) Lipidomics: a tool for studies of atherosclerosis. *Curr. Atheroscler. Rep.*, **12**, 273–281.
69 Plump, A.S., Smith, J.D., Hayek, T., Aalto-Setala, K., Walsh, A., Verstuyft, J.G., Rubin, E.M., and Breslow, J.L. (1992) Severe hypercholesterolemia and atherosclerosis in apolipoprotein E-deficient mice created by homologous recombination in ES cells. *Cell*, **71**, 343–353.
70 Zhang, S.H., Reddick, R.L., Piedrahita, J.A., and Maeda, N. (1992) Spontaneous hypercholesterolemia and arterial lesions in mice lacking apolipoprotein E. *Science*, **258**, 468–471.
71 Nakashima, Y., Plump, A.S., Raines, E.W., Breslow, J.L., and Ross, R. (1994) ApoE-deficient mice develop lesions of all phases

of atherosclerosis throughout the arterial tree. *Arterioscler. Thromb.*, **14**, 133–140.

72 Rosenfeld, M.E., Polinsky, P., Virmani, R., Kauser, K., Rubanyi, G., and Schwartz, S. M. (2000) Advanced atherosclerotic lesions in the innominate artery of the ApoE knockout mouse. *Arterioscler. Thromb. Vasc. Biol.*, **20**, 2587–2592.

73 Kobayashi, T., Ito, T., and Shiomi, M. (2011) Roles of the WHHL rabbit in translational research on hypercholesterolemia and cardiovascular diseases. *J. Biomed. Biotechnol.*, **2011**, 406473.

74 Zhang, F., Jia, Z., Gao, P., Kong, H., Li, X., Chen, J., Yang, Q., Yin, P., Wang, J., Lu, X., Li, F., Wu, Y., and Xu, G. (2009) Metabonomics study of atherosclerosis rats by ultra fast liquid chromatography coupled with ion trap-time of flight mass spectrometry. *Talanta*, **79**, 836–844.

75 Zha, W.AJ., Wang, G., Yan, B., Gu, S., Zhu, X., Hao, H., Huang, Q., Sun, J., Zhang, Y., and Cao, B., and Ren, H. (2009) Metabonomic characterization of early atherosclerosis in hamsters with induced cholesterol. *Biomarkers*, **14**, 372–380.

76 Manicke, N.E., Nefliu, M., Wu, C., Woods, J.W., Reiser, V., Hendrickson, R.C., and Cooks, R.G. (2009) Imaging of lipids in atheroma by desorption electrospray ionization mass spectrometry. *Anal. Chem.*, **81**, 8702–8707.

77 Meikle, P.J., Wong, G., Tsorotes, D., Barlow, C.K., Weir, J.M., Christopher, M. J., MacIntosh, G.L., Goudey, B., Stern, L., Kowalczyk, A., Haviv, I., White, A.J., Dart, A.M., Duffy, S.J., Jennings, G.L., and Kingwell, B.A. (2011) Plasma lipidomic analysis of stable and unstable coronary artery disease. *Arterioscler. Thromb. Vasc. Biol.*, **31**, 2723–2732.

78 Ford, D.A. (2010) Lipid oxidation by hypochlorous acid: chlorinated lipids in atherosclerosis and myocardial ischemia. *Clin. Lipidol.*, **5**, 835–852.

79 Laaksonen, R., Katajamaa, M., Paiva, H., Sysi-Aho, M., Saarinen, L., Junni, P., Lutjohann, D., Smet, J., Van Coster, R., Seppanen-Laakso, T., Lehtimaki, T., Soini, J., and Oresic, M. (2006) A systems biology strategy reveals biological pathways and plasma biomarker candidates for potentially toxic statin-induced changes in muscle. *PLoS ONE*, **1**, e97.

80 Bergheanu, S.C., Reijmers, T., Zwinderman, A.H., Bobeldijk, I., Ramaker, R., Liem, A.H., van der Greef, J., Hankemeier, T., and Jukema, J.W. (2008) Lipidomic approach to evaluate rosuvastatin and atorvastatin at various dosages: investigating differential effects among statins. *Curr. Med. Res. Opin.*, **24**, 2477–2487.

81 Kaddurah-Daouk, R., Baillie, R.A., Zhu, H., Zeng, Z.B., Wiest, M.M., Nguyen, U. T., Watkins, S.M., and Krauss, R.M. (2010) Lipidomic analysis of variation in response to simvastatin in the cholesterol and pharmacogenetics study. *Metabolomics*, **6**, 191–201.

82 Suhre, K., Romisch-Margl, W., de Angelis, M.H., Adamski, J., Luippold, G., and Augustin, R. (2011) Identification of a potential biomarker for FABP4 inhibition: the power of lipidomics in preclinical drug testing. *J. Biomol. Screen*, **16**, 467–475.

83 Makowski, L., Boord, J.B., Maeda, K., Babaev, V.R., Uysal, K.T., Morgan, M.A., Parker, R.A., Suttles, J., Fazio, S., Hotamisligil, G.S., and Linton, M.F. (2001) Lack of macrophage fatty-acid-binding protein aP2 protects mice deficient in apolipoprotein E against atherosclerosis. *Nat. Med.*, **7**, 699–705.

84 Sergent, O., Ekroos, K., Lefeuvre-Orfila, L., Rissel, M., Forsberg, G.B., Oscarsson, J., Andersson, T.B., and Lagadic-Gossmann, D. (2009) Ximelagatran increases membrane fluidity and changes membrane lipid composition in primary human hepatocytes. *Toxicol. In Vitro*, **23**, 1305–1310.

85 Thomas, M.C., Mitchell, T.W., Harman, D.G., Deeley, J.M., Nealon, J.R., and Blanksby, S.J. (2008) Ozone-induced dissociation: elucidation of double bond position within mass-selected lipid ions. *Anal. Chem.*, **80**, 303–311.

86 Malmberg, P., Borner, K., Chen, Y., Friberg, P., Hagenhoff, B., Mansson, J.E., and Nygren, H. (2007) Localization of lipids in the aortic wall with imaging TOF-SIMS. *Biochim. Biophys. Acta*, **1771**, 185–195.

14
Lipid Metabolism in Neurodegenerative Diseases

Lynette Lim, Guanghou Shui, and Markus R. Wenk

14.1
Introduction

Neurodegeneration is a broad term for diseases that encompass progressive loss of neuronal structure and function, eventually leading to cell death. The most common types of neurodegenerative diseases are Alzheimer's (AD), Parkinson's (PD), and Huntington's (HD). A different brain region is affected in each of these diseases resulting in distinct clinical symptoms. Genetically, with the exception of HD, most incidents of neurodegenerative diseases have unknown etiology. In both PD and AD, genetically linked mutations represent only about 5–20% of the total cases. Yet, these rare genetic mutations have led to many insights into the underlying molecular events preceding disease. Interestingly, despite the strikingly different clinical features of PD, HD, and AD, at the cellular and molecular levels, these diseases share many similarities, including the synaptic dysfunction [1], axonal transport deficit [2], deranged calcium signaling [3] and mitochondrial functions [4], and amyloidogenic protein self-aggregation [5], all of which precede the actual loss of neurons.

Lipids account for about 50% of the brain's dry weight. One of their main functions has been attributed to electrical insulation in myelin [6]. However, it is now clear that lipids play many other roles in brain function, including modulation of cellular calcium (Ca^{2+}) signaling and other signal transduction cascades and targeting of proteins to membranes [7–10]. Thus, alterations in the levels and distribution of various classes of lipids could, therefore, influence biological membrane properties such as membrane fluidity, the clustering of certain receptors, and so on. Ultimately, this could lead to changes in synaptic fidelity [11]. Thus, it is not too surprising that alterations in lipid metabolism have been linked to various neurodegenerative diseases.

In the context of AD, which is characterized by a progressive loss of memory and cognition, alterations in both sterol and glycerophospholipid (GPL) metabolism have been reported. Brains from AD patients are marked with senile plaques of

Lipidomics, First Edition. Edited by Kim Ekroos.

beta-amyloid peptides (Aβ), neurofibrillary tangles (NFT), and lipid aggregates. Loss of synaptic function precedes neuronal death. Interestingly, the loss of pertinent membrane lipid composition and architecture appears to be an early metabolic event along with the loss of synapses and the formation of lipid–protein aggregates/plaques.

PD, on the other hand, is classified as a movement disorder and marked by the selective degeneration of dopaminergic neurons in the nigrostriatal pathway [12–14]. Compared to AD, apparent links of (aberrant) lipid metabolism with onset of PD are less obvious. In clinical studies, low-density lipoprotein cholesterol (LDL-C), associated with higher risk of PD [13, 14], and higher serum levels of total cholesterol are associated with a significantly decreased risk of Parkinson's disease [15]. In the brains of post-mortem PD patients, elevated levels of polyunsaturated fatty acids (PUFA) were detected [16], suggesting misregulation of lipid could play a role in the disease.

Synaptic defects, misregulation of autophagy or mitophagy, protein misfolding, and formation of protein–lipid aggregates appear to be common events in all three diseases mentioned above. In HD, the mutated trinucleotide expansion of the HTT gene (mHTT) can self-aggregate to form a nuclear inclusion. In PD, Lewy bodies are formed by the protein α-synuclein. Beta-amyloid plaques are formed by self-association of Aβ. More intriguingly, the monomeric form of all three peptides, mHTT, Aβ, and α-synuclein, is intrinsically unstructured, whereby the secondary structures are unlikely to spontaneously fold into well-organized globular structures [17]. In the case of both Aβ and α-synuclein, the influence of different types of lipids has been shown to accelerate or stabilize the formation of fibrils like beta-sheets [18–25].

Thus, could (subtle) changes in specific lipid classes or particular lipid species either directly or in concert with other changes and factors contribute to brain defects in mitochondrial dysfunction and synaptic transmission? Do alterations in cellular lipids promote aggregation of structures associated with neurodegenerative diseases?

In this chapter, we will first briefly highlight some of the major lipids found in brain and provide an overview of two general mass spectrometry-based ("targeted" and "nontargeted") approaches for their detection. We will discuss how misregulation in lipid metabolism could contribute to neurological disease (AD and PD). We hope to bring in some new perspectives on current challenges for lipidomics in these contexts. Emphasis will be placed on how complementary observations (in particular, genetics) could be used in combined approaches.

14.1.1
Brain Lipids

GPL, sphingolipids, and sterols comprise the bulk of lipid mass in the brain and in proportions comparable to those found in mammalian cell membranes. This distribution varies somewhat between nerve cell bodies (gray matter), axons (white

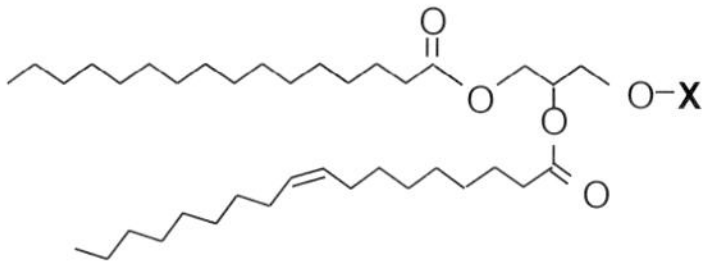

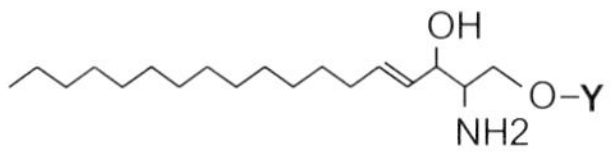

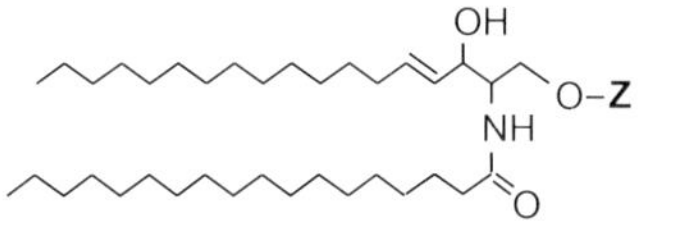

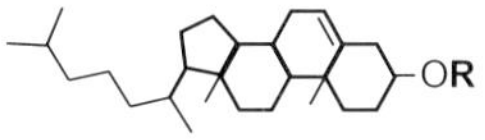

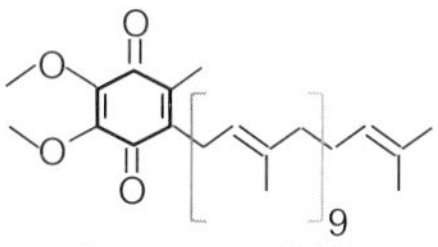

	Head group	Lipid	Mass spectrometry
X:	P (Phosphate)	PA	ESI-
	P-choline	PC	ESI+/-
	P-ethanolamine	PE	ESI-/+
	P-serine	PS	ESI-
	P-inositol	PI	ESI-
	P-glycerol	PG	ESI-
	P-PG	CL	ESI-
	Fatty acyl	TAG	ESI+; APCI+
	Hydrogen	DAG	ESI+; APCI+
Y:	Hydrogen	Sph	ESI+
	P (Phosphate)	S1P	ESI-/+
Z:	Hydrogen	Cer	ESI-/+
	P (Phosphate)	Cer-1-P	ESI-
	P-ethanolamine	PE-Cer	ESI-/+
	P-choline	SM	ESI+/-
	Glucose	GluCer	ESI-/+
	Galactose	GalCer	ESI+/-
	Lactose	LacCer	ESI+
	Sulfated galactoside	ST	ESI-
	aNeu5Ac(2-3)bDGalp (1-4)bDGlcp(1-1)	GM3	ESI-
R:	Hydrogen	Cho	APCI+
	Fatty acyl	CE	ESI+; APCI+
	Sulfate	CS	ESI-
	Other modifications	Oxysterols	APCI+; GCMS
	Prenol lipids	Quinones	APCI+
	Examples	Isoprenoids, etc	

Figure 14.1 Molecular structures of some of the lipids discussed in this chapter.

matter), and myelin (Figure 14.1), in particular with respect to sphingomyelin that is enriched in myelin [26].

Free cholesterol (FC, i.e., nonesterified) is an essential structural component of the plasma membrane of a cell. In humans, the average level of free cholesterol in the central nervous system (CNS) is higher than in any other tissue. The cholesterol required for growth and various CNS functions comes from *de novo* synthesis rather than from the diet and transport via the blood and blood–brain barrier. Regulation of the sterol excretory process is very important to maintain cholesterol homeostasis when the rate of synthesis exceeds the need for new structural sterols.

GPL are key components of cellular membranes and the major classes in brain include phosphatidylcholine (PC), phosphatidylethanolamine (PE),

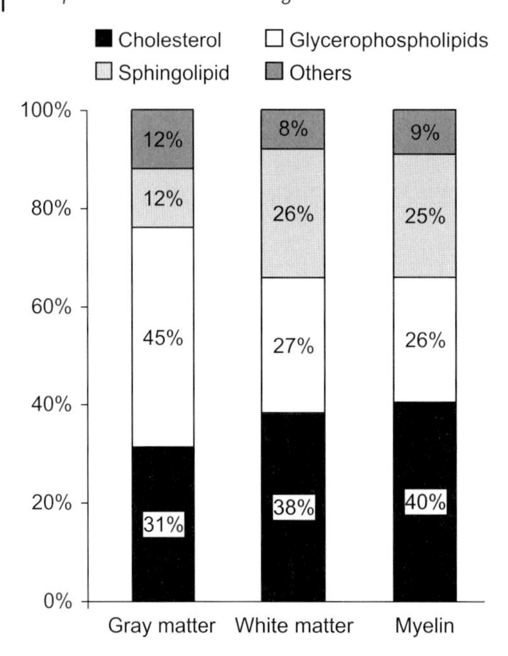

Figure 14.2 Distribution of various classes of lipids in gray matter, white matter, and myelin. Graphical plot of the percentage of the major lipid classes in gray matter, white matter, and myelin from the frontal lobes of human brains. Values were calculated using reported data from Ref. [26].

phosphatidylinositol (PI), phosphatidylserine (PS), as well as phosphatidic acid (PA), phosphatidylglycerol (PG), and cardiolipin (CL) (Figure 14.2). Brain is particularly rich in phosphoinositides (phosphorylated derivatives of PI) that play important roles in signaling and have been reviewed extensively [27, 28].

Sphingolipids are a complex family sharing a sphingoid base backbone that is synthesized *de novo* primarily from serine and a long-chain (C16) fatty acyl-CoA. Sphingolipids play important roles in various cellular functions including growth arrest, apoptosis, proliferation, differentiation, and cell recognition. The main composition of human brain sphingolipids are sphingomyelins (SM), ceramides (Cer), and glycosylated forms of ceramides (glucosylceramide (GluCer), galactosylceramide (GalCer), and gangliosides GM3, GM2, GM1, GD1, GD2, and GD3). In addition to three major classes of lipids mentioned above, brain tissue contains other, less abundant lipids, such as prenol lipids and fatty acyl-derived mediators.

14.1.2
Mass Spectrometry of Brain Lipids

The workflow for analyzing brain tissue lipids is illustrated in Figure 14.3. Brain is relatively a soft tissue (as opposed to muscle, for example), thus

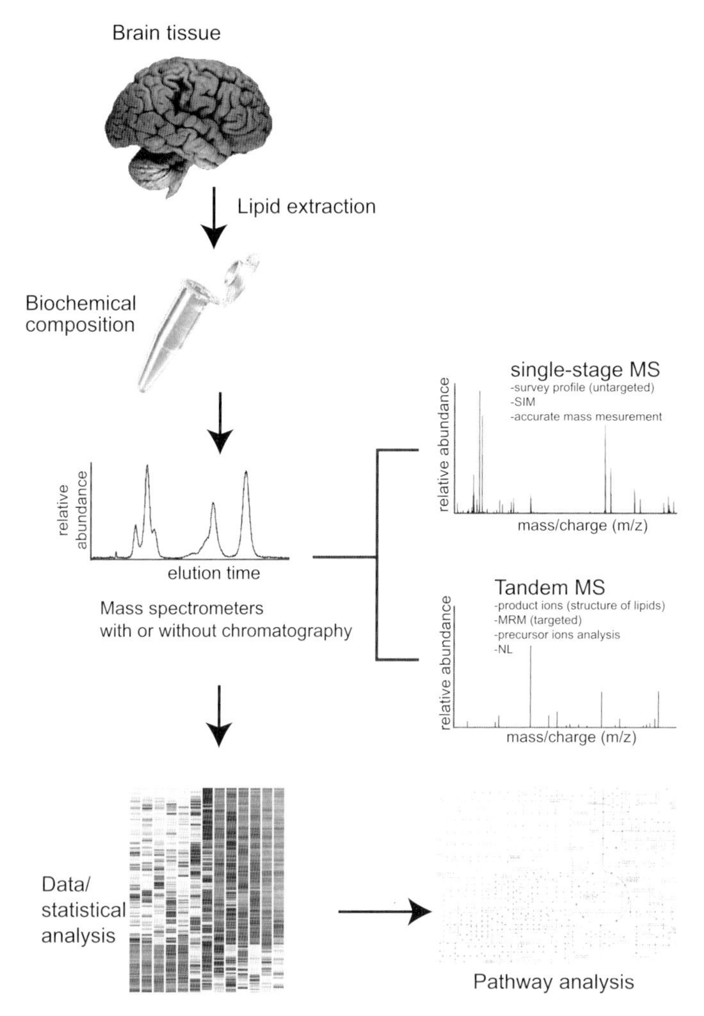

Figure 14.3 Generalized workflow for mass spectrometric measurement of lipids in extracts from brain.

simplifying homogenization that is either done in frozen form (powder formation under liquid nitrogen or, less optimal, on ice). Selection of a robust and efficient extraction protocol for quantitative recovery of lipids from the tissue is in any case the crucial (and often ignored) first step. Most widely used extraction protocols, which are either based on or modified from Bligh and Dyer or Folch methods, have proven to be reliable for qualitative lipid analysis and comparative studies. Typically, crude lipid extracts are directly introduced into a mass spectrometer without (shotgun lipidomics) or with prior chromatographic separation (LC-MS). Shotgun approaches have the advantage of simple experimental setup and high capacity (high-throughput analysis). Coupling chromatography to MS significantly reduces ion suppression and enables detection of minor lipid species.

Lipids are preferably detected using mass spectrometers with soft ionization techniques such as electrospray ionization (ESI) and atmospheric pressure chemical ionization (APCI). Most phospholipid and sphingolipids are well ionized in ESI modes, while cholesterol and coenzyme Q9 are well ionized in APCI mode. ESI/MS has been widely used to measure lipids in various applications. While many cellular phospholipids (PS, PA, PG, PI, and PE) are effectively ionized and detected by ESI-MS in the negative ionization mode, PC and SM are more sensitively detected in positive ionization mode. Both phospholipids and sphingolipids are effectively ionized and detected by ESI-MS in the negative ionization mode, with or without a chemical modifier or signal enhancer. Thus, negative ESI/MS could be directly used to obtain full-scan MS spectra of the crude lipid extracts for nontargeted lipidomics. For instance, an MS scan ranging from 400 to 1200 amu will include various polar lipids such as lysophospholipids, phospholipids (PC, PE, PI, PS, PG, and PA), and sphingolipids. Furthermore, other abundant sphingolipids such as sulfatides and GM3 are measured in negative ESI mode. Minor neutral lipid species, such as TAG and cholesteryl ester species, can be monitored as adduct ions in positive ESI mode [29, 30].

Present lipidomic strategies can be classified into two broad categories according to the nature of the MS data collected, (1) nontargeted approaches and (2) full MS scan mode, although alternative naming according to the particular single or tandem MS operation is equally valid and often more precise in describing the technical details of the approaches taken (Figure 14.3). For simplicity, we shall use "nontargeted" for single-stage MS and "targeted" for tandem MS based on multiple reaction monitoring, MRM, approaches. Nontargeted lipidomic approaches aim to cover the lipidome as broadly as possible, while targeted approaches focus only on lipid species of interest. Nontargeted approaches are important tools in discovering novel lipids or unexpected lipid metabolites between paired sample sets. Targeted approaches, on the other hand, are usually more specific, sensitive, and confer relative ease in terms of data processing. Inevitably, loss of information due to the restrictions imposed by the targeting lists of lipids to be measured is a limitation.

The selection of a suitable combination of mass analyzers (quadrupole, time of flight, ion trap, Orbitrap) represents the next critical step in order to ensure an effective analytical scheme for both qualitative investigation (i.e., mass detection based on m/z values) and quantitative (i.e., measuring the intensity of an ion of interest) analyses. In the case of nontargeted lipidomic approaches, a high-resolution (TOF or Orbitrap) mass analyzer is desirable to provide accurate mass data (m/z values) at sufficiently high spectral resolution. In targeted lipidomic approaches, quadrupole mass analyzers are popular due to the relative ease of operation (e.g., both MRM and SIM, precursor ion, and neutral loss scans fall into this category to a certain degree).

14.2 Alzheimer's Disease

AD, first described by the German psychiatrist Alois Alzheimer in 1906, is now one of the most common forms of incurable neurodegenerative diseases. In a short paper, Alzheimer identified three main features in the brain of Mrs Deter, who suffered from advanced dementia in her 50s and died at the age of 56. These main features are (1) striking changes in neurofibrils (better known as neurofibrillary tangles, NFT), (2) plaques or foci visible without staining (beta-amyloid plaques), and (3) lipoid inclusion granules in glia [31–33] (Table 14.1). Since then researchers have made substantial progress toward biochemical understanding of these immunohistological features.

Table 14.1 Summary of lipids that have been implicated in Alzheimer's disease.

Lipid name	Common abbreviation	Role in disease	References
Arachidonic acid	AA	AA is increased upon Aβ production and elevated AMPAR and excitotoxcity	[34, 35]
Cholesterol and cholesteryl-ester	FC and CE	In AD models, hyperactive ACAT1 led to increased CE; ACAT1 inhibitor are protective in AD mouse models	[36–41]
Docosahexanoic acid/neuroprotectin D1	DHA/NPD1	DHA and NPD1 are survival factors in Aβ-induced toxicity. They promote neurite outgrowth and reduce Aβ-42 formation	[42–45]
Diacylglycerol	DAG	DAG increases α-secretase activity, thereby reducing Aβ formation	[46]
Phosphatidylinositol-4,5-bisphosphate	PI(4,5)P2	Reduction of PI(4,5)P2 is observed in AD models due to hyperactive PLC activity; increase in PI(4,5)P2 is protective in AD model	[27, 47, 48]
Plasmalogen phosphatidylethanolamine	pPE	Decreased levels of pPE were observed in gray matter of AD brain. pPE may act as a buffer for oxidative stress. AGPS, the enzyme, which catalyzes the formation of pPE, can be modulated by APP	[49–51]
Phosphatidylcholines	PC	Decreased levels of PC were observed in AD brain	[52]
Sulfatides		Depleted levels of sulfatide in brain and CSF of AD patients. Sulfatide levels are regulated by ApoE alleles	[53–55]

For example, the "changes in neurofibrils" are now considered hyperphosphorylation of the microtubule-associated protein tau. The "foci" or "plaque" are now "β-amyloid plaques" composed of beta-amyloid peptides of 36–43 amino acids. The precursor protein APP is cleaved by either beta (BACE1) or gamma-secretase (consisting of presenilin (PS1, nicastrin, APH-1, and PEN-2) to generate these beta-amyloid peptides of varying length. Various components of this pathway, such as APP and PS1, have been identified to cause early-onset Alzheimer's disease, possibly due to the changes in APP cleavage [56–58].

14.2.1 Cholesterol and Cholesterol Esters

However, the third feature, that is, the "lipoid inclusion," was largely ignored until the early 1990s when gene-mapping studies identified carriers of the apolipoprotein (APOE) e4 allele with increased risk for AD [59, 60]. Interestingly, the APOE e2 allele appears to be associated with lowered risk of the disease [60]. In addition, APOE e4 allele is the only known major genetic risk factor that accounts for 95% of both sporadic and familial form of late-onset AD [61]. In the nervous system, APOE is produced by astrocytes, which allows for the transport of cholesterol into neurons. While APOE is particularly important in the brain, other apolipoproteins are involved in sterol/lipid transport in the body periphery. Although correlations between hypercholesterolemia in mid-life as a risk factor for AD have been identified in epidemiological studies [62, 63], the precise details of how peripheral sterol/lipid levels affect the brain remain poorly understood.

APP and Aβ are found to be associated with cholesterol-rich microdomains. More specifically, Puglielli *et al.* showed that elevation of cholesteryl-esters (CE) but not free cholesterol results in increased Aβ production [36, 37]. There are (at least) two major interchangeable pools of cellular cholesterol, (1) FC in membranes and (2) CE in cytoplasmic lipid droplets. Acetyl-coenzyme A acetyltransferase (ACAT) regulates this dynamic equilibrium, and is thus likely to be an important function in the brain that does not rely on bulk fat storage in the form of lipid droplets.

Inhibition of ACAT that leads to reduced levels of CEs also lowers Aβ generation [36, 38–40, 64]. Furthermore, in fibroblasts from AD patients ACAT-1 mRNA levels are increased significantly. Recent genetic evidence associates a single nucleotide polymorphism (SNP, codon 405 isoleucine to valine (V405)) of the cholesterol ester transfer protein (CETP) with the lowered risk of dementia [65, 66]. Similar to what was first noted by Alzheimer, ultrastructural studies on autopsied brain tissue from Alzheimer's disease patients using immunocytochemistry with an antibody that recognizes amyloid-beta peptides revealed cytosolic clusters of lipid droplets in immunopositive areas and there are cytosolic granules [67]. One possibility is that sterols could modulate processing and/or accumulation of amyloid beta.

Supporting this view, cognitive defects seen in mice expressing the human form of the Swedish APP mutant (hAPPsw) are ameliorated in ACTA1 gene ablated background (ACAT1−/−, [41]. In addition, ACAT1 inhibitor CP-113818 [68] and CI-1011 [69] showed some effectiveness in preclinical models. ACAT1−/− animals also exhibit an increase in 24-hydroxycholesterol in the endoplasmic reticulum and decreased rate of sterol synthesis in the brain [41], thus an overall effect on cholesterol metabolism. Indeed, it has been widely proposed to test the effects of statins as potential therapeutics for AD treatment. Statins, which inhibit HMG-CoA reductase, have been used in clinical trials for AD with inconsistent results within the prospective cohorts. However, it remains to be seen how effectively statins actually get transported to the brain and influence sterol biosynthesis in this organ.

14.2.2 Sulfatides

Sulfatides, a class of sulfated galactosphingolipids, are found in high abundance in the CNS. They are synthesized by oligodendrocytes and misregulation in their distribution has been shown in AD brains. One of the first evidence of aberrant sphingolipid metabolism in AD was the observation that within degenerating neurons from both the cortex and the hippocampus of AD cases, there were high levels of immunoreactivity for a monoclonal antibody raised against gangliosides, A2B5 [70]. It was later shown that A2B5 also binds to sulfatides [71].

Using an unbiased lipidomics approach, Han *et al.* found that at very early stages of AD, sulfatides are substantially and specifically depleted both in brains and in cerebrospinal fluid (CSF) of individuals with AD [54, 55]. To identify the mechanism(s) of sulfatide loss concurrent with AD onset, the same group analyzed the sulfatide content in the cortex and hippocampus from animals either lacking ApoE or expressing the human ApoE4 allele. Interestingly, they found that the ApoE4-expressing mice had approximately 60% less sulfatides than those found in wild-type mice of the same age [53]. In contrast, the sulfatide levels in hippocampus and cortex of ApoE knockout mice were, respectively, 61 and 114% higher than in wild-type mice, suggesting that ApoE and the ApoE alleles play important role in regulating sulfatide levels. A separate study of whole-brain extracts from mice with various ApoE knock-in alleles did not find major changes in lipids [30], which may be attributed to masking effects (whole brain versus selected regions). Thus, one possible model proposed by Han is that ApoE plays a role in carrying sulfatides from oligodendrocytes (where they are synthesized) to neurons [72], similar to what has been shown for trafficking of cholesterol [73]. Factors that disrupt this process lead to alteration in sulfatide levels in the brain, serum, or CSF.

14.2.3 Plasmalogen Ethanolamines

Plasmalogen phosphatidyl ethanolamine (pPE) is one of the most abundant lipids in neuronal cell membranes, representing about 30% of total phospholipids [72]. Using unbiased approaches, a number of groups have showed that in post-mortem temporal cerebral and other gray matter regions, there is a significant reduction of pPE [72]. Interestingly, the magnitude of the deficiencies is correlated with severity of AD progression. In patients' serum, levels of pPE were observed to be significantly decreased in pathologically diagnosed AD subjects at all stages of dementia compared to age-matched controls. Again, the severity of this decrease correlated with the severity of dementia [49]. In disease models such as transgenic mice carrying the APP Swedish mutation, pPE is decreased in the cerebral cortices but not the cerebellum, suggesting that pPE decreased as a result of AD [51].

More recently, the intercellular domain of APP was found to decrease the expression of alkyl-dihydroxyacetonephosphate synthase (AGPS), the rate-limiting enzyme in pPE biosynthesis [50]. In addition to modulating AGPS levels, AD pathogenesis could affect pPE levels in another way. Since Aβ has been shown to induce oxidative stress and pPE lipids are sensitive to oxidization due to the vinyl ether bond [74], it is likely that Aβ induces decrease in pPE by oxidation of pPE. While this would suggest that pPE level decreases as a result of Aβ-mediated oxidative stress and regulation of AGPS, it remains unclear whether pPE decrease could influence or accelerate the progression of AD. Particularly, as the largest risk factor for AD is age, and levels of pPE decrease with age, it is possible that a reduction in plasmalogens could, in fact, influence AD progression.

14.2.4 Phospholipases

Phospholipases (C, D, A) as well as the levels of their substrates (GPLs) and products (PAs, DAGs, lyso-GPLs, and fatty acyls) have long been implicated with inflammation in the brain and AD.

14.2.4.1 Phospholipase A2

The role of phospholipase A2 (PLA2) in Aβ-dependent cognitive deficit and toxicity has been well studied, but this also revealed a lot of complexity. In general, PLA2 catalyzes the reaction of phospholipids to lysophospholipids and fatty acids. In humans, PLA2 can be classified into 12 different groups, with more than 19 different isoforms [75], though most of the studies on the central nervous system have focused on three groups, group IV cytosolic (cPLA2), group II secretory (sPLA), and group VI – Ca^{2+}-independent PLA2 (GVI-PLA2). cPLA2 is regulated by intracellular Ca^{2+}, found at high levels in the hippocampus, and has strongest substrate specificity for arachidonic acid (AA). GVI-PLA2 appears to not have particular substrate preference but is the determinant enzyme to control docosahexaenoic acid (DHA) in the brain [76].

Using a lipidomics approach to profile various fatty acids in brain tissue of an AD model transgenic mice, researchers have found an increase in arachidonic acids and its metabolites [35]. AA levels, the main products of cPLA2, have been implicated in various roles relevant to AD such as inflammatory response, synaptic transmission, and oxidative stress. Several studies have shown that there is likely a feedback loop between AA production and synaptic strength, long-term potentiation, as well as NMDAR and AMPAR levels in dendritic spines [34, 77, 78]. How AA inhibition mediates neuroprotection has not been completely elucidated, however.

Hypoactivity in GIV-PLA2 in the hippocampus impairs long- and short-term memory [79]. Moreover, in a cohort of AD patients, GIV-PLA2 activity was found to have decreased [80]. The human isoform for GIV-PLA2, PLA2G6, is also mutated in neurodegenerative disorders with high brain iron including AD, PD, and neuroaxonal dystrophies [81, 82]. Exactly why reduction of GIV-PLA2 activity and its substrate, DHA, levels are involved in AD is unclear. Unlike AA, which has been shown to be involved in synaptic functions, there is little evidence of the role of DHA in neural transmission. Most of the studies on DHA point toward its importance in brain development and neuronal survival [43]. Increasing DHA levels by modification of the diet modestly improves cognition in AD animal models [42]. Several epidemiological studies have reported that reduced levels of DHA are associated with higher risk of neurodegeneration. The mechanisms of these observations of neuroprotection are unclear. One model suggests that DHA functions as an antioxidant to counterbalance the inflammatory response during AD [44].

Evidence supporting this model comes from studies on the DHA-derived messenger, neuroprotectin D1 (NPD1). Using mass spectrometry to detect DHA and NPD1 in a variety of brain and retinal degenerative models, Bazan's group has found that NPD1 exerts protective effects and antiinflammatory bioactivity [83]. In cells overexpressing beta-amyloid precursor protein (βAPP), the NPD1 suppressed Aβ42 formation by downregulating BACE1 while activating the α-secretase. This is effective in changing the amyloidogenic pathway into a nonamyloidogenic, neurotrophic pathway [84]. NPD1 also mediates survival of retinal epithelial cells via the PI3K/Akt pathway [85]. Whether or not similar mechanisms are at work in the hippocampal or frontal cortex remains to be elucidated.

14.2.4.2 **Phospholipase C and Phospholipase D**

It has been shown that Aβ42 oligomers promote the hydrolysis of phosphatidylinositol-4,5-bisphosphate ($PI(4,5)P_2$) via the phospholipase C (PLC) pathway. Blocking this pathway has also been shown to protect against the synapse-impairing actions of Aβ [47]. Activation of PLC results in the production of diacylglycerol (DAG) and inositol-1,4,5-trisphosphate (IP3) (Figure 14.4). One hypothesis is that excessive IP3 increases release of Ca^{2+} that may induce synaptic dysfunction; thus, blocking this event would prevent synaptic loss. On the other hand, DAG, the membrane-bound product of PLC action, has been shown to increase alpha-secretase activity via protein kinase C (PKC) [46]. Increase in alpha-secretase could result in lowered levels of beta-amyloid peptides, which would expect to be protective. Interestingly, genetic ablation of synaptojanin 1 (SYNJ1), a phosphatase that converts PI(4,5)P2 to PI(4)P, increases PI(4,5)P2 levels and also protects against Aβ toxicity [47], suggesting that PI(4,5)P2 levels could also play an important in neuroprotection and that phospholipid metabolism is closely linked to susceptibility of Aβ insults.

Phospholipase D (PLD) catalyzes the reaction of phosphatidylcholine to phosphatidic acid (Figure 14.4). PA is a bioactive lipid and also represents a major cross-road in lipid biosynthesis. While PA can be quickly converted to DAG in a

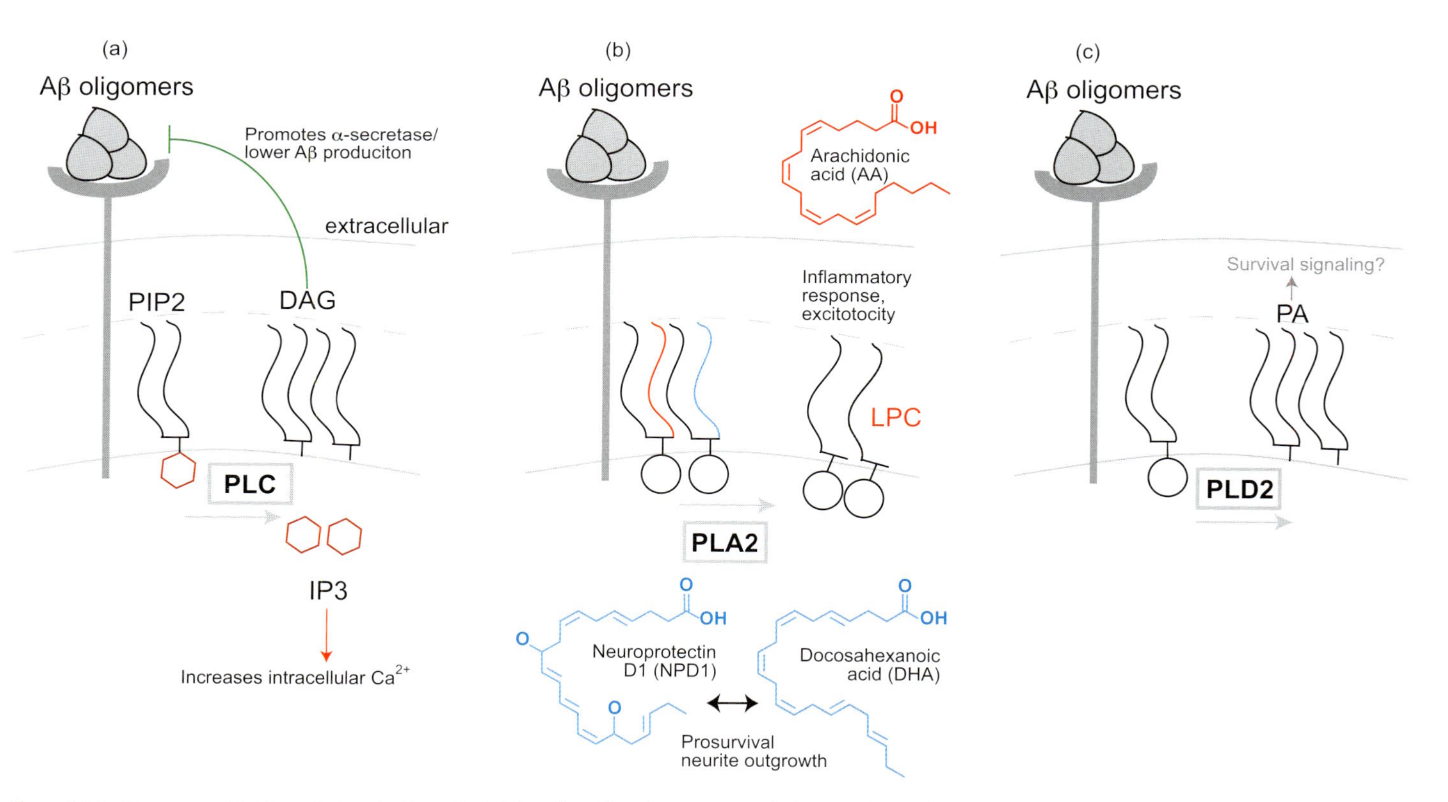

Figure 14.4 Summary of lipids and phospholipase in Aβ signaling. Signaling through Aβ creates various complex cascades that could be either protective or deleterious to neuronal survival. Figure 14.4a shows activation of PLC, which results in hydrolysis of PIP2 and generation of IP3 and DAG. While DAG can lower Aβ production, IP3 increases intracellular Ca^{2+} levels. Figure 14.4b shows activation of PLA2, which could result in hydrolysis of phosphatidylcholine (PC) to arachidonic or docosahexanoic acid (AA or DHA, respectively) and lysophosphatidylcholine (LPC). Both AA and LPC induce inflammatory response and excitotoxicity. However, DHA, which can be converted further to NPD1, elicits prosurvival signals. Figure 14.4c shows activation of PLD2, resulting in the hydrolysis of PC to PA and choline.

cell, there appear to be distinct "pools" of PA and DAG signals. An attractive model proposes fatty acyl tails to be distinguishing factors: DAG signaling consists of polyunsaturated fatty acids whereas PLD-derived PA is monounsaturated or saturated [86]. Mass spectrometry approaches have thus enabled the distinction not only of lipid classes but also of the saturation state. In AD mouse models, genetic ablation of PLD2 ameliorates cognitive deficits and synaptic dysfunction. Two PA species, 32:1 and 38:4, are consistently increased in the PLD2+/+ animals compared to the PLD2−/− [87]. In an independent study, it was found that Spo14 regulatory factor 1 (Srf1) activates PLD and this is essential to protect neural cells from sustained exposure to Aβ [88]. These results indicate that a fine balance between phosphatidylinositol-4,5-bisphosphate (PIP2)/DAG/PA is required, but how this is regulated at the molecular level of microdomains remains to be determined.

14.3 Parkinson's Disease

Parkinson's Disease is a movement disorder marked by selective degeneration of dopaminergic neurons in the nigrostriatum pathway [12]. It affects about 1% of the population over 60 years old and 4% of people over 80 years old. Named after the clinician who first described the disease in 1817, James Parkinson, this is at present the second most common age-related neurodegenerative disorder [89].

The clinical characterization of the disease is extremely accurate. With experienced clinicians, PD is diagnosed with 98% accuracy [90], compared to about 83% for AD [91]. Most patients with PD exhibit numerous deficits in movement with obvious symptoms such as rigidity or stiffness of limbs and/or neck, tremor, bradykinesia, and reduction of movement. Other less apparent symptoms include depression, dementia or confusion, uncontrolled drooling, speech impairment, swallowing difficulty, and constipation.

The main clinical features of PD are caused by the selective loss of dopaminergic neurons in the nigrostriatal pathway, which is also a hallmark of the disease. However, it is important to note that neuronal cell death is also detected in other regions of the brain such as the cerebellum and cortex. Furthermore, cytoplasmic aggregates called Lewy bodies, comprising alpha-synuclein, are present in brains of PD patients. Clearly, the most severely affected region is the nigrostriatal pathway, which regulates fine voluntary movements. Thus, the main motor deficits in PD patients are most likely attributed to this pathway, while the other less common and nonmotor symptoms such as confusion and dementia are most likely due to impairments in other brain regions.

Despite the clear clinical diagnosis, only 5–10% of PD cases are linked to known genes (Table 14.2). Furthermore, the pathological mechanisms responsible for onset and progression of the disease are still unknown. The consensus is that PD is a multifactorial disease, whereby both extrinsic factors, such as exposure to environmental toxins, and intrinsic factors, such as genetic

Table 14.2 Gene and locus linked to Parkinson's disease.

	Mode of inheritance	Age of onset	Locus/gene/protein	Role in pathogenesis and membrane function	References
PARK1/4	Dominant	40s	α-Synuclein	Knockout animals have defects in synaptic vesicle release	[12, 92, 93]
PARK2	Recessive	20s	Parkin	Transgenic animals with parkin truncated mutation showed loss of dopaminergic neurons at 16 mo	[94, 95]
PARK3	Dominant	60s	2p13		
PARK5	Dominant	50s	UCHL1	Knockout animals showed axonopathy	[12]
PARK6	Recessive	30s	PINK1	PINK1 interacts with E3 ligase and controls mitophagy	[96, 97]
PARK7	Recessive	30s	DJ-1	DJ-1 knockout animals have defects in calcium signaling and increased oxidative stress	[98]
PARK8	Dominant	Variable	LRRK2	Lrrk2 regulates synaptic vesicle endocytosis	[99, 100]
PARK9	Recessive	Variable	ATP13A2		[102]
PARK10	Recessive	Variable	1p32		
PARK11	Dominant	Variable	GIGYF2	GIgYF2 (+/−) has motor dysfuncitons	[103]
PARK12			Xq21-q25		
PARK13			HTRA2	Knockout mouse model of PARK13 showed abnormality of lower motor neuron function	[104]
PARK14			PLA2G6	The loss of function mutation has severe motor dysfunction and models infantile neuroaxonal dystrophy	[105]
PARK15	Recessive		FBXO7		[84]
PARK16			1q32		

background, are important for pathogenesis. More importantly, there is a growing list of components associated with PD etiology such as oxidative stress, ER protein misfolding, mitochondria dysfunction, transcription factor changes, epigenetic changes, calcium toxicity, and cholesterol misregulation. It is not known if all these factors are related or are independent factors for disease progression.

14.3.1 Cerebrosides

Unlike AD, for which a dominant risk factor (APOE) was identified a few decades ago, there has been much less evidence for a direct and specific implication of lipids in PD. This started to change around 2004 when the first clinical study identified an association of parkinsonism with mutation in the glucocerebrosidase gene (GBA) in Ashkenazi Jews [106]. Many other studies on different patients cohorts also found similar associations (see for review Ref. [107]). To date, GBA variants are the most common risk factor associated with parkinsonism, and PD patients are five times more likely to carry a GBA mutation than control individuals [108].

While the clinical link of GBA and PD has been well established, less is known about the underlying mechanisms. Along with GBA–PD linkage studies, a number of groups have observed neuroprotective effects of gangliosides in models for PD [109, 110]. Short-term clinical trials have also demonstrated some benefits to neurological functions in PD patients [109, 111, 112]. Studies on an animal cell line with inhibited glycosyl-ceramide synthase led to various features of lysosomal pathology, including compromised lysosomal activity, enhanced lysosomal membrane permeabilization, and increased cytotoxicity [113], suggesting a connection between ceramide metabolism and lysosomal functions. Glucosylceramide (GluCer), the substrate of glucocerebrosidase, stabilizes oligomeric forms of α-syn and enhances amyloid formation. At the same time, α-syn species lead to depletion of lysosomes. Loss of glucocerebrosidase activity causes accumulation of α-synuclein but not that of tau, another aggregation-prone protein [114].

Studies in GBA models suggest that α-syn and lipids can coregulate each other. However, lipidomic data to support these findings remain rather limiting. For example, untargeted lipidomics in transgenic mice with human α-syn or deleted endogenous α-syn reveal differences between ages and gender and smaller changes associated with α-syn genotypes [115], thus providing little guidance on how changes in lipids may affect progression of PD. Recently, a multicenter study presented lipidomic analyses of human PD visual cortex (a region devoid of obvious pathology), amygdala and anterior cingulated cortex (regions with Lewy bodies but lacking degeneration). No significant differences were detected between the amygdala and the anterior cingulated cortex of controls and patients. In contrast, levels of approximately 70 lipids were different in the visual cortex of controls versus patients [29].

Table 14.3 Summary of lipids that exacerbate or ameliorate degeneration in Parkinson's disease.

Lipid name	Abbreviation	Role/mechanism in neuroprotection	References
Anandamide		Anandamide levels are increased in PD patients' CSF. Exogenous anadamide activates the indirect pathway in nigrostriatal synapse	[118, 119]
Coenzyme Q	CoQ	CoQ is protective in mouse/primate model of PD	[120–124]
Glucosylceramide and ganglioside	GluCer	GluCer stabilizes oligomeric form of α-syn. Increased GluCer is observed in α-syn PD mouse model. Gangliosides are neuroprotective in PD models	[111, 112, 114]
Lanosterol		Lanosterol levels are decreased in mouse model of PD. Exogenous lanosterol uncouples neuronal mitochondria	[125]
Methylene blue	MB	MB reroutes electrons and bypasses complex I, II, III, and IV, similar to uncouplers	[126]

14.3.2
Coenzyme Q

In the list of rare genetic PD cases, many genes point toward mitochondrial dysfunction (Table 14.3). In addition, in patients with idiopathic PD, the catalytic activity of brain mitochondrial complex I is compromised [116]. A number of environmental toxins that directly affect mitochondrial function induce parkinsonism. Perhaps, the best example of toxin-induced PD is MPTP (or its active metabolite MPP+). MPP+ selectively enters into dopaminergic neurons via the dopamine transporter (DAT), inhibits complex I of mitochondria, and thereby induces clinical symptoms that are reminiscent of those found in patients with PD [117].

Because MPTP/MPP+ toxicity emulates PD symptoms, it has been widely used in animal and cellular models to study neuronal cell death and to screen for neuroprotective agents. Many of the identified neuroprotective agents are found to modulate mitochondrial function, including L-carnitine, creatine, and the endogenous lipid electron carrier, coenzyme Q10 (CoQ) [127]. In MPTP rodent and primate animal models and some cohorts of patients, the supplementation of CoQ is partially protective or beneficial [120–124]. Similar to CoQ, methylene blue, an FDA-approved drug and an antidote to cyanide poisoning, is neuroprotective in PD models and acts by rerouting electrons similar to known uncouplers [126].

Understanding the modulation of mitochondrial redox states has gained a lot of interest in the context of PD research in recent years. Uncoupling proteins (UCPs), which depend on fatty acids, are neuroprotective in the MPTP model of PD [128–130], and their expression is downregulated in mice lacking DJ-1, a gene linked to early onset of PD [98]. Using a targeted approach based on GC-MS for measurement of cholesterol (and some of its precursors and oxidized derivatives), we discovered that lanosterol is substantially reduced in the affected brain regions of mice treated with MPTP. We further uncovered a novel role of lanosterol in neuroprotection and showed that it induces uncoupling, thus providing a link between sterol biosynthesis and mitochondrial function [125].

14.3.3
Endocannabinoids

Another lipid class that can ameliorate movement deficits in PD is that of endocannabinoids. Using a targeted approach (via GC-MS), Giuffrida *et al.* found that motor activity stimulated the release of anandamide but no other species of endogenous cannabinoids in the brain [131]. This discovery became an important starting point for subsequent studies that link PD to anandamide signaling. In a rodent model of Parkinson's disease (induced by unilateral lesion with 6-hydroxydopamine), the striatal levels of anandamide, but not that of the other endocannabinoid, were affected [132]. Anandamides were also detected at higher levels in the MPTP mouse model than in control animals. Finally, anandamide is twice as abundant in the CSF of PD patients than it is in age-matched controls [119], though levels do not seem to correlate with disease severity.

One question arising from these observations is whether an increase in anandamide is part of the pathology of PD or a protective mechanism upon loss of dopaminergic neurons. The answer to this question came only after several studies elucidated the role of endocannabinoids in the basal ganglia circuitry. Fine movement in the mammalian system depends largely on basal ganglia, which integrates inputs from dopaminergic, cortical, and thalamic neurons via the striatum (Figure 14.5). Striatal projection signals to the thalamic neurons via a direct or indirect pathway. Imbalances by hypoactivity in the direct and overactivity in the indirect pathways have been proposed to underlie the motor deficits in Parkinson's disease [133]. In models of Parkinson's disease, indeed, the indirect pathway endocannabinoid-mediated depression is absent causing the overactivity, which can be rescued by inhibitors of endocannabinoid degradation. Administration of these drugs *in vivo* reduces parkinsonian motor deficits, suggesting that endocannabinoid-mediated activation of the indirect pathway synapses has a critical role in the control of movement [118].

While it is quite clear that endocannabinoid signaling is linked to neurotransmission in the basal ganglia, it is important to note that each species of

(a)

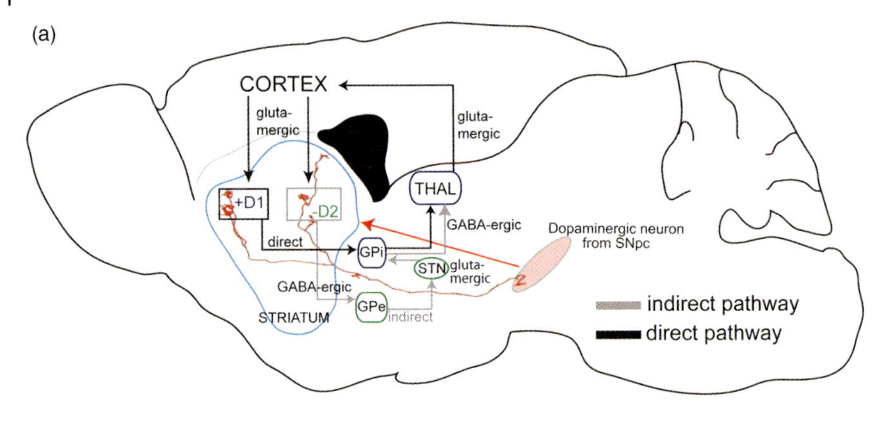
CORTEX
gluta-
mergic
gluta-
mergic
THAL
+D1
-D2
GABA-ergic
Dopaminergic neuron
from SNpc
direct
GPi
STN
gluta-
mergic
GABA-ergic
GPe
indirect
STRIATUM
indirect pathway
direct pathway

(b)

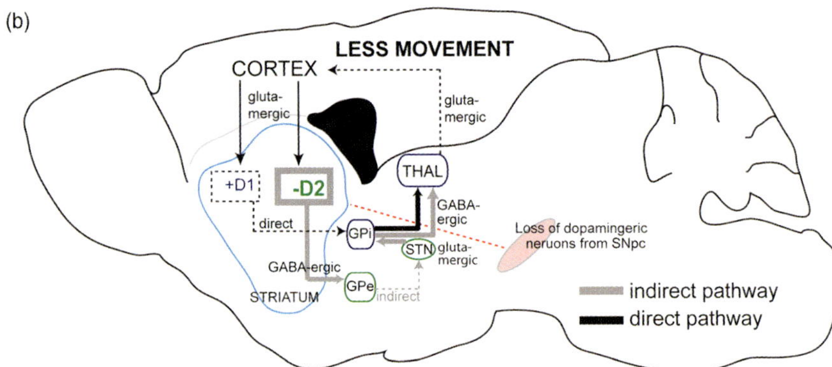
LESS MOVEMENT
CORTEX
gluta-
mergic
gluta-
mergic
THAL
+D1
-D2
GABA-
ergic
direct
GPi
Loss of dopamingeric
neruons from SNpc
STN
gluta-
mergic
GABA-ergic
GPe
indirect
STRIATUM
indirect pathway
direct pathway

(c)

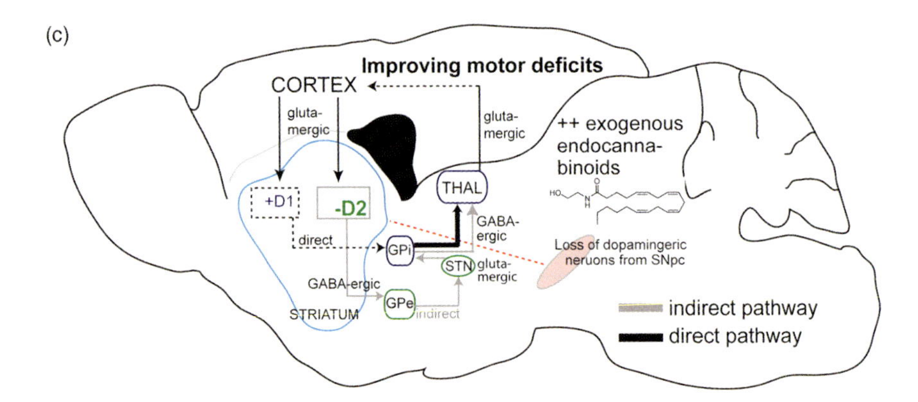
Improving motor deficits
CORTEX
gluta-
mergic
gluta-
mergic
++ exogenous
endocanna-
binoids
THAL
+D1
-D2
GABA-
ergic
direct
GPi
Loss of dopamingeric
neruons from SNpc
STN
gluta-
mergic
GABA-ergic
GPe
indirect
STRIATUM
indirect pathway
direct pathway

endocannabinoid may play different physiological roles in different neuronal types. For example, reuptake of GABA in striatal neurons is modulated by 2-arachidonoyl-glycerol (2-AG) but not anandamides [134]. In addition, exogenous anandamide produces either inhibitory or excitatory effects on subthalmic nucleus (STN) neurons depending on its localization in the STN [135]. Thus, the challenge for neuronal lipidomics is to understand cell type-specific lipid classes. It is still unclear if there is a class of neurons that synthesize and secrete endocannabinoid analogous to neurons that secrete and produce dopamine. Future advances in mass spectrometry with single-cell resolutions could answer some of these questions.

14.4 Conclusions

Mass spectrometry of lipids is an outstanding new tool for lipid research. It is most promising when combined with complementary approaches. In this short chapter, we have provided examples on how such integration could be guided in the case of two major neurodegenerative diseases.

Acknowledgments

Work in our laboratory is supported by the National University of Singapore and by grants from the Singapore National Research Foundation under CRP Award No. 2007-04 and from the Biomedical Research Council of Singapore (R-183-000-234-305).

Figure 14.5 The nigrostriatal pathway in healthy and disease states. Graphical drawing of the sagittal plane of a rodent brain (left represents anterior and right represents posterior). (a) In a healthy state, dopaminergic neurons in the substantial nigra par compacta (SNpc) send excitatory and inhibitory signals to two classes of GABAergic neurons of the caudate putamen (striatum), consisting of D1 and D2 receptors, respectively. In the direct pathway, D1 GABAergic neurons synapse onto GABAergic neurons of the globus pallidus internal segment (GPi). GPi GABAergic neurons synapse onto gluatmergic neurons of the thalamus (THAL), sending signal to the motor cortex. In the indirect pathway, D2 GABAergic neurons synapse onto GABAergic neurons of the globus pallidus external segment (GPe). GPe GABAergic neurons synapse onto glutamatergic neurons of the subthalmic nucleus (STN). From STN, glutamatergic neurons synapse onto GPi GABAerigc neurons. The balance between this direct and indirect pathway results in fine motor coordination such as speech. (b) In PD, upon the loss of dopaminergic neurons, D1 (direct) pathway is less active wheras D2 (indirect) is hyperactive. The final result is overinhibition of thalamic neurons, reducing glutaminergic synapses from the thalamus to motor cortex. Movement deficits are thus in patients with defects in the nigrostriatal circuit. (c) In PD model, upon loss of dopaminergic neurons, exogenous addition of endocannbinoids causes partial defects in the indirect pathway and improves motor deficits.

References

1 Nimmrich, V. and Ebert, U. (2009) Is Alzheimer's disease a result of presynaptic failure? Synaptic dysfunctions induced by oligomeric beta-amyloid. *Rev. Neurosci.*, **20**, 1–12.

2 Hirokawa, N., Niwa, S., and Tanaka, Y. (2011) Molecular motors in neurons: transport mechanisms and roles in brain function, development, and disease. *Neuron*, **68**, 610–638.

3 Bezprozvanny, I. and Mattson, M.P. (2008) Neuronal calcium mishandling and the pathogenesis of Alzheimer's disease. *Trends Neurosci.*, **31**, 454–463.

4 Vives-Bauza, C. and Przedborski, S. (2011) Mitophagy: the latest problem for Parkinson's disease. *Trends Mol. Med.*, **17**, 158–165.

5 Goedert, M. and Spillantini, M.G. (2006) A century of Alzheimer's disease. *Science*, **314**, 777–781.

6 Piomelli, D., Astarita, G., and Rapaka, R. (2007) A neuroscientist's guide to lipidomics. *Nat. Rev. Neurosci.*, **8**, 743–754.

7 DiNitto, J.P., Cronin, T.C., and Lambright, D.G. (2003) Membrane recognition and targeting by lipid-binding domains. *Sci. STKE*, **2003**, re16.

8 Fukuda, M., Kojima, T., and Mikoshiba, K. (1996) Phospholipid composition dependence of Ca^{2+}-dependent phospholipid binding to the C2A domain of synaptotagmin IV. *J. Biol. Chem.*, **271**, 8430–8434.

9 Fukuda, M., Kojima, T., and Mikoshiba, K. (1997) Regulation by bivalent cations of phospholipid binding to the C2A domain of synaptotagmin III. *Biochem. J.*, **323** (Pt 2), 421–425.

10 Wenk, M.R., Lucast, L., Di Paolo, G., Romanelli, A.J., Suchy, S.F., Nussbaum, R.L., Cline, G.W., Shulman, G.I., McMurray, W., and De Camilli, P. (2003) Phosphoinositide profiling in complex lipid mixtures using electrospray ionization mass spectrometry. *Nat. Biotechnol.*, **21**, 813–817.

11 Lim, L. and Wenk, M. (2009) Neuronal membrane lipids: their role in the synaptic vesicle cycle, in *Handbook of Neurochemistry and Molecular Neurobiology*, 3rd edn, Springer Science, New York, pp. 224–238.

12 Dauer, W. and Przedborski, S. (2003) Parkinson's disease: mechanisms and models. *Neuron*, **39**, 889–909.

13 Huang, X., Abbott, R.D., Petrovitch, H., Mailman, R.B., and Ross, G.W. (2008) Low LDL cholesterol and increased risk of Parkinson's disease: prospective results from Honolulu-Asia Aging Study. *Mov. Disord.*, **23**, 1013–1018.

14 Huang, X., Chen, H., Miller, W.C., Mailman, R.B., Woodard, J.L., Chen, P.C., Xiang, D., Murrow, R.W., Wang, Y. Z., and Poole, C. (2007) Lower low-density lipoprotein cholesterol levels are associated with Parkinson's disease. *Mov. Disord.*, **22**, 377–381.

15 de Lau, L.M., Koudstaal, P.J., Hofman, A., and Breteler, M.M. (2006) Serum cholesterol levels and the risk of Parkinson's disease. *Am. J. Epidemiol.*, **164**, 998–1002.

16 Sharon, R., Bar-Joseph, I., Mirick, G.E., Serhan, C.N., and Selkoe, D.J. (2003) Altered fatty acid composition of dopaminergic neurons expressing alpha-synuclein and human brains with alpha-synucleinopathies. *J. Biol. Chem.*, **278**, 49874–49881.

17 Dyson, H.J. and Wright, P.E. (2005) Intrinsically unstructured proteins and their functions. *Nat. Rev. Mol. Cell Biol.*, **6**, 197–208.

18 Koob, A.O., Ubhi, K., Paulsson, J.F., Kelly, J., Rockenstein, E., Mante, M., Adame, A., and Masliah, E. (2010) Lovastatin ameliorates alpha-synuclein accumulation and oxidation in transgenic mouse models of alpha-synucleinopathies. *Exp. Neurol.*, **221**, 267–274.

19 Liu, J.P., Tang, Y., Zhou, S., Toh, B.H., McLean, C., and Li, H. (2010) Cholesterol involvement in the pathogenesis of neurodegenerative diseases. *Mol. Cell Neurosci.*, **43**, 33–42.

20 Madine, J., Doig, A.J., and Middleton, D. A. (2006) A study of the regional effects of alpha-synuclein on the organization

and stability of phospholipid bilayers. *Biochemistry*, **45**, 5783–5792.

21 Naito, A. and Kawamura, I. (2007) Solid-state NMR as a method to reveal structure and membrane-interaction of amyloidogenic proteins and peptides. *Biochim. Biophys. Acta*, **1768**, 1900–1912.

22 Pfefferkorn, C.M. and Lee, J.C. (2011) Tryptophan probes at the alpha-synuclein and membrane interface. *J. Phys. Chem.*, **114**, 4615–4622.

23 Rantham Prabhakara, J.P., Feist, G., Thomasson, S., Thompson, A., Schommer, E., and Ghribi, O. (2008) Differential effects of 24-hydroxycholesterol and 27-hydroxycholesterol on tyrosine hydroxylase and alpha-synuclein in human neuroblastoma SH-SY5Y cells. *J. Neurochem.*, **107**, 1722–1729.

24 Ryan, T.M., Griffin, M.D., Teoh, C.L., Ooi, J., and Howlett, G.J. (2011) High-affinity amphipathic modulators of amyloid fibril nucleation and elongation. *J. Mol. Biol.*, **406**, 416–429.

25 Sheikh, A.M. and Nagai, A. (2011) Lysophosphatidylcholine modulates fibril formation of amyloid beta peptide. *FEBS J.*, **278**, 634–642.

26 O'Brien, J.S. and Sampson, E.L. (1965) Fatty acid and fatty aldehyde composition of the major brain lipids in normal human gray matter, white matter, and myelin. *J. Lipid Res.*, **6**, 545–551.

27 Di Paolo, G. and De Camilli, P. (2006) Phosphoinositides in cell regulation and membrane dynamics. *Nature*, **443**, 651–657.

28 Wenk, M.R. (2005) The emerging field of lipidomics. *Nat. Rev. Drug Discov.*, **4**, 594–610.

29 Cheng, D., Jenner, A.M., Shui, G., Cheong, W.F., Mitchell, T.W., Nealon, J.R., Kim, W.S., McCann, H., Wenk, M. R., Halliday, G.M., and Garner, B. (2011) Lipid pathway alterations in Parkinson's disease primary visual cortex. *PLoS One*, **6**, e17299.

30 Sharman, M.J., Shui, G., Fernandis, A. Z., Lim, W.L., Berger, T., Hone, E., Taddei, K., Martins, I.J., Ghiso, J., Buxbaum, J.D., Gandy, S., Wenk, M.R., and Martins, R.N. (2010) Profiling brain and plasma lipids in human APOE epsilon2, epsilon3, and epsilon4 knock-in mice using electrospray ionization mass spectrometry. *J. Alzheimer's Dis.*, **20**, 105–111.

31 Foley, P. (2010) Lipids in Alzheimer's disease: a century-old story. *Biochim. Biophys. Acta*, **1801**, 750–753.

32 Graeber, M.B., Kosel, S., Egensperger, R., Banati, R.B., Muller, U., Bise, K., Hoff, P., Moller, H.J., Fujisawa, K., and Mehraein, P. (1997) Rediscovery of the case described by Alois Alzheimer in 1911: historical, histological and molecular genetic analysis. *Neurogenetics*, **1**, 73–80.

33 Graeber, M.B. and Mehraein, P. (1999) Reanalysis of the first case of Alzheimer's disease. *Eur. Arch. Psy. Clin. N.*, **249** (Suppl. 3), 10–13.

34 Nishikawa, T., Tomori, Y., Yamashita, S., and Shimizu, S. (1989) Inhibition of Na^+,K^+-ATPase activity by phospholipase A2 and several lysophospholipids: possible role of phospholipase A2 in noradrenaline release from cerebral cortical synaptosomes. *J. Pharm. Pharmacol.*, **41**, 450–458.

35 Sanchez-Mejia, R.O., Newman, J.W., Toh, S., Yu, G.Q., Zhou, Y., Halabisky, B., Cisse, M., Scearce-Levie, K., Cheng, I.H., Gan, L., Palop, J.J., Bonventre, J.V., and Mucke, L. (2008) Phospholipase A2 reduction ameliorates cognitive deficits in a mouse model of Alzheimer's disease. *Nat. Neurosci.*, **11**, 1311–1318.

36 Puglielli, L., Konopka, G., Pack-Chung, E., Ingano, L.A., Berezovska, O., Hyman, B.T., Chang, T.Y., Tanzi, R.E., and Kovacs, D.M. (2001) Acyl-coenzyme A: cholesterol acyltransferase modulates the generation of the amyloid beta-peptide. *Nat. Cell Biol.*, **3**, 905–912.

37 Puglielli, L., Tanzi, R.E., and Kovacs, D. M. (2003) Alzheimer's disease: the cholesterol connection. *Nat. Neurosci.*, **6**, 345–351.

38 Huttunen, H.J., Greco, C., and Kovacs, D.M. (2007) Knockdown of ACAT-1 reduces amyloidogenic processing of APP. *FEBS Lett.*, **581**, 1688–1692.

39 Huttunen, H.J., Peach, C., Bhattacharyya, R., Barren, C.,

Pettingell, W., Hutter-Paier, B., Windisch, M., Berezovska, O., and Kovacs, D.M. (2009) Inhibition of acyl-coenzyme A: cholesterol acyl transferase modulates amyloid precursor protein trafficking in the early secretory pathway. *FASEB J.*, **23**, 3819–3828.

40 Huttunen, H.J., Puglielli, L., Ellis, B.C., MacKenzie Ingano, L.A., and Kovacs, D. M. (2009) Novel N-terminal cleavage of APP precludes Abeta generation in ACAT-defective AC29 cells. *J. Mol. Neurosci.*, **37**, 6–15.

41 Bryleva, E.Y., Rogers, M.A., Chang, C.C., Buen, F., Harris, B.T., Rousselet, E., Seidah, N.G., Oddo, S., LaFerla, F.M., Spencer, T.A., Hickey, W.F., and Chang, T.Y. (2010) ACAT1 gene ablation increases 24(*S*)-hydroxycholesterol content in the brain and ameliorates amyloid pathology in mice with AD. *Proc. Natl. Acad. Sci. USA*, **107**, 3081–3086.

42 Arsenault, D., Julien, C., Tremblay, C., and Calon, F. (2011) DHA improves cognition and prevents dysfunction of entorhinal cortex neurons in 3xTg-AD mice. *PLoS One*, **6**, e17397.

43 Cole, G.M., Ma, Q.L., and Frautschy, S.A. (2009) Omega-3 fatty acids and dementia. *Prostaglandins Leukot. Essent. Fatty Acids*, **81**, 213–221.

44 Pomponi, M., Di Gioia, A., Bria, P., and Pomponi, M.F. (2008) Fatty aspirin: a new perspective in the prevention of dementia of Alzheimer's type? *Curr. Alzheimer Res.*, **5**, 422–431.

45 Zhao, Y., Calon, F., Julien, C., Winkler, J.W., Petasis, N.A., Lukiw, W.J., and Bazan, N.G. (2011) Docosahexaenoic acid-derived neuroprotectin D1 induces neuronal survival via secretase – and PPARgamma-mediated mechanisms in Alzheimer's disease models. *PLoS One*, **6**, e15816.

46 Lee, J., Kang, J.H., Han, K.C., Kim, Y., Kim, S.Y., Youn, H.S., Mook-Jung, I., Kim, H., Lo Han, J.H., Ha, H.J., Kim, Y. H., Marquez, V.E., Lewin, N.E., Pearce, L. V., Lundberg, D.J., and Blumberg, P.M. (2006) Branched diacylglycerol-lactones as potent protein kinase C ligands and alpha-secretase activators. *J. Med. Chem.*, **49**, 2028–2036.

47 Berman, D.E., Dall'armi, C., Voronov, S.V., McIntire, L.B., Zhang, H., Moore, A.Z., Staniszewski, A., Arancio, O., Kim, T.W., and Di Paolo, G. (2008) Oligomeric amyloid-beta peptide disrupts phosphatidylinositol-4,5-bisphosphate metabolism. *Nat. Neurosci.*, **11**, 547–554.

48 Di Paolo, G., Sankaranarayanan, S., Wenk, M.R., Daniell, L., Perucco, E., Caldarone, B.J., Flavell, R., Picciotto, M.R., Ryan, T.A., Cremona, O., and De Camilli, P. (2002) Decreased synaptic vesicle recycling efficiency and cognitive deficits in amphiphysin 1 knockout mice. *Neuron*, **33**, 789–804.

49 Goodenowe, D.B., Cook, L.L., Liu, J., Lu, Y., Jayasinghe, D.A., Ahiahonu, P.W., Heath, D., Yamazaki, Y., Flax, J., Krenitsky, K.F., Sparks, D.L., Lerner, A., Friedland, R.P., Kudo, T., Kamino, K., Morihara, T., Takeda, M., and Wood, P.L. (2007) Peripheral ethanolamine plasmalogen deficiency: a logical causative factor in Alzheimer's disease and dementia. *J. Lipid Res.*, **48**, 2485–2498.

50 Grimm, M.O., Kuchenbecker, J., Rothhaar, T.L., Grosgen, S., Hundsdorfer, B., Burg, V.K., Friess, P., Muller, U., Grimm, H.S., Riemenschneider, M., and Hartmann, T. (2011) Plasmalogen synthesis is regulated via alkyl-dihydroxyacetonephosphate-synthase by amyloid precursor protein processing and is affected in Alzheimer's disease. *J. Neurochem.*, **116**, 916–925.

51 Han, X., Holtzman, D.M., and McKeel, D.W., Jr. (2001) Plasmalogen deficiency in early Alzheimer's disease subjects and in animal models: molecular characterization using electrospray ionization mass spectrometry. *J. Neurochem.*, **77**, 1168–1180.

52 Svennerholm, L. and Gottfries, C.G. (1994) Membrane lipids, selectively diminished in Alzheimer brains, suggest synapse loss as a primary event in early-onset form (type I) and demyelination in late-onset form (type II). *J. Neurochem.*, **62**, 1039–1047.

53 Han, X., Cheng, H., Fryer, J.D., Fagan, A. M., and Holtzman, D.M. (2003) Novel

role for apolipoprotein E in the central nervous system. Modulation of sulfatide content. *J. Biol. Chem.*, **278**, 8043–8051.

54 Han, X.D.M.H., McKeel, D.W., Jr., Kelley, J., and Morris, J.C. (2002) Substantial sulfatide deficiency and ceramide elevation in very early Alzheimer's disease: potential role in disease pathogenesis. *J. Neurochem.*, **82**, 809–818.

55 Han, X., Fagan, A.M., Cheng, H., Morris, J.C., Xiong, C., and Holtzman, D.M. (2003) Cerebrospinal fluid sulfatide is decreased in subjects with incipient dementia. *Ann. Neurol.*, **54**, 115–119.

56 Chartier-Harlin, M.C., Crawford, F., Houlden, H., Warren, A., Hughes, D., Fidani, L., Goate, A., Rossor, M., Roques, P., Hardy, J. *et al.* (1991) Early-onset Alzheimer's disease caused by mutations at codon 717 of the beta-amyloid precursor protein gene. *Nature*, **353**, 844–846.

57 Laudon, H., Winblad, B., and Naslund, J. (2007) The Alzheimer's disease-associated gamma-secretase complex: functional domains in the presenilin 1 protein. *Physiol. Behav.*, **92**, 115–120.

58 Murrell, J., Farlow, M., Ghetti, B., and Benson, M.D. (1991) A mutation in the amyloid precursor protein associated with hereditary Alzheimer's disease. *Science*, **254**, 97–99.

59 Corder, E.H., Saunders, A.M., Strittmatter, W.J., Schmechel, D.E., Gaskell, P.C., Small, G.W., Roses, A.D., Haines, J.L., and Pericak-Vance, M.A. (1993) Gene dose of apolipoprotein E type 4 allele and the risk of Alzheimer's disease in late onset families. *Science*, **261**, 921–923.

60 Pericak-Vance, M.A. and Haines, J.L. (1995) Genetic susceptibility to Alzheimer disease. *Trends Genet.*, **11**, 504–508.

61 Strittmatter, W.J. (2000) Apolipoprotein E and Alzheimer's disease. *Ann. N. Y. Acad. Sci.*, **924**, 91–92.

62 Anstey, K.J., Lipnicki, D.M., and Low, L. F. (2008) Cholesterol as a risk factor for dementia and cognitive decline: a systematic review of prospective studies with meta-analysis. *Am. J. Geriatr. Psychiatry*, **16**, 343–354.

63 Notkola, I.L., Sulkava, R., Pekkanen, J., Erkinjuntti, T., Ehnholm, C., Kivinen, P., Tuomilehto, J., and Nissinen, A. (1998) Serum total cholesterol, apolipoprotein E epsilon 4 allele, and Alzheimer's disease. *Neuroepidemiology*, **17**, 14–20.

64 Bryleva, E.Y., Rogers, M.A., Chang, C.C., Buen, F., Harris, B.T., Rousselet, E., Seidah, N.G., Oddo, S., LaFerla, F.M., Spencer, T.A., Hickey, W.F., and Chang, T.Y. (2011) ACAT1 gene ablation increases 24(*S*)-hydroxycholesterol content in the brain and ameliorates amyloid pathology in mice with AD. *Proc. Natl. Acad. Sci. USA*, **107**, 3081–3086.

65 Rodriguez, E., Mateo, I., Infante, J., Llorca, J., Berciano, J., and Combarros, O. (2006) Cholesteryl ester transfer protein (CETP) polymorphism modifies the Alzheimer's disease risk associated with APOE epsilon4 allele. *J. Neurol.*, **253**, 181–185.

66 Sanders, A.E., Wang, C., Katz, M., Derby, C.A., Barzilai, N., Ozelius, L., and Lipton, R.B. (2010) Association of a functional polymorphism in the cholesteryl ester transfer protein (CETP) gene with memory decline and incidence of dementia. *JAMA*, **303**, 150–158.

67 Gomez-Ramos, P. and Asuncion Moran, M. (2007) Ultrastructural localization of intraneuronal Abeta-peptide in Alzheimer disease brains. *J. Alzheimer's Dis.*, **11**, 53–59.

68 Hutter-Paier, B., Huttunen, H.J., Puglielli, L., Eckman, C.B., Kim, D.Y., Hofmeister, A., Moir, R.D., Domnitz, S.B., Frosch, M.P., Windisch, M., and Kovacs, D.M. (2004) The ACAT inhibitor CP-113,818 markedly reduces amyloid pathology in a mouse model of Alzheimer's disease. *Neuron*, **44**, 227–238.

69 Huttunen, H.J., Havas, D., Peach, C., Barren, C., Duller, S., Xia, W., Frosch, M.P., Hutter-Paier, B., Windisch, M., and Kovacs, D.M. (2010) The acyl-coenzyme A: cholesterol acyltransferase inhibitor CI-1011 reverses diffuse brain amyloid pathology in aged amyloid precursor protein transgenic mice. *J. Neuropathol. Exp. Neurol.*, **69**, 777–788.

70 Emory, C.R., Ala, T.A., and Frey, W.H., 2nd (1987) Ganglioside monoclonal antibody (A2B5) labels Alzheimer's neurofibrillary tangles. *Neurology*, **37**, 768–772.

71 Majocha, R.E., Jungalwala, F.B., Rodenrys, A., and Marotta, C.A. (1989) Monoclonal antibody to embryonic CNS antigen A2B5 provides evidence for the involvement of membrane components at sites of Alzheimer degeneration and detects sulfatides as well as gangliosides. *J. Neurochem.*, **53**, 953–961.

72 Han, X. (2010) The pathogenic implication of abnormal interaction between apolipoprotein E isoforms, amyloid-beta peptides, and sulfatides in Alzheimer's disease. *Mol. Neurobiol.*, **41**, 97–106.

73 Vance, J.E., Karten, B., and Hayashi, H. (2006) Lipid dynamics in neurons. *Biochem. Soc. Trans.*, **34**, 399–403.

74 Zoeller, R.A., Lake, A.C., Nagan, N., Gaposchkin, D.P., Legner, M.A., and Lieberthal, W. (1999) Plasmalogens as endogenous antioxidants: somatic cell mutants reveal the importance of the vinyl ether. *Biochem. J.*, **338** (Pt 3), 769–776.

75 Sun, G.Y., Xu, J., Jensen, M.D., and Simonyi, A. (2004) Phospholipase A2 in the central nervous system: implications for neurodegenerative diseases. *J. Lipid Res.*, **45**, 205–213.

76 Green, J.T., Orr, S.K., and Bazinet, R.P. (2008) The emerging role of group VI calcium-independent phospholipase A2 in releasing docosahexaenoic acid from brain phospholipids. *J. Lipid Res.*, **49**, 939–944.

77 Bernard, J., Lahsaini, A., and Massicotte, G. (1994) Potassium-induced long-term potentiation in area CA1 of the hippocampus involves phospholipase activation. *Hippocampus*, **4**, 447–453.

78 Miller, B., Sarantis, M., Traynelis, S.F., and Attwell, D. (1992) Potentiation of NMDA receptor currents by arachidonic acid. *Nature*, **355**, 722–725.

79 Schaeffer, E.L. and Gattaz, W.F. (2005) Inhibition of calcium-independent phospholipase A2 activity in rat hippocampus impairs acquisition of short- and long-term memory. *Psychopharmacology*, **181**, 392–400.

80 Talbot, K., Young, R.A., Jolly-Tornetta, C., Lee, V.M., Trojanowski, J.Q., and Wolf, B.A. (2000) A frontal variant of Alzheimer's disease exhibits decreased calcium-independent phospholipase A2 activity in the prefrontal cortex. *Neurochem. Int.*, **37**, 17–31.

81 Gregory, A., Westaway, S.K., Holm, I.E., Kotzbauer, P.T., Hogarth, P., Sonek, S., Coryell, J.C., Nguyen, T.M., Nardocci, N., Zorzi, G., Rodriguez, D., Desguerre, I., Bertini, E., Simonati, A., Levinson, B., Dias, C., Barbot, C., Carrilho, I., Santos, M., Malik, I., Gitschier, J., and Hayflick, S.J. (2008) Neurodegeneration associated with genetic defects in phospholipase A (2). *Neurology*, **71**, 1402–1409.

82 Morgan, N.V., Westaway, S.K., Morton, J.E., Gregory, A., Gissen, P., Sonek, S., Cangul, H., Coryell, J., Canham, N., Nardocci, N., Zorzi, G., Pasha, S., Rodriguez, D., Desguerre, I., Mubaidin, A., Bertini, E., Trembath, R.C., Simonati, A., Schanen, C., Johnson, C.A., Levinson, B., Woods, C.G., Wilmot, B., Kramer, P., Gitschier, J., Maher, E.R., and Hayflick, S.J. (2006) PLA2G6, encoding a phospholipase A2, is mutated in neurodegenerative disorders with high brain iron. *Nat. Genet.*, **38**752–754.

83 Bazan, N.G. (2009) Cellular and molecular events mediated by docosahexaenoic acid-derived neuroprotectin D1 signaling in photoreceptor cell survival and brain protection. *Prostaglandins Leukot. Essent. Fatty Acids*, **81**, 205–211.

84 Zhao, T., De Graaff, E., Breedveld, G.J., Loda, A., Severijnen, L.A., Wouters, C. H., Verheijen, F.W., Dekker, M.C., Montagna, P., Willemsen, R., Oostra, B. A., and Bonifati, V. (2011) Loss of nuclear activity of the FBXO7 protein in patients with parkinsonian-pyramidal syndrome (PARK15). *PLoS One*, **6**, e16983.

85 Halapin, N.A. and Bazan, N.G. (2010) NPD1 induction of retinal pigment

epithelial cell survival involves PI3K/Akt phosphorylation signaling. *Neurochem. Res.*, **35**, 1944–1947.

86 Hodgkin, M.N., Pettitt, T.R., Martin, A., Michell, R.H., Pemberton, A.J., and Wakelam, M.J. (1998) Diacylglycerols and phosphatidates: which molecular species are intracellular messengers? *Trends Biochem. Sci.*, **23**, 200–204.

87 Oliveira, T.G., Chan, R.B., Tian, H., Laredo, M., Shui, G., Staniszewski, A., Zhang, H., Wang, L., Kim, T.W., Duff, K. E., Wenk, M.R., Arancio, O., and Di Paolo, G. (2010) Phospholipase d2 ablation ameliorates Alzheimer's disease-linked synaptic dysfunction and cognitive deficits. *J. Neurosci.*, **30**, 16419–16428.

88 Kennedy, M.A., Kabbani, N., Lambert, J. P., Swayne, L.A., Ahmed, F., Figeys, D., Bennett, S.A., Bryan, J., and Baetz, K. (2010) Srf1 is a novel regulator of phospholipase D activity and is essential to buffer the toxic effects of C16:0 platelet activating factor. *PLoS Genet.*, **7**, e1001299.

89 Elbaz, A. and Moisan, F. (2008) Update in the epidemiology of Parkinson's disease. *Curr. Opin. Neurol.*, **21**, 454–460.

90 de Lau, L.M. and Breteler, M.M. (2006) Epidemiology of Parkinson's disease. *Lancet Neurol.*, **5**, 525–535.

91 Lim, A., Tsuang, D., Kukull, W., Nochlin, D., Leverenz, J., McCormick, W., Bowen, J., Teri, L., Thompson, J., Peskind, E.R., Raskind, M., and Larson, E.B. (1999) Clinico-neuropathological correlation of Alzheimer's disease in a community-based case series. *J. Am. Geriatr. Soc.*, **47**, 564–569.

92 Larsen, K., Hedegaard, C., Bertelsen, M. F., and Bendixen, C. (2009) Threonine 53 in alpha-synuclein is conserved in long-living non-primate animals. *Biochem. Biophys. Res. Commun.*, **387**, 602–605.

93 Lo Bianco, C., Ridet, J.L., Schneider, B.L., Deglon, N., and Aebischer, P. (2002) Alpha-synucleinopathy and selective dopaminergic neuron loss in a rat lentiviral-based model of Parkinson's disease. *Proc. Natl. Acad. Sci. USA*, **99**, 10813–10818.

94 Frank-Cannon, T.C., Tran, T., Ruhn, K.A., Martinez, T.N., Hong, J., Marvin, M., Hartley, M., Trevino, I., O'Brien, D.E., Casey, B., Goldberg, M.S., and Tansey, M.G. (2008) Parkin deficiency increases vulnerability to inflammation-related nigral degeneration. *J. Neurosci.*, **28**, 10825–10834.

95 Lu, X.H., Fleming, S.M., Meurers, B., Ackerson, L.C., Mortazavi, F., Lo, V., Hernandez, D., Sulzer, D., Jackson, G.R., Maidment, N.T., Chesselet, M.F., and Yang, X.W. (2009) Bacterial artificial chromosome transgenic mice expressing a truncated mutant parkin exhibit age-dependent hypokinetic motor deficits, dopaminergic neuron degeneration, and accumulation of proteinase K-resistant alpha-synuclein. *J. Neurosci.*, **29**, 1962–1976.

96 Geisler, S., Holmstrom, K.M., Skujat, D., Fiesel, F.C., Rothfuss, O.C., Kahle, P.J., and Springer, W. (2010) PINK1/Parkin-mediated mitophagy is dependent on VDAC1 and p62/SQSTM1. *Nat. Cell Biol.*, **12**, 119–131.

97 Narendra, D.P., Jin, S.M., Tanaka, A., Suen, D.F., Gautier, C.A., Shen, J., Cookson, M.R., and Youle, R.J. (2010) PINK1 is selectively stabilized on impaired mitochondria to activate Parkin. *PLoS Biology*, **8**, e1000298.

98 Guzman, J.N., Sanchez-Padilla, J., Wokosin, D., Kondapalli, J., Ilijic, E., Schumacker, P.T., and Surmeier, D.J. (2010) Oxidant stress evoked by pacemaking in dopaminergic neurons is attenuated by DJ-1. *Nature*, **468**, 696–700.

99 Li, Y., Liu, W., Oo, T.F., Wang, L., Tang, Y., Jackson-Lewis, V., Zhou, C., Geghman, K., Bogdanov, M., Przedborski, S., Beal, M.F., Burke, R.E., and Li, C. (2009) Mutant LRRK2 (R1441G) BAC transgenic mice recapitulate cardinal features of Parkinson's disease. *Nat. Neurosci.*, **12**, 826–828.

100 Ramonet, D., Daher, J.P., Lin, B.M., Stafa, K., Kim, J., Banerjee, R., Westerlund, M., Pletnikova, O., Glauser, L., Yang, L., Liu, Y., Swing, D.A., Beal, M. F., Troncoso, J.C., McCaffery, J.M., Jenkins, N.A., Copeland, N.G., Galter, D.,

Thomas, B., Lee, M.K., Dawson, T.M., Dawson, V.L., and Moore, D.J. (2011) Dopaminergic neuronal loss, reduced neurite complexity and autophagic abnormalities in transgenic mice expressing G2019S mutant LRRK2. *PLoS One*, **6**, e18568.

101 Shin, N., Jeong, H., Kwon, J., Heo, H.Y., Kwon, J.J., Yun, H.J., Kim, C.H., Han, B. S., Tong, Y., Shen, J., Hatano, T., Hattori, N., Kim, K.S., Chang, S., and Seol, W. (2008) LRRK2 regulates synaptic vesicle endocytosis. *Exp. Cell Res.*, **314**, 2055–2065.

102 Di Fonzo, A., Chien, H.F., Socal, M., Giraudo, S., Tassorelli, C., Iliceto, G., Fabbrini, G., Marconi, R., Fincati, E., Abbruzzese, G., Marini, P., Squitieri, F., Horstink, M.W., Montagna, P., Libera, A. D., Stocchi, F., Goldwurm, S., Ferreira, J. J., Meco, G., Martignoni, E., Lopiano, L., Jardim, L.B., Oostra, B.A., Barbosa, E.R., and Bonifati, V. (2007) ATP13A2 missense mutations in juvenile parkinsonism and young onset Parkinson disease. *Neurology*, **68**, 1557–1562.

103 Giovannone, B., Tsiaras, W.G., de la Monte, S., Klysik, J., Lautier, C., Karashchuk, G., Goldwurm, S., and Smith, R.J. (2009) GIGYF2 gene disruption in mice results in neurodegeneration and altered insulin-like growth factor signaling. *Hum. Mol. Genet.*, **18**, 4629–4639.

104 Jones, J.M., Albin, R.L., Feldman, E.L., Simin, K., Schuster, T.G., Dunnick, W.A., Collins, J.T., Chrisp, C.E., Taylor, B.A., and Meisler, M.H. (1993) mnd2: a new mouse model of inherited motor neuron disease. *Genomics*, **16**, 669–677.

105 Wada, H., Yasuda, T., Miura, I., Watabe, K., Sawa, C., Kamijuku, H., Kojo, S., Taniguchi, M., Nishino, I., Wakana, S., Yoshida, H., and Seino, K. (2009) Establishment of an improved mouse model for infantile neuroaxonal dystrophy that shows early disease onset and bears a point mutation in Pla2g6. *Am. J. Pathol.*, **175**, 2257–2263.

106 Aharon-Peretz, J., Rosenbaum, H., and Gershoni-Baruch, R. (2004) Mutations in the glucocerebrosidase gene and Parkinson's disease in Ashkenazi Jews. *N. Engl. J. Med.*, **351**, 1972–1977.

107 DePaolo, J., Goker-Alpan, O., Samaddar, T., Lopez, G., and Sidransky, E. (2009) The association between mutations in the lysosomal protein glucocerebrosidase and parkinsonism. *Mov. Disord.*, **24**, 1571–1578.

108 Velayati, A., Yu, W.H., and Sidransky, E. (2010) The role of glucocerebrosidase mutations in Parkinson disease and Lewy body disorders. *Curr. Neurol. Neurosci. Rep.*, **10**, 190–198.

109 Schneider, J.S., Roeltgen, D.P., Mancall, E.L., Chapas-Crilly, J., Rothblat, D.S., and Tatarian, G.T. (1998) Parkinson's disease: improved function with GM1 ganglioside treatment in a randomized placebo-controlled study. *Neurology*, **50**, 1630–1636.

110 Wei, J., Fujita, M., Sekigawa, A., Sekiyama, K., Waragai, M., and Hashimoto, M. (2009) Gangliosides' protection against lysosomal pathology of synucleinopathies. *Autophagy*, **5**, 860–861.

111 Schneider, J.S. (1998) GM1 ganglioside in the treatment of Parkinson's disease. *Ann. N. Y. Acad. Sci.*, **845**, 363–373.

112 Schneider, J.S., Sendek, S., Daskalakis, C., and Cambi, F. (2010) GM1 ganglioside in Parkinson's disease: results of a five year open study. *J. Neurol. Sci.*, **292**, 45–51.

113 Wei, J., Fujita, M., Nakai, M., Waragai, M., Sekigawa, A., Sugama, S., Takenouchi, T., Masliah, E., and Hashimoto, M. (2009) Protective role of endogenous gangliosides for lysosomal pathology in a cellular model of synucleinopathies. *Am. J. Pathol.*, **174**, 1891–1909.

114 Mazzulli, J.R., Xu, Y.H., Sun, Y., Knight, A.L., McLean, P.J., Caldwell, G.A., Sidransky, E., Grabowski, G.A., and Krainc, D. (2011) Gaucher disease glucocerebrosidase and alpha-synuclein form a bidirectional pathogenic loop in synucleinopathies. *Cell*, **146**, 37–52.

115 Rappley, I., Myers, D.S., Milne, S.B., Ivanova, P.T., Lavoie, M.J., Brown, H.A., and Selkoe, D.J. (2009) Lipidomic profiling in mouse brain reveals

differences between ages and genders, with smaller changes associated with alpha-synuclein genotype. *J. Neurochem.*, **111**, 15–25.

116 Keeney, P.M., Xie, J., Capaldi, R.A., and Bennett, J.P., Jr. (2006) Parkinson's disease brain mitochondrial complex I has oxidatively damaged subunits and is functionally impaired and misassembled. *J. Neurosci.*, **26**, 5256–5264.

117 Watanabe, Y., Himeda, T., and Araki, T. (2005) Mechanisms of MPTP toxicity and their implications for therapy of Parkinson's disease. *Med. Sci. Monit.*, **11**, RA17–RA23.

118 Kreitzer, A.C. and Malenka, R.C. (2007) Endocannabinoid-mediated rescue of striatal LTD and motor deficits in Parkinson's disease models. *Nature*, **445**, 643–647.

119 Pisani, A., Fezza, F., Galati, S., Battista, N., Napolitano, S., Finazzi-Agro, A., Bernardi, G., Brusa, L., Pierantozzi, M., Stanzione, P., and Maccarrone, M. (2005) High endogenous cannabinoid levels in the cerebrospinal fluid of untreated Parkinson's disease patients. *Ann. Neurol.*, **57**, 777–779.

120 Cleren, C., Yang, L., Lorenzo, B., Calingasan, N.Y., Schomer, A., Sireci, A., Wille, E.J., and Beal, M.F. (2008) Therapeutic effects of coenzyme Q10 (CoQ10) and reduced CoQ10 in the MPTP model of parkinsonism. *J. Neurochem.*, **104**, 1613–1621.

121 Galpern, W.R. and Cudkowicz, M.E. (2007) Coenzyme Q treatment of neurodegenerative diseases of aging. *Mitochondrion*, (7 Suppl.), S146–S153.

122 Horvath, T.L., Diano, S., Leranth, C., Garcia-Segura, L.M., Cowley, M.A., Shanabrough, M., Elsworth, J.D., Sotonyi, P., Roth, R.H., Dietrich, E.H., Matthews, R.T., Barnstable, C.J., and Redmond, D.E., Jr. (2003) Coenzyme Q induces nigral mitochondrial uncoupling and prevents dopamine cell loss in a primate model of Parkinson's disease. *Endocrinology*, **144**, 2757–2760.

123 Muller, T., Buttner, T., Gholipour, A.F., and Kuhn, W. (2003) Coenzyme Q10 supplementation provides mild symptomatic benefit in patients with Parkinson's disease. *Neurosci. Lett.*, **341**, 201–204.

124 Yang, L., Calingasan, N.Y., Wille, E.J., Cormier, K., Smith, K., Ferrante, R.J., and Beal, M.F. (2009) Combination therapy with coenzyme Q10 and creatine produces additive neuroprotective effects in models of Parkinson's and Huntington's diseases. *J. Neurochem.*, **109**, 1427–1439.

125 Lim, L., Jackson-Lewis, V., Wong, L.C., Shui, G.H., Goh, A.X., Kesavapany, S., Jenner, A.M., Fivaz, M., Przedborski, S., and Wenk, M.R. (2011) Lanosterol induces mitochondrial uncoupling and protects dopaminergic neurons from cell death in a model for Parkinson's disease. *Cell Death Differ.* DOI: 10.1038/cdd.2011.105.

126 Wen, Y., Li, W., Poteet, E.C., Xie, L., Tan, C., Yan, L.J., Ju, X., Liu, R., Qian, H., Marvin, M.A., Goldberg, M.S., She, H., Mao, Z., Simpkins, J.W., and Yang, S.H. (2011) Alternative mitochondrial electron transfer as a novel strategy for neuroprotection. *J. Biol. Chem.*, **286**, 16504–16515.

127 Virmani, A., Gaetani, F., and Binienda, Z. (2005) Effects of metabolic modifiers such as carnitines, coenzyme Q10, and PUFAs against different forms of neurotoxic insults: metabolic inhibitors, MPTP, and methamphetamine. *Ann. N. Y. Acad. Sci.*, **1053**, 183–191.

128 Andrews, Z.B., Diano, S., and Horvath, T.L. (2005) Mitochondrial uncoupling proteins in the CNS: in support of function and survival. *Nat. Rev. Neurosci.*, **6**, 829–840.

129 Andrews, Z.B., Horvath, B., Barnstable, C.J., Elsworth, J., Yang, L., Beal, M.F., Roth, R.H., Matthews, R.T., and Horvath, T.L. (2005) Uncoupling protein-2 is critical for nigral dopamine cell survival in a mouse model of Parkinson's disease. *J. Neurosci.*, **25**, 184–191.

130 Conti, B., Sugama, S., Lucero, J., Winsky-Sommerer, R., Wirz, S.A., Maher, P., Andrews, Z., Barr, A.M., Morale, M.C., Paneda, C., Pemberton, J., Gaidarova, S., Behrens, M.M., Beal, F., Sanna, P.P., Horvath, T., and Bartfai, T. (2005)

Uncoupling protein 2 protects dopaminergic neurons from acute 1,2,3,6-methyl-phenyl-tetrahydropyridine toxicity. *J. Neurochem.*, **93**, 493–501.

131 Giuffrida, A., Parsons, L.H., Kerr, T.M., Rodriguez de Fonseca, F., Navarro, M., and Piomelli, D. (1999) Dopamine activation of endogenous cannabinoid signaling in dorsal striatum. *Nat. Neurosci.*, **2**, 358–363.

132 Gubellini, P., Picconi, B., Bari, M., Battista, N., Calabresi, P., Centonze, D., Bernardi, G., Finazzi-Agro, A., and Maccarrone, M. (2002) Experimental parkinsonism alters endocannabinoid degradation: implications for striatal glutamatergic transmission. *J. Neurosci.*, **22**, 6900–6907.

133 Kravitz, A.V., Freeze, B.S., Parker, P.R., Kay, K., Thwin, M.T., Deisseroth, K., and Kreitzer, A.C. (2010) Regulation of parkinsonian motor behaviours by optogenetic control of basal ganglia circuitry. *Nature*, **466**, 622–626.

134 Venderova, K., Brown, T.M., and Brotchie, J.M. (2005) Differential effects of endocannabinoids on [(3)H]-GABA uptake in the rat globus pallidus. *Exp. Neurol.*, **194**, 284–287.

135 Morera-Herreras, T., Ruiz-Ortega, J.A., Linazasoro, G., and Ugedo, L. (2011) Nigrostriatal denervation changes the effect of cannabinoids on subthalamic neuronal activity in rats. *Psychopharmacology*, **214**, 379–389.

15
The Tumor Mitochondrial Lipidome and Respiratory Bioenergetic Insufficiency

Thomas N. Seyfried, Jeffrey H. Chuang, Lu Zhang, Xianlin Han, and Michael A. Kiebish

15.1
Introduction

The lipidome comprises all lipids in a tissue, cell, or cellular organelle. The content, composition, and diversity of lipid classes and their molecular species can provide insight into the metabolic status of biological specimens. Although the genome and proteome have been the focus of much attention in tumorigenesis, little attention has been given to the lipidome as a potential origin of the tumorigenic phenotype. Lipids maintain the integrity of the outer and inner mitochondrial membranes. Abnormalities in lipids can compromise mitochondrial function and the efficiency of oxidative phosphorylation (OxPhos), as the functions of electron transport chain (ETC) proteins depend to a considerable degree on the lipid composition of the inner mitochondrial membrane. Recent advancements in mass spectrometric and high-throughput lipidomic platforms can accurately identify and quantify a diversity of lipid molecular species [1, 2]. Multidimensional mass spectrometry-based shotgun lipidomics (MDMS-SL) technology for analysis of the tumor mitochondrial lipidome has provided insight into the metabolic derangements of cancer [2, 3]. Hence, high-throughput platforms that can decipher the tumor mitochondrial lipidome can provide novel information on the metabolic abnormalities in tumor cells.

Otto Warburg originally proposed that all cancers arise from protracted damage to cellular respiration [4, 5]. In order for cells with insufficient respiration to remain viable, they would require a compensatory source of energy. This compensatory energy source in tumor cells mostly involves fermentation through glycolysis. Tumor cells differ from normal cells in that they continue to ferment glucose even in the presence of oxygen. Normal cells respire rather than ferment in oxygen. Although respiration may still function in some low malignancy tumor cells, no tumor cell is known with a respiratory capacity comparable to that in its normal cell counterpart [6]. The persistent fermentation of glucose in oxygen is known as the "Warburg effect" and arises from any number of gene and

Lipidomics, First Edition. Edited by Kim Ekroos.

environment insults that diminish cellular OxPhos [3, 7]. As the mitochondrial lipidome is essential for maintaining mitochondrial function, and ultimately cellular energy homeostasis, any disturbance of the mitochondrial lipidome is expected to alter cellular energy homeostasis. The displacement of OxPhos with fermentation is the bioenergetic signature of cancer leading eventually to nuclear genomic instability and all hallmarks of cancer [7]. Little is known, however, on how disturbances in the mitochondrial lipidome might contribute to the altered energy metabolism in tumor cells.

Lipid analysis in tumor cells has progressed over the decades from lipid class content and fatty acid composition to molecular species architecture using soft ionization electrospray mass spectrometry in high-throughput platforms. These experimental methods have linked abnormalities in the tumor lipidome with abnormalities in biological function. It is difficult to know if lipidomic changes seen in cultured tumor cells were also present in tumor cells grown *in vivo*. A comparative analysis of the lipidome in tumor cells grown *in vivo* with that from tumor cells grown *in vitro* can help identify those changes likely to have importance in the natural environment. To this end, we focused our attention on the mitochondrial lipidome of tumors and of their tissue of origin. We employed a novel bioinformatics approach to integrate lipidomic dynamics and energy metabolism in the mitochondria of normal cells and tumor cells. This approach has helped link structural changes in the tumor mitochondrial lipidome in tumor cells with respiratory insufficiency. We did not detect PIP3 in purified brain mitochondria from either normal brain cells or tumor cells. We also found very low levels of lyso-phospholipids in mitochondrial membranes. While there is a large literature on PIP3 and lyso-phospholipids in tumor cells and cancer, we are not able to comment on the role of these lipids in mitochondria from our brain tumor models.

15.1.1 Lipidomic Abnormalities in Tumor Mitochondria

Pederson reviewed numerous studies showing that mitochondrial lipid abnormalities are common in all tumors examined [8]. It is important to see from Figure 15.1 that cholesterol is generally more enriched in the outer than in the inner mitochondrial membrane of normal cells. In contrast to normal cells, Feo *et al.* showed that the cholesterol/phospholipid ratio was significantly higher in mitochondria from hepatomas than in mitochondria from normal liver cells [9, 10]. As cholesterol reduces membrane fluidity, elevated levels of cholesterol are expected to reduce the fluidity properties of mitochondrial membranes. Mitochondrial phospholipids in normal tissues contain an abundance of long-chain polyunsaturated fatty acids, whereas mitochondria in tumor cells contain phospholipids with higher amounts of short-chain saturated or monounsaturated species [8]. We have confirmed these findings in mouse brain tumors [3]. Most importantly, we found several abnormalities in the structure of cardiolipin (CL), the major lipid of the inner mitochondrial membrane.

15.2 Cardiolipin and Electron Transport Chain Abnormalities in Mouse Brain Tumor Mitochondria

Cardiolipin (1,3-diphosphatidyl-*sn*-glycerol) is a complex mitochondrial specific phospholipid that regulates numerous enzyme activities especially those related to OxPhos and coupled respiration [12–17]. Several studies have shown that CL is essential for efficient oxidative energy production and mitochondrial function [12, 14, 17–31]. Alterations in the content or composition of CL can alter cellular respiration. Before describing evidence linking CL abnormalities to insufficient respiration in brain tumors, it would be good to first briefly review information about the unique properties of this mitochondrial lipid.

CL contains two phosphate head groups, three glycerol moieties, and four fatty acyl chains and is primarily enriched in the inner mitochondrial membrane (Figures 15.1 and 15.2). Enrichment in the inner mitochondrial membrane makes cardiolipin a pivotal molecule for regulating cristae structure and OxPhos [32]. CL binds complex I, III, IV, and V and stabilizes the super complexes (I/III/IV, I/III, and III/IV) demonstrating an absolute requirement of CL for catalytic activity of these respiratory enzyme complexes [3, 14, 15, 33, 34]. CL restricts pumped protons within its head group domain, thus providing the structural basis for mitochondrial membrane potential and in supplying protons to the ATP synthase [13, 17].

Some ETC proteins that interact with CL have evolved to form hydrophobic grooves on their surface [33, 35]. These grooves accommodate the fatty acid chains

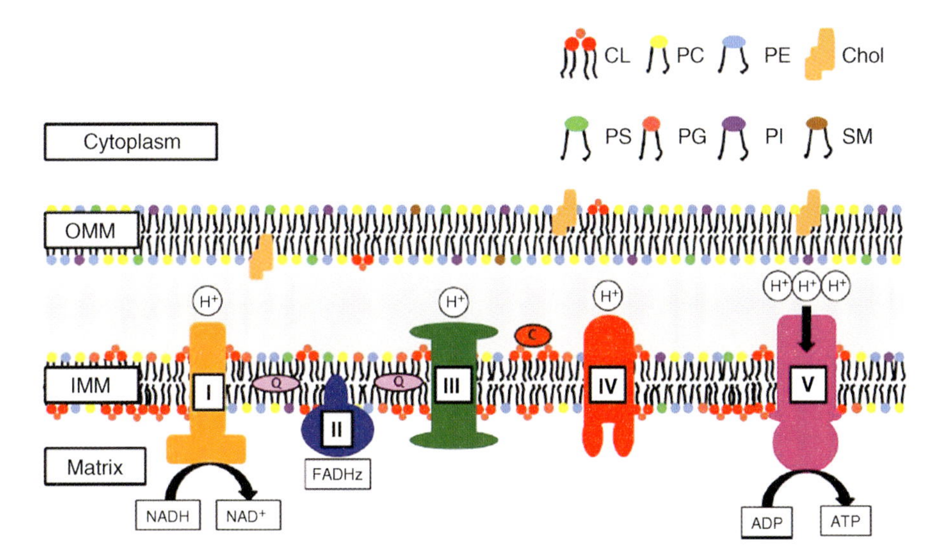

Figure 15.1 Topology of lipid distribution in mitochondrial membranes. Cardiolipin (red) is enriched primarily in the inner mitochondrial membrane and plays an important role in maintaining the proton motive gradient and efficiency of the electron transport chain. Reproduced with permission from Ref. [11].

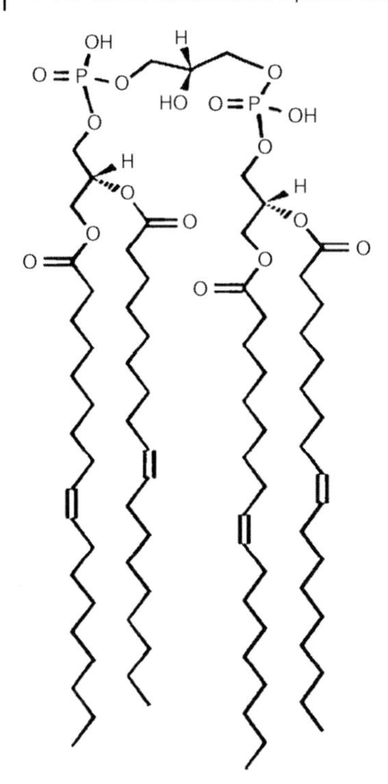

Figure 15.2 Structure of cardiolipin (1, 1′, 2, 2′-tetraoleyl cardiolipin). Cardiolipin is a complex mitochondrial specific phospholipid that regulates numerous enzyme activities, especially those related to oxidative phosphorylation and coupled respiration (see text for details). Reproduced with permission from Ref. [3].

of CL (Figure 15.3). Since long-chain carbon molecules appeared earlier in evolution than did membrane proteins, it is likely that the grooves evolved to accommodate already existing fatty acids. While the amino acid sequence of electron transport proteins is highly conserved across species, considerable variability occurs for the fatty acid sequences of CL. Although the respiratory protein structure is largely invariant, the fatty acid composition of CL can be modulated through changes in nutrition and physiological environment (Seyfried and Ta, unpublished). CL can modulate ETC activities without altering the primary sequence of amino acids. Hence, changes in CL content and composition can influence electron transport and ultimately the efficiency of OxPhos.

The activity of respiratory enzymes in complex I and complex III and their linked activities are directly related to CL content [14, 27, 36]. The activities of the respiratory enzyme complexes are also dependent on the composition of CL molecular species [13]. Indeed, the degree of CL unsaturation is related to the respiratory state [12, 22]. Respiratory efficiency depends on the degree of CL remodeling. Remodeling is a complex process where immature CL is remodeled to form mature CL.

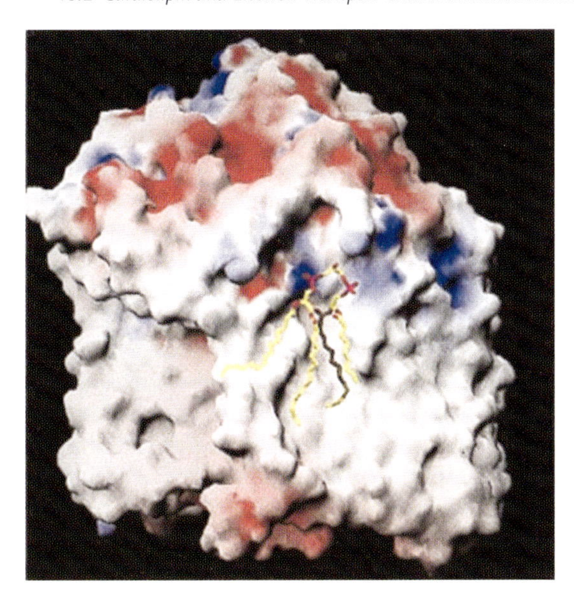

Figure 15.3 Interaction of cardiolipin with protein amino acid sequence. The amino acid sequence of highly conserved proteins of the electron transport chain evolved to generate hydrophobic regions to selectively mold to the structural diversity of cardiolipin molecular species. This relationship between protein and lipid interactions both generates functional regulation of enzymatic efficiency and emphasizes the importance of lipidomic organization of the mitochondrial membrane, thus linking cardiolipin structure with enzymatic functionality. Reproduced with permission from Ref. [35].

This process involves the replacement of shorter chain and less unsaturated fatty acids in immature CL with longer chain and more complex (polyunsaturated) fatty acids in mature CL. In general, remodeling produces longer chain unsaturated fatty acid species characteristic of CL in differentiated cells. The respiratory energy efficiency in tissues is, therefore, dependent to a large extent on the expression of mature CL.

Almost 100 molecular species of CL were recently detected in the mitochondria from mammalian brain [37, 38]. Moreover, these molecular species form a symmetric pattern consisting of seven major groups when arranged according to fatty acid chain length and degree of unsaturation (Figure 15.4) [37]. This unique fatty acid pattern is expressed in CL analyzed from both synaptic (Syn) mitochondria (enriched in neurons) and nonsynaptic (NS) mitochondria (mostly enriched in cell bodies of neurons and glia) in mature mouse brain. CL analyzed from nonneural cells contains mostly tetra 18:2, that is, four 18-carbon chains with each chain containing two double bonds.

We recently showed that the lipid composition and/or content in mouse brain tumor mitochondria differed markedly from that in mitochondria derived from the normal syngeneic host brain tissue [3]. These brain tumors covered a spectrum of growth behaviors seen in most human malignant brain cancers. Two of the tumors

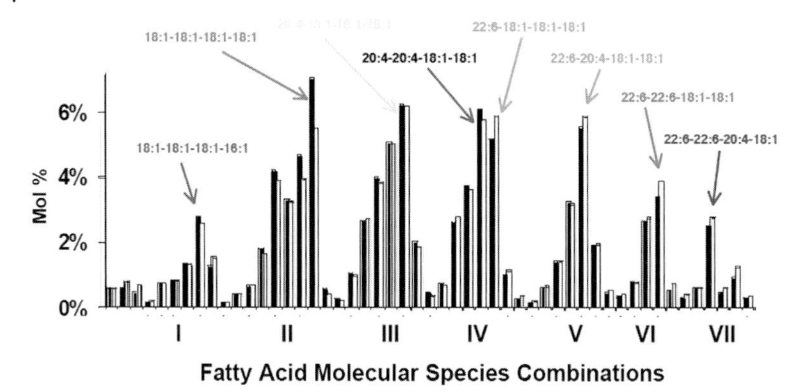

Figure 15.4 Distribution of CL molecular species in nonsynaptic (black bar) and synaptic (white bar) mitochondria of mouse brain. CL molecular species were arranged according to the mass to charge ratio based on percentage distribution. CL molecular species were subdivided into seven groups, which contained a predominance of oleic, arachidonic, and/or docosahexanoic fatty acids in varying concentrations. Corresponding mass content of molecular species in nonsynaptic (NS) and synaptic (Syn) mitochondria were as we described [37]. All values are expressed as the mean of three independent samples ($n = 3$), where six mouse cerebral cortexes were pooled for each sample. Adapted from Ref. [37].

evaluated, an ependymoblastoma (EPEN) and an astrocytoma (CT-2A), were derived from implantation of 20-methylcholantherene into the brains of inbred C57BL/6J mice [39–41]. Three of the tumors evaluated, VM-M2, VM-M3, and VM-NM1, arose spontaneously in the brains of inbred VM mice. The VM mouse strain is unique in developing a relatively high incidence of brain tumors [42]. The VM-M2 and VM-M3 tumors express multiple properties of myeloid/mesenchymal cells and display the invasive growth behavior of human glioblastoma multiforme [43–45]. The VM-NM1 is rapidly growing, but is neither invasive in the brain nor metastatic when grown outside the brain [44]. We produced clonal cell lines from each of the five brain tumors. Mitochondria were isolated from each tumor that was grown subcutaneously in the syngeneic mouse host [3]. We employed both Ficoll and sucrose gradients to obtain highly purified mitochondria from normal brain tissue and from brain tumor tissue [3, 37]. This isolation procedure maintained both the structure and the function of the purified mitochondria [37]. Besides expressing multiple abnormalities in the major phospholipids (phosphatidylcholine and phosphatidylethanolamine), we found that the content and composition of CL differed markedly between normal brain tissue and tumor tissue.

CL content was significantly lower in the mitochondria from the CT-2A and the EPEN tumors than in the mitochondria from the normal control B6 mouse brain (Figure 15.5). In contrast to the B6 mouse brain, which contains about 100 molecular fatty acid species of CL symmetrically distributed over seven major groups, the VM mouse brain is unique in having only about 45 major CL molecular species. Moreover, the molecular species in groups IV, V, and VII are not seen in the mitochondria isolated from VM brain (Figure 15.6a and b). We addressed the importance of the CL changes in the VM mice in relationship to the inheritance of

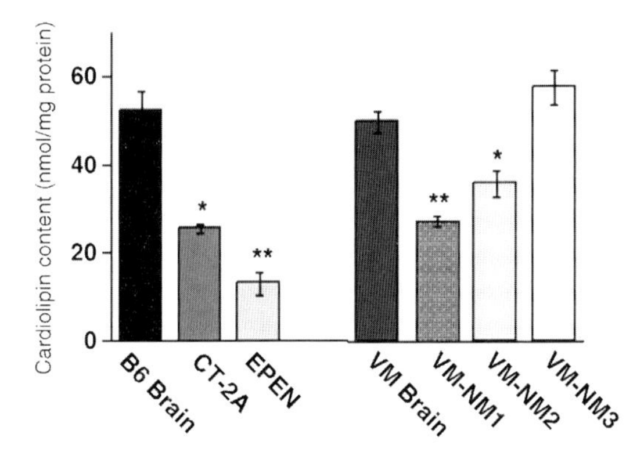

Figure 15.5 Cardiolipin content in mitochondria isolated from normal mouse brain and mouse brain tumors. Reduction in CL content suggests fewer mitochondria or mitochondria with reduced amounts of inner membrane. Mitochondria were isolated as we described [37]. Values are represented as the mean ± standard deviation (SD) of three independent mitochondrial preparations from brain or tumor tissue. Asterisks indicate that the tumor values differ significantly from the B6 or the VM normal brain values at the $^{*}p < 0.01$ or $^{**}p < 0.001$ levels as determined by the two-tailed t-test. Reproduced with permission from Ref. [3].

brain tumors in this strain [46]. CL content was also significantly lower in the mitochondria from the VM-NM1 and the VM-M2 tumors than in the mitochondria from the control VM mouse brain (Figure 15.5). Interestingly, CL content in the VM-M3 tumor was not significantly reduced. None of the tumors had a CL fatty acid composition similar to that of the normal parental brain tissue. We also found that the CL abnormalities in these tumors were associated with significant reductions in electron transport chain (ETC) activities consistent with the pivotal role of CL in maintaining the structural integrity of the inner mitochondrial membrane [3, 12, 13, 19].

The activities of the ETC complexes I, I/III, and II/III were significantly lower in the brain tumors than in their normal syngeneic brain tissue [3]. As mitochondrial ETC activities depend on the content and the composition of CL, we used a bioinformatics approach to model ETC activities as a function of CL content and composition in the five mouse brain tumors. The two main variables included (1) total CL content and (2) the distribution of CL molecular species in mitochondria. The information about the molecular species distribution was simplified into a single number, which described the degree of relationship of the CL composition of the tumor mitochondria with that of brain mitochondria from the host mouse strain [3]. This number was generated as a Pearson product moment correlation. We used the correlation coefficient to assess the degree of "compositional similarity" of CL from the host mouse brain mitochondria with that of tumor mitochondria.

Utilizing a computational approach, we analyzed the relationship of cardiolipin content, molecular species composition, and enzyme activity (Figure 15.7). We showed that a low coefficient indicated dissimilarity between the host brain mitochondria and the tumor mitochondria for CL fatty acid molecular species composition. A high correlation indicated that CL molecular species composition was similar between the host brain mitochondria and the tumor mitochondria. The ETC activities in each tumor were then measured using standard enzymatic procedures, whereas MDMS-SL was used to measure the content and composition of CL [37]. A two-dimensional linear regression was used to fit the measured ETC activity values with CL composition. The best-fit relationship for each complex with the content and composition of CL was expressed as a quadratic surface (Figure 15.7). We compared the data for the CT-2A and the EPEN tumors with their B6 host strain and compared the data for the VM-NM1, VM-M2, and VM-M3 tumors with their VM host strain. This analysis demonstrated a direct relationship between

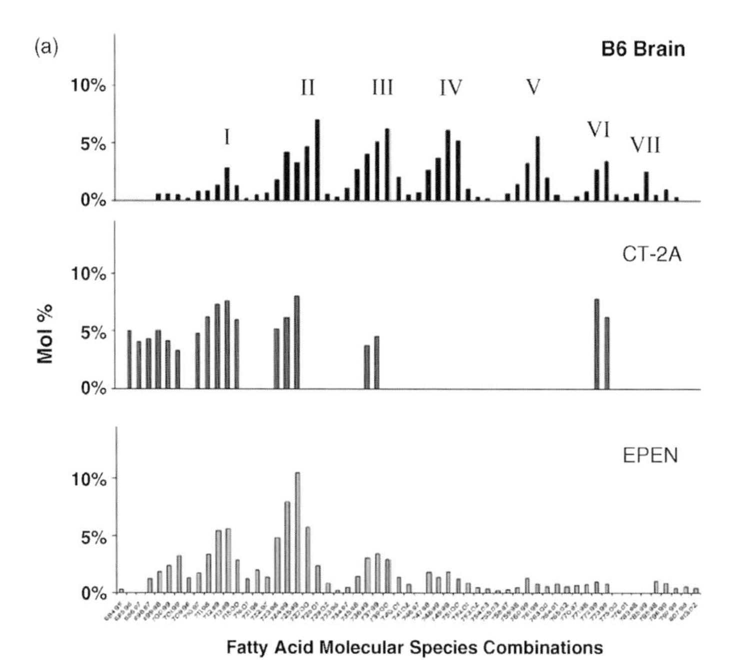

Figure 15.6 Distribution of cardiolipin molecular species in mitochondria isolated and purified from normal mouse brain and brain tumor mitochondria. (a) Distribution in a syngeneic C57BL/6J (B6) mouse brain and in the CT-2A and the EPEN tumors. (b) Distribution in syngeneic VM mouse brain and the VM-NM1, the VM-M2, and the VM-M3 tumors. Cardiolipin fatty acid molecular species are plotted on the abscissa and arranged according to the mass to charge ratio based on percentage distribution. The molecular species are subdivided into seven major groups (I–VII) as described in Figures 14.5–14.10. Corresponding mass content of molecular species in normal brain and tumor mitochondria can be found in Table 1 of our previous study [3]. It is clear that fatty acid molecular species composition differs markedly between tumors and their syngeneic mouse host's brain tissue. As CL composition influences ETC activities and mitochondrial energy production, these findings indicate that mitochondrial energy efficiency differs between normal brain tissue and brain tumor tissue. All values are expressed as the mean of three independent mitochondrial preparations, where tissues from six brain cortexes or tumors were pooled for each preparation. Reproduced with permission from Ref. [3].

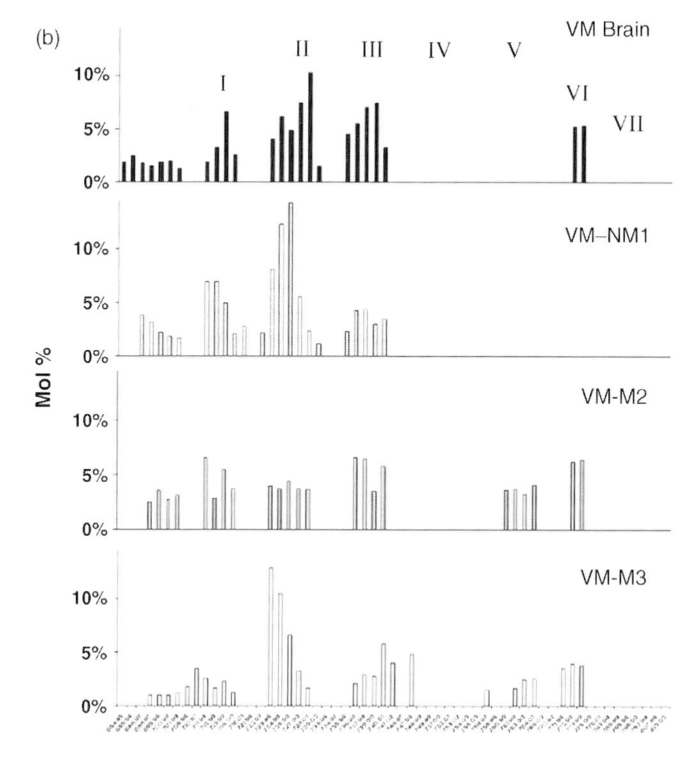

Figure 15.6 (*Continued*)

ETC activity, CL content, and the distribution of molecular species. It was clear from our studies that abnormalities in the content and composition of CL could underlie the abnormal energy metabolism of these diverse brain tumors. This is the type of connection that Sidney Weinhouse considered essential for establishing the credibility of the Warburg theory [47]. Hence, our lipidomic studies in mouse brain tumors provide credibility to Warburg's original theory.

Our findings are also consistent with earlier studies in rat hepatomas showing an increase in shorter chain saturated fatty acid content (palmitic and stearic) characteristic of immature CL [48, 49]. Our studies are consistent with more recent findings in rhabdomyosarcoma, a type of muscle tumor, showing that the reduction in complex I activity was associated with CL abnormalities [50]. Continued expression of immature CL would reduce respiratory energy production. In light of what we know about CL structure and respiratory function, it is difficult to conceive how mitochondrial OxPhos could function normally in tumors that express CL abnormalities.

How might abnormalities in the content and composition of CL arise? Would CL abnormalities be related to the cause or effect of tumor formation? We proposed that CL abnormalities could arise either from inherited cancer risk factors as seen in the VM mice or from numerous epigenetic and environmental cancer risk factors including inflammation, viruses, hypoxia, radiation, and so on (Figure 15.8)

Complex I

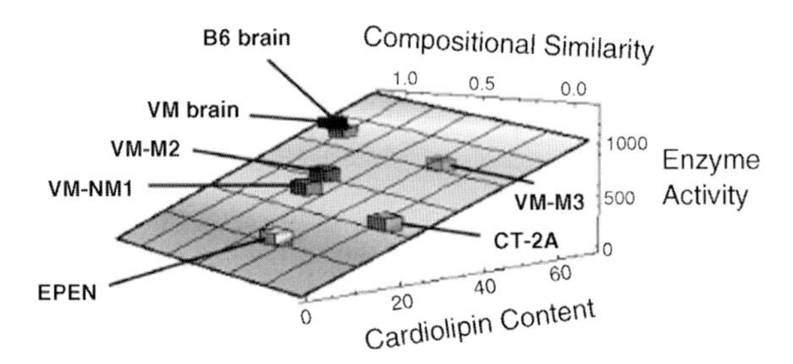

Complex I / III

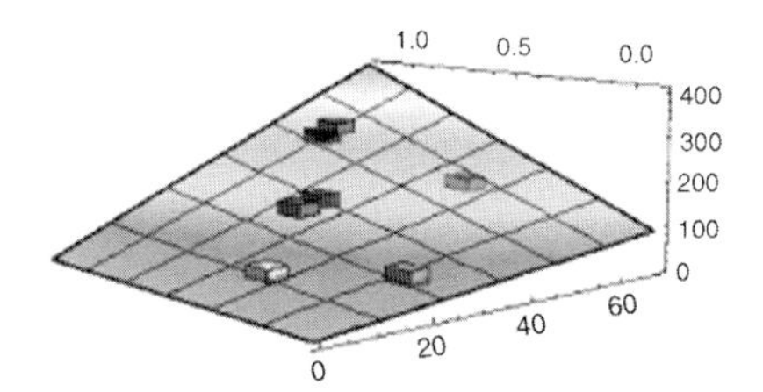

Complex II / III

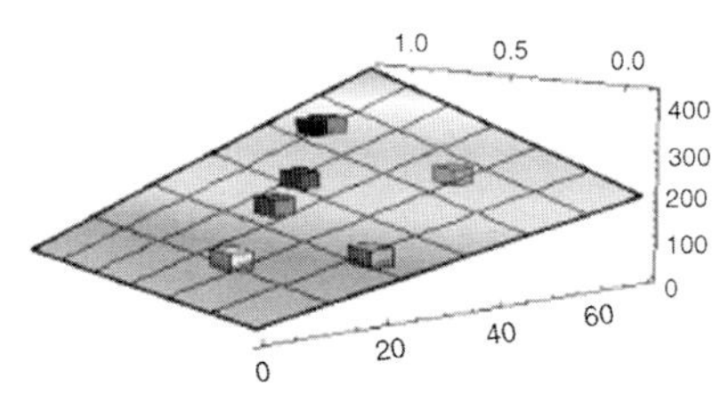

Figure 15.7 Relationship of cardiolipin abnormalities with electron transport chain activities in the B6 and the VM mouse brain tumors. The data are expressed on the best-fit three-dimensional quadratic surface for each electron transport chain complex as we recently described [3]. In order to illustrate the position of all tumors on the same graph relative to their host strain, the data for the VM strain and tumors were fit to the B6-fit quadratic surface as described [3]. The data indicate that changes in tumor ETC activities can be directly related to changes in CL content and fatty acid molecular species composition. The data show that ETC complex activity differs markedly between tumors and their syngeneic B6 and VM hosts and that these differences are linked to abnormalities in the content and composition of CL. The results suggest that CL abnormalities are associated with reduced efficiency of respiration. Reproduced with permission from Ref. [3].

[3, 7, 46]. Indeed, gamma radiation is known to induce free radical CL fragmentation, which would compromise respiratory function [51]. Based on these and other observations, we suggested that most tumors regardless of cell origin would contain abnormalities in CL composition and/or content [3]. Regardless of whether the CL abnormalities are related to the cause of the tumor or arise during tumor

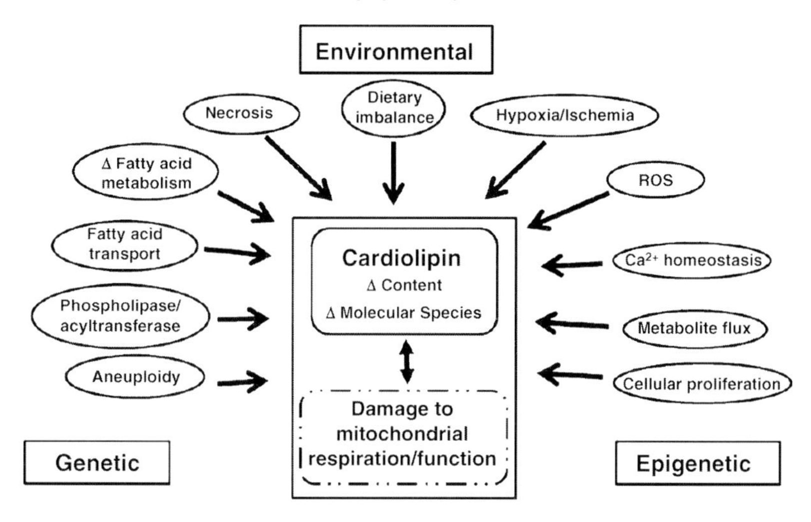

Figure 15.8 Relationship of genetic, epigenetic, and environmental factors to dysfunctional respiration associated with abnormalities in cardiolipin content and composition. ROS: reactive oxygen species. Reproduced with permission from Ref. [3].

progression, the CL abnormalities will significantly reduce the efficiency of mitochondrial OxPhos.

Our findings of immature CL molecular species in tumor mitochondria together with associated abnormalities in ETC activity are consistent with the recent findings of Roman Eliseev and colleagues who showed enhanced replication of immature mitochondria in malignant osteosarcoma cells compared to normal osteoblasts [52]. Mitochondrial immaturity would predict insufficient energy production through OxPhos. The association of CL abnormalities with impaired respiratory function is expected based on the localization and role of CL in the inner mitochondrial membrane.

15.3 Complicating Influence of the *in vitro* Growth Environment on Cardiolipin Composition and Energy Metabolism

It is important to recognize that numerous studies of energy metabolism in tumor cells are conducted on the cells grown in tissue culture. We recently showed for the first time that the *in vitro* growth environment produces lipidomic and electron transport abnormalities in mitochondria from both nontumorigenic cells and tumor cells [53]. The implications of this observation are profound. How is it possible to fully describe the metabolic abnormalities of cancer cells if the environment in which the cells are grown alters energy metabolism?

Using MDMS-SL analysis, we found that the mitochondrial lipidome of the CT-2A and EPEN brain tumor cells grown in tissue culture differed markedly from the mitochondrial lipidome of these same brain tumor cells when they were grown in their natural host. The difference is clearly seen by comparing the CL molecular species distribution of the CT-2A and EPEN tumors grown *in vivo* (Figure 15.6a) with the species distribution of these same tumors grown *in vitro* (Figure 15.9). Moreover, the CL molecular species distribution of the nontumorigenic astrocytes was more similar to those of the cultured tumor CT-2A and EPEN cells than to those of the normal brain (Figures 15.4 and 15.9). Both the astrocytes and the normal brain tissue were from the same genetic background (C57BL/6). Hence, growth environment significantly alters CL composition and ultimately respiratory function. The CL composition of the cultured cells was largely immature containing a preponderance of shorter chain saturated or monounsaturated species indicative of failed remodeling [53]. These findings indicate that the *in vitro* growth

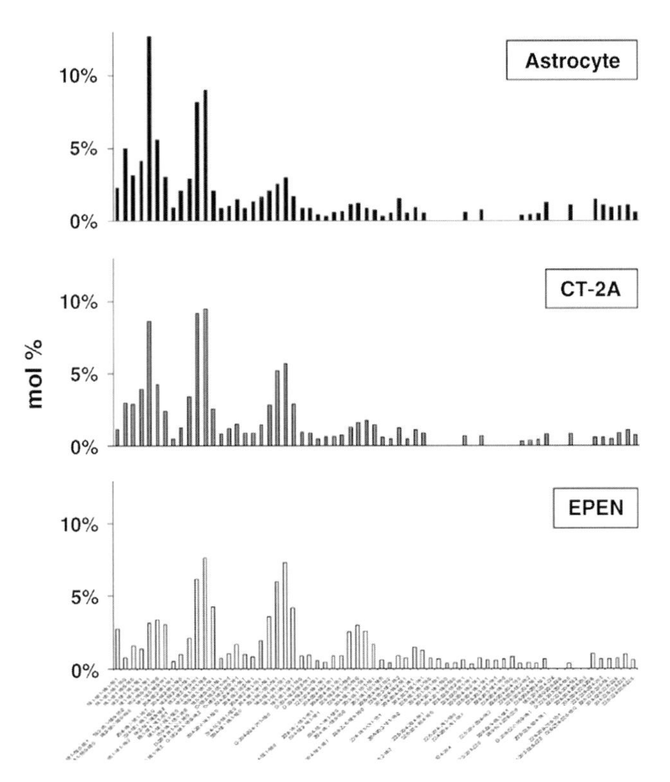

Figure 15.9 Distribution of cardiolipin molecular species in mitochondria isolated from astrocytes and brain tumor cells grown *in vitro*. In contrast to the unique differences seen for CL composition between the normal brain and the CT-2A and EPEN brain tumors grown *in vivo* (Figure 15.6a and b), no major differences are seen between nontumorigenic astrocytes and the CT-2A and EPEN tumors when grown as cultured cells. The fatty acid composition and mass content of each molecular species presented on the *x*-axis can be found in supplementary Table S3 at http://www.asnneuro.org/an/001/an001e011.add.htm [53]. It appears that growth in the cell culture environment alters CL composition. Reproduced with permission from Ref. [53].

environment produced abnormalities in CL remodeling. A failure to remodel CL will prevent efficient energy production through OxPhos. Our findings in the mouse brain tumors are consistent with numerous reports indicating that the content and composition of CL is essential for normal respiratory function [12, 14, 17–31, 50]. Hence, brain tumors with CL abnormalities will require an alternative energy source to OxPhos in order to maintain viability. According to the Warburg theory, this alternative energy source comes from glucose fermentation [5].

A failure to remodel CL will alter the activities of respiratory enzyme activities. That such alterations occur in the cultured cells is clearly illustrated in Figure 15.10. These findings connect abnormal CL to abnormal ETC enzyme activities. The activity of complex I was especially reduced in the cultured cells. Complex I activity is essential for the initiation of electron transport. The linked complex I/III activities were also significantly reduced in the cultured cells. These findings

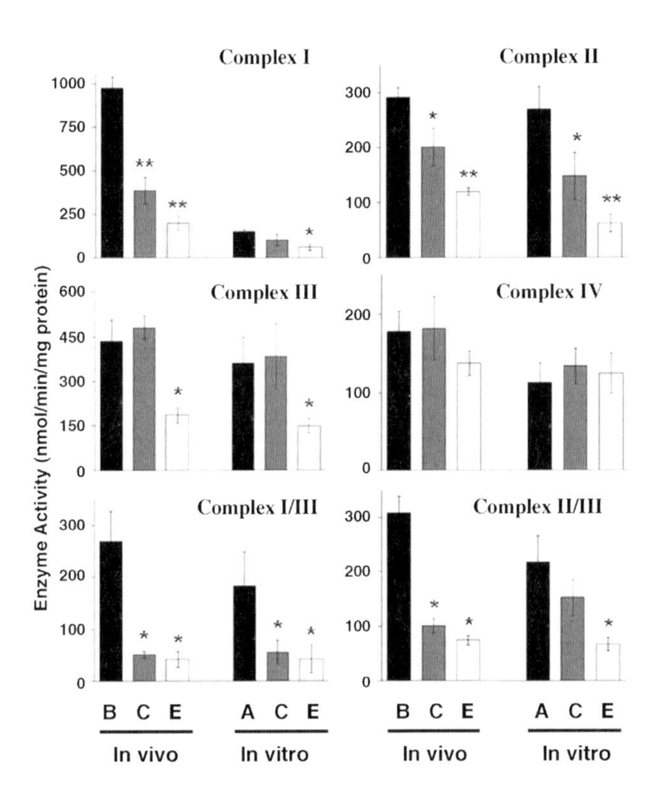

Figure 15.10 Electron transport chain enzyme activities in purified mitochondria from mouse brain, brain tumors, and cultured tumor cells. Enzyme activities are expressed as nmol/min/mg protein as described [53]. B, C, E, and A represent enzyme activities in mitochondria isolated from normal brain, CT-2A, EPEN, and astrocytes (nontumorigenic), respectively. Other conditions are as described [53]. Asterisks indicate that the activities in the brain tumor samples differ from those of the control samples (either mouse brain or astrocytes) at the $^{*}p < 0.03$ or $^{**}p < 0.005$ levels as determined by the two-tailed t-test. Reprinted with permission from Ref. [53].

suggest that growth in the *in vitro* environment reduces electron transport and energy production through OxPhos. If energy through OxPhos is compromised in cultured tumor cells, how do these cells maintain their viability?

High levels of glucose and other metabolites in culture media can increase glycolysis and inhibit OxPhos. This effect was first described by Herbert Crabtree in the late 1920s and is referred to as the "Crabtree effect" [53–56]. We do not exclude the possibility that the lipidomic and ETC abnormalities observed in the cultured nontumorigenic astrocytes and brain tumors cells could arise in part from the Crabtree effect. It is also interesting that several lipidomic differences found between the brain tumor mitochondria and the normal mitochondria in the *in vivo* environment are not seen between the brain tumor cells and the nontumorigenic astrocytes in the *in vitro* environment. These findings indicate that the *in vitro* growth environment obscures lipidomic differences related to tumorigenesis. It is important to mention that the growth environment for cultured cells usually includes high glucose, DMEM with high glutamine, and 10% fetal calf serum. While this culture environment is ideal for the growth and survival of transformed cells and tumor cells, it differs markedly from the natural *in vivo* growth environment. The unnatural conditions of the cell culture environment will negatively impact mitochondrial lipid composition, membrane fluidity, and ultimately respiratory energy metabolism. It should also not be surprising to find differences in lipid metabolism between primary cell cultures and cultured cells adapted to the *in vitro* environment. Others have also reported similar phenomena [57]. A failure to recognize these facts could confound data interpretation.

Further support for respiratory energy inefficiency in the cultured cells comes from our findings that lactic acid production is high in the CT-2A tumor cells and the nontumorigenic astrocytes when grown under identical *in vitro* growth conditions indicating that aerobic glucose fermentation is enhanced in these cells [53]. Recent studies in rhabdomyosarcoma support our findings in the brain tumors [50]. Freyssenet and coworkers showed that reduced CL content in rhabdomyosarcoma was linked to mitochondrial energy dysfunction requiring compensatory energy production through glycolysis [50].

We suggest that cell proliferation *in vitro* and the Crabtree effect could obscure or mask lipidomic abnormalities between normal cells and tumor cells due to tumorigenesis [53]. Warburg considered highly malignant tumor cells as having irreversibly damaged respiration. This respiratory damage would prevent differentiation and facilitate continuous cell proliferation through enhanced fermentation [7]. In addition to enhancing aerobic fermentation, an impaired mitochondrial lipidome could also enhance energy production through substrate-level phosphorylation in the TCA cycle itself as we recently reported [58]. It is well documented that glutamine is a necessary energy metabolite for many cells grown in culture. TCA cycle substrate-level phosphorylation together with glycolysis could compensate for the energy lost through respiration in order to preserve cell viability [58–61]. This could explain why proliferating cultured cells, either tumorigenic or nontumorigenic, rely heavily on glutaminolysis and glycolysis for viability.

As most metazoan cells did not evolve to grow as microorganisms, growth in culture will produce a physiological state different from that of the intact tissue environment. Viewed collectively, our findings indicate that the *in vitro* growth environment produces lipidomic and ETC abnormalities in nontumorigenic astrocytes and in brain tumor cells, which would disrupt energy production through OxPhos, thus confounding the relationship of altered energy metabolism to tumorigenesis. It is surprising that many researchers in the cancer field are unaware of this fact [62].

15.4 Bioinformatic Methods to Interpret Alterations in the Mitochondrial Lipidome

High-throughput lipid molecular species data generated from lipidomic experiments, particularly for tumor samples, provide novel information that can be used for bioinformatic analysis of lipids. From a mechanistic perspective, CL remodeling in tumors is one lipid system that has been most thoroughly analyzed computationally [3, 63]. Similar methods have also been used for mechanistic analysis of nontumor dynamic lipid systems [63–66]. After *de novo* synthesis, the four acyl chains of CL are replaced or exchanged with those from other lipids. The MDMS-SL approach can detect more than 100 mass peaks of CL, each containing tens to hundreds of possible CL isomers with different combinations of chains at the four acyl positions in CL.

Remarkably, a relatively simple process controls CL remodeling in normal mouse brain. The four CL chain positions (*sn*-1, *sn*-1′; *sn*-2, *sn*-2′) are independently and identically remodeled. This behavior was demonstrated by determining whether the B6 brain CL data could be fit to data generated by an independent and identical distribution (IID) model [63]. In this model, the relative prevalence of a CL isomer with acyl chain α_i at position *sn*-1, α_j at position *sn*-1′, α_k at position *sn*-2, and α_l at position *sn*-2′ is predicted to be proportional to the product of probabilities determined by the fatty acid chains individually, that is,

$$P(\alpha_i\,\alpha_j\,\alpha_k\,\alpha_l) = P(\alpha_i)P(\alpha_j)P(\alpha_k)P(\alpha_l),$$

where the $P(\alpha)$ refers to the relative prevalence of individual fatty acids, and the four CL positions are assumed to have equal access to fatty acids from the remodeling pool. Statistical inference methods can be used to find the set of parameters, that is, the $P(\alpha)$ values, that provide the closest fit between the experimental and the predicted CL distributions. As shown in Figure 15.11, this model can account for the distribution of B6 brain mitochondrial CL species. This is a remarkable fit given that the number of possible fatty acids is much smaller than the number of peaks in the CL species distribution.

The IID model provides a good first approximation to the behavior of CL distributions in tumors as well, though tumors also exhibit signs of more specific remodeling behaviors. Out of 14 mouse brain samples of varying tumor type, most showed a good correlation between the IID fit and the observed data (13/14

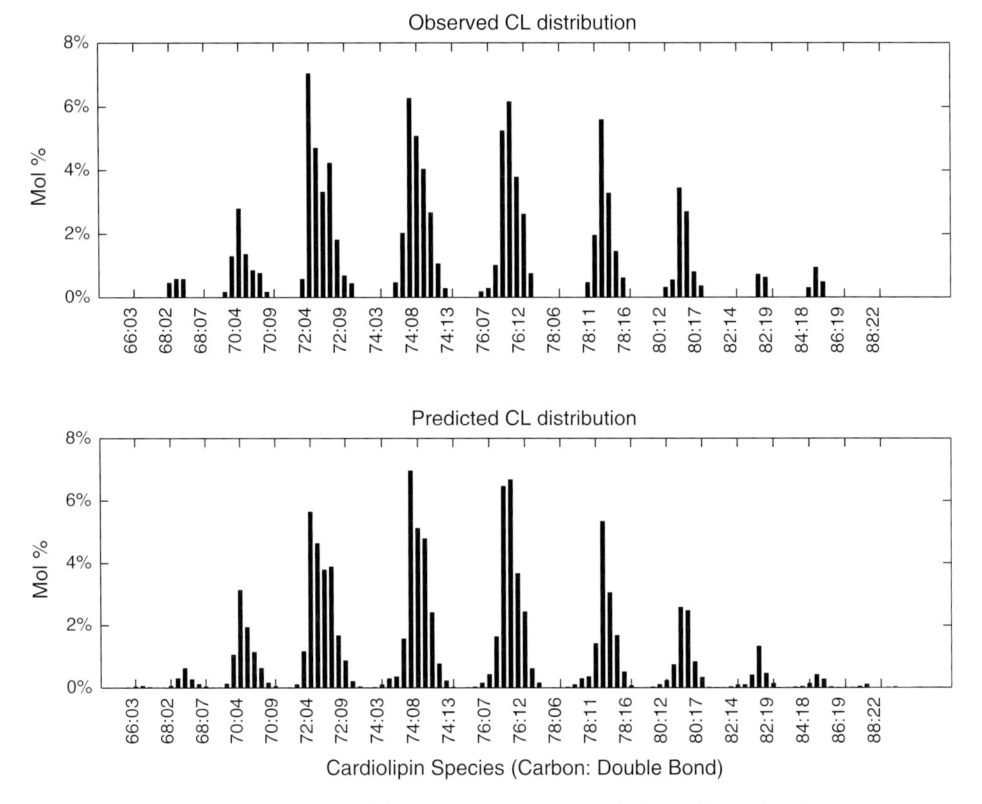

Figure 15.11 Comparison of the IID model prediction and observed CL molecular species distribution for B6 mouse brain.

samples exhibited a Pearson correlation >0.7). The predominant fatty acid in mouse brain is 18:1. However, 18:1 is reduced in brain tumor samples, especially in VM-M2 and VM-M3 vitro and VM-M3 vivo. The relative concentrations of different CL fatty acids in each tumor are shown in Figure 15.12, with comparison to the concentrations in normal tissues of C57BL/6 mice.

Deviations from IID behavior also exhibit certain regularities. For example, the CL peaks at 80 carbons:14 double bonds and 80:15 are unusually high in the non-tumorigenic BV2 cells grown *in vitro* (BV2 vitro). Similar findings were obtained for the VM M3 tumor grown *in vitro* and for the CT-2A tumor grown *in vivo*, and are consistently underestimated by the IID model. These findings indicate that position-specific or cooperative remodeling mechanisms are active in some tumors. In brain tumors, the pool of chains in $P(\alpha)$ appears to derive from processes dependent on the head group of the molecule donating acyl chains (PC, PE, PG, etc.), but with relatively little sensitivity to the actual chain being transferred [63].

Lipidomics has also been valuable for direct determination of diagnostic markers. In one important study, lipidomic measurements were performed on 267

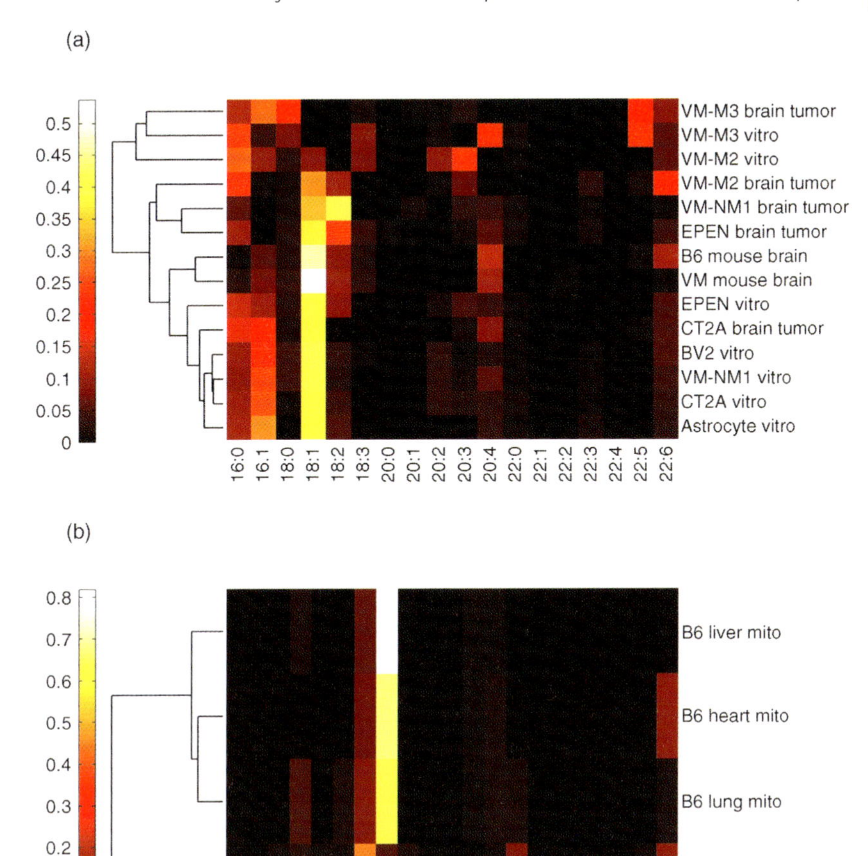

Figure 15.12 Comparison of CL FA compositions between 14 mouse brain samples (a) and 4 B6 mouse tissues (b). 18:1 is predominant in brain (~40%), while its percentage is reduced in some tumor samples (e.g., VM-M2 vitro, VM-M3 vivo VM-M3 vitro). 18:2 is predominant in heart, lung, and liver (~70%). Reproduced with permission from Ref. [63].

human breast tissues to identify lipid abnormalities associated with the cancer phenotype [67]. These studies were conducted on whole tissues rather than purified mitochondria as we did. Products of *de novo* fatty acid synthesis, such as phosphatidylcholine containing palmitate, were increased in tumors, and experiments to silence lipid metabolism-regulating genes reduced the viability of breast cancer cells. These findings indicate that information on the tumor lipidome has therapeutic potential. Studies in human tumor samples will be needed to test further our predictions.

15.5 Conclusions

We provide new information showing how abnormalities in CL content or composition can reduce ATP production through OxPhos leading to respiratory insufficiency in tumor cells. Alterations in CL can arise through the process of tumorigenesis and from the growth of mammalian cells in the *in vitro* environment. A take home message from these studies is that caution should be used in comparing energy metabolism in nontumorigenic cells and tumorigenic cells grown in tissue culture environments that do not replicate the growth conditions of the *in vivo* environment. In light of our findings and those of Peter Pedersen [8], it is surprising that many investigators would consider OxPhos as normal in cancer cells [62, 68]. Information on the tumor lipidome can provide insight into underlying metabolic abnormalities, which might be exploited for therapy [69]. As cancer is primarily a metabolic disease, further studies of CL in tumor cells will provide insight into the bioenergetic abnormalities of cancer.

Acknowledgments

This work was supported, in part, by the National Institutes of Health [68] (grant numbers HD-39722, NS- 55195 and CA-102135, and NIA/NIDDK R01 AG31675) and by a grant from the American Institute of Cancer Research and the Boston College Expense Fund.

References

1 Han, X. and Gross, R.W. (2005) Shotgun lipidomics: electrospray ionization mass spectrometric analysis and quantitation of cellular lipidomes directly from crude extracts of biological samples. *Mass Spectrom. Rev.*, **24**, 367–412.
2 Han, X., Yang, K., and Gross, R.W. (2011) Multi-dimensional mass spectrometry-based shotgun lipidomics and novel strategies for lipidomic analyses. *Mass Spectrom. Rev.*, **31**, 134–178.
3 Kiebish, M.A., Han, X., Cheng, H., Chuang, J.H., and Seyfried, T.N. (2008) Cardiolipin and electron transport chain abnormalities in mouse brain tumor mitochondria: lipidomic evidence supporting the Warburg theory of cancer. *J. Lipid Res.*, **49**, 2545–2556.
4 Warburg, O. (1931) *The Metabolism of Tumours*, Richard R. Smith, New York.
5 Warburg, O. (1956) On the origin of cancer cells. *Science (New York)*, **123**, 309–314.
6 Warburg, O. (1969) *The prime cause of cancer and prevention. Part 2.* Annual meeting of Nobelists at Lindau, Germany, http://www.hopeforcancer.com/OxyPlus.htm.
7 Seyfried, T.N. and Shelton, L.M. (2010) Cancer as a metabolic disease. *Nutr. Metab.*, **7**, 7.
8 Pedersen, P.L. (1978) Tumor mitochondria and the bioenergetics of cancer cells. *Prog. Exp. Tumor Res.*, **22**, 190–274.
9 Feo, F., Canuto, R.A., Bertone, G., Garcea, R., and Pani, P. (1973) Cholesterol and phospholipid composition of

mitochondria and microsomes isolated from Morris hepatoma 5123 and rat liver. *FEBS Lett.*, **33**, 229–232.

10 Feo, F., Canuto, R.A., Garcea, R., and Gabriel, L. (1975) Effect of cholesterol content on some physical and functional properties of mitochondria isolated from adult rat liver, fetal liver, cholesterol-enriched liver and hepatomas AH-130,3924A and 5123. *Biochim. Biophys. Acta*, **413**, 116–134.

11 Kiebish, M.A. (2008) *Mitochondrial lipidome and genome alterations in mouse brain and experimental murine brain tumors*. Ph.D. thesis, Boston College, Chestnut Hill.

12 Hoch, F.L. (1992) Cardiolipins and biomembrane function. *Biochim. Biophys. Acta*, **1113**, 71–133.

13 Chicco, A.J. and Sparagna, G.C. (2007) Role of cardiolipin alterations in mitochondrial dysfunction and disease. *Am. J. Physiol. Cell Physiol.*, **292**, C33–C44.

14 Fry, M. and Green, D.E. (1981) Cardiolipin requirement for electron transfer in complex I and III of the mitochondrial respiratory chain. *J. Biol. Chem.*, **256**, 1874–1880.

15 Fry, M. and Green, D.E. (1980) Cardiolipin requirement by cytochrome oxidase and the catalytic role of phospholipid. *Biochem. Biophys. Res. Commun.*, **93**, 1238–1246.

16 Fry, M., Blondin, G.A., and Green, D.E. (1980) The localization of tightly bound cardiolipin in cytochrome oxidase. *J. Biol. Chem.*, **255**, 9967–9970.

17 Haines, T.H. and Dencher, N.A. (2002) Cardiolipin: a proton trap for oxidative phosphorylation. *FEBS Lett.*, **528**, 35–39.

18 Houtkooper, R.H. and Vaz, F.M. (2008) Cardiolipin, the heart of mitochondrial metabolism. *Cell Mol. Life Sci.*, **65**, 2493–2506.

19 Ordys, B.B., Launay, S., Deighton, R.F., and McCulloch, J., and Whittle, I.R. (2010) The role of mitochondria in glioma pathophysiology. *Mol. Neurobiol.*, **42**, 64–75.

20 Schagger, H. (2002) Respiratory chain supercomplexes of mitochondria and bacteria. *Biochim. Biophys. Acta*, **1555**, 154–159.

21 Koshkin, V. and Greenberg, M.L. (2002) Cardiolipin prevents rate-dependent uncoupling and provides osmotic stability in yeast mitochondria. *Biochem. J.*, **364**, 317–322.

22 Hoch, F.L. (1998) Cardiolipins and mitochondrial proton-selective leakage. *J. Bioenerg. Biomembr.*, **30**, 511–532.

23 Eilers, M., Endo, T., and Schatz, G. (1989) Adriamycin, a drug interacting with acidic phospholipids, blocks import of precursor proteins by isolated yeast mitochondria. *J. Biol. Chem.*, **264**, 2945–2950.

24 Mileykovskaya, E., Zhang, M., and Dowhan, W. (2005) Cardiolipin in energy transducing membranes. *Biochemistry (Moscow)*, **70**, 154–158.

25 Zhang, M., Mileykovskaya, E., and Dowhan, W. (2005) Cardiolipin is essential for organization of complexes III and IV into a supercomplex in intact yeast mitochondria. *J. Biol. Chem.*, **280**, 29403–29408.

26 Shidoji, Y., Hayashi, K., Komura, S., Ohishi, N., and Yagi, K. (1999) Loss of molecular interaction between cytochrome c and cardiolipin due to lipid peroxidation. *Biochem. Biophys. Res. Commun.*, **264**, 343–347.

27 Pfeiffer, K., Gohil, V., Stuart, R.A. *et al.* (2003) Cardiolipin stabilizes respiratory chain supercomplexes. *J. Biol. Chem.*, **278**, 52873–52880.

28 Ostrander, D.B., Zhang, M., Mileykovskaya, E. *et al.* (2001) Lack of mitochondrial anionic phospholipids causes an inhibition of translation of protein components of the electron transport chain. A yeast genetic model system for the study of anionic phospholipid function in mitochondria. *J. Biol. Chem.*, **276**, 25262–25272.

29 Gohil, V.M., Hayes, P., Matsuyama, S. *et al.* (2004) Cardiolipin biosynthesis and mitochondrial respiratory chain function are interdependent. *J. Biol. Chem.*, **279**, 42612–42618.

30 Kagan, V.E., Tyurina, Y.Y., Bayir, H. *et al.* (2006) The "pro-apoptotic genies" get out of mitochondria: oxidative lipidomics and redox activity of cytochrome c/cardiolipin complexes. *Chem. Biol. Interact.*, **163**, 15–28.

31 Gold, V.A., Robson, A., Bao, H. *et al.* (2010) The action of cardiolipin on the

bacterial translocon. *Proc. Natl. Acad. Sci. USA*, **107**, 10044–10049.

32 Alirol, E. and Martinou, J.C. (2006) Mitochondria and cancer: is there a morphological connection? *Oncogene*, **25**, 4706–4716.

33 Shinzawa-Itoh, K., Aoyama, H., Muramoto, K. *et al.* (2007) Structures and physiological roles of 13 integral lipids of bovine heart cytochrome c oxidase. *EMBO J.*, **26**, 1713–1725.

34 McKenzie, M., Lazarou, M., Thorburn, D. R., and Ryan, M.T. (2006) Mitochondrial respiratory chain supercomplexes are destabilized in Barth Syndrome patients. *J. Mol. Biol.*, **361**, 462–469.

35 McAuley, K.E., Fyfe, P.K., Ridge, J.P. *et al.* (1999) Structural details of an interaction between cardiolipin and an integral membrane protein. *Proc. Natl. Acad. Sci. USA*, **96**, 14706–14711.

36 Zhang, M., Mileykovskaya, E., and Dowhan, W. (2002) Gluing the respiratory chain together. Cardiolipin is required for supercomplex formation in the inner mitochondrial membrane. *J. Biol. Chem.*, **277**, 43553–43556.

37 Kiebish, M.A., Han, X., Cheng, H. *et al.* (2008) Lipidomic analysis and electron transport chain activities in C57BL/6J mouse brain mitochondria. *J. Neurochem.*, **106**, 299–312.

38 Cheng, H., Mancuso, D.J., Jiang, X. *et al.* (2008) Shotgun lipidomics reveals the temporally dependent, highly diversified cardiolipin profile in the mammalian brain: temporally coordinated postnatal diversification of cardiolipin molecular species with neuronal remodeling. *Biochemistry*, **47**, 5869–5880.

39 Mukherjee, P., Abate, L.E., and Seyfried, T.N. (2004) Antiangiogenic and proapoptotic effects of dietary restriction on experimental mouse and human brain tumors. *Clin. Cancer Res.*, **10**, 5622–5629.

40 Mukherjee, P., El-Abbadi, M.M., Kasperzyk, J.L. *et al.* (2002) Dietary restriction reduces angiogenesis and growth in an orthotopic mouse brain tumour model. *Br. J. Cancer*, **86**, 1615–1621.

41 Seyfried, T.N., el-Abbadi, M., and Roy, M. L. (1992) Ganglioside distribution in murine neural tumors. *Mol. Chem. Neuropathol.*, **17**, 147–167.

42 Fraser, H. (1986) Brain tumours in mice, with particular reference to astrocytoma. *Food Chem. Toxicol.*, **24**, 105–111.

43 Huysentruyt, L.C. and Seyfried, T.N. (2010) Perspectives on the mesenchymal origin of metastatic cancer. *Cancer Metastasis Rev.*, **29**, 695–707.

44 Huysentruyt, L.C., Mukherjee, P., Banerjee, D. *et al.* (2008) Metastatic cancer cells with macrophage properties: evidence from a new murine tumor model. *Int. J. Cancer*, **123**, 73–84.

45 Shelton, L.M., Mukherjee, P., Huysentruyt, L.C. *et al.* (2010) A novel pre-clinical *in vivo* mouse model for malignant brain tumor growth and invasion. *J. Neurooncol.*, **99**, 165–176.

46 Kiebish, M.A., Han, X., Cheng, H. *et al.* (2008) Brain mitochondrial lipid abnormalities in mice susceptible to spontaneous gliomas. *Lipids*, **43**, 951–959.

47 Weinhouse, S. (1976) The Warburg hypothesis fifty years later. *Z. Krebsforsch. Klin. Onkol. (Cancer Res. Clin. Oncol.)*, **87**, 115–126.

48 Hartz, J.W., Morton, R.E., Waite, M.M., and Morris, H.P. (1982) Correlation of fatty acyl composition of mitochondrial and microsomal phospholipid with growth rate of rat hepatomas. *Lab. Invest.*, **46**, 73–78.

49 Canuto, R.A., Biocca, M.E., Muzio, G., and Dianzani, M.U. (1989) Fatty acid composition of phospholipids in mitochondria and microsomes during diethylnitrosamine carcinogenesis in rat liver. *Cell Biochem. Funct.*, **7**, 11–19.

50 Jahnke, V.E., Sabido, O., Defour, A. *et al.* (2010) Evidence for mitochondrial respiratory deficiency in rat rhabdomyosarcoma cells. *PLoS One*, **5**, e8637.

51 Shadyro, O.I., Yurkova, I.L., Kisel, M.A. *et al.* (2004) Radiation-induced fragmentation of cardiolipin in a model membrane. *Int. J. Radiat. Biol.*, **80**, 239–245.

52 Shapovalov, Y., Hoffman, D., Zuch, D. *et al.* (2011) Mitochondrial dysfunction in cancer cells due to aberrant mitochondrial replication. *J. Biol. Chem.*, **286** (25), 22331–22338.

53 Kiebish, M.A., Han, X., Cheng, H., and Seyfried, T.N. (2009) In vitro growth environment produces lipidomic and electron transport chain abnormalities in mitochondria from non-tumorigenic astrocytes and brain tumours. *ASN Neuro*, **1** (3), 125–138.

54 Frezza, C. and Gottlieb, E. (2009) Mitochondria in cancer: not just innocent bystanders. *Semin. Cancer Biol.*, **19**, 4–11.

55 Guppy, M., Greiner, E., and Brand, K. (1993) The role of the Crabtree effect and an endogenous fuel in the energy metabolism of resting and proliferating thymocytes. *Eur. J. Biochem.*, **212**, 95–99.

56 Crabtree, H.G. (1929) Observations on the carbohydrate metabolism of tumors. *Biochem. J.*, **23**, 536–545.

57 Sergent, O., Ekroos, K., Lefeuvre-Orfila, L. *et al.* (2009) Ximelagatran increases membrane fluidity and changes membrane lipid composition in primary human hepatocytes. *Toxicol. In Vitro*, **23**, 1305–1310.

58 Shelton, L.M., Strelko, C.L., Roberts, M.F., and Seyfried, N.T. (2010) Krebs cycle substrate-level phosphorylation drives metastatic cancer cells, in Proceedings of the 101st Annual Meeting of the American Association for Cancer Research, Washington, D.C.

59 Weinberg, J.M., Venkatachalam, M.A., Roeser, N.F., and Nissim, I. (2000) Mitochondrial dysfunction during hypoxia/reoxygenation and its correction by anaerobic metabolism of citric acid cycle intermediates. *Proc. Natl. Acad. Sci. USA*, **97**, 2826–2831.

60 Phillips, D., Aponte, A.M., French, S.A. *et al.* (2009) Succinyl-CoA synthetase is a phosphate target for the activation of mitochondrial metabolism. *Biochemistry*, **48**, 7140–7149.

61 Schwimmer, C., Lefebvre-Legendre, L., Rak, M. *et al.* (2005) Increasing mitochondrial substrate-level phosphorylation can rescue respiratory growth of an ATP synthase-deficient yeast. *J. Biol. Chem.*, **280**, 30751–30759.

62 Jose, C., Bellance, N., and Rossignol, R. (2010) Choosing between glycolysis and oxidative phosphorylation: a tumor's dilemma? *Biochim. Biophys. Acta*, **1807** (6), 552–561.

63 Zhang, L., Bell, R.J., Kiebish, M.A. *et al.* (2011) A mathematical model for the determination of steady-state cardiolipin remodeling mechanisms using lipidomic data. *PLoS one*, **6**, e21170.

64 Kiebish, M.A., Bell, R., Yang, K. *et al.* (2010) Dynamic simulation of cardiolipin remodeling: greasing the wheels for an interpretative approach to lipidomics. *J. Lipid Res.*, **51**, 2153–2170.

65 Gupta, S., Maurya, M.R., Stephens, D.L. *et al.* (2009) An integrated model of eicosanoid metabolism and signaling based on lipidomics flux analysis. *Biophys. J.*, **96**, 4542–4551.

66 Kainu, V., Hermansson, M., and Somerharju, P. (2008) Electrospray ionization mass spectrometry and exogenous heavy isotope-labeled lipid species provide detailed information on aminophospholipid acyl chain remodeling. *J. Biol. Chem.*, **283**, 3676–3687.

67 Hilvo, M., Denkert, C., Lehtinen, L. *et al.* (2011) Novel theranostic opportunities offered by characterization of altered membrane lipid metabolism in breast cancer progression. *Cancer Res.*, **71**, 3236–3245.

68 Koppenol, W.H., Bounds, P.L., and Dang, C.V. (2011) Otto Warburg's contributions to current concepts of cancer metabolism. *Nat. Rev.*, **11**, 325–337.

69 Seyfried, T.N., Kiebish, M.A., Marsh, J. *et al.* (2010) Metabolic management of brain cancer. *Biochim. Biophys. Acta*, **1807**, 577–594.

16
Lipidomics for Pharmaceutical Research

Yoshinori Satomi

16.1
Introduction

Lipids represent the major class of molecules in human body. In addition to being used as the building blocks for cells, they also have several biological functions such as to maintain homeostasis of life. When a drug is administered into a body, it generally produces its efficacy by altering molecular profiles and by modifying cellular functions and morphologies. Some drugs can also directly or indirectly influence the biosynthesis or metabolism of lipids, thus leading to efficacies for multiple diseases. For a better understanding of the mechanism of drug actions, the interest in lipidomics has grown extensively in drug development and proven as an essential tool-kit.

Traditional approach based on ELISA or RIA can analyze only single lipid molecular species at a time in drug mechanism study. Recent advancement in technology such as mass spectrometry (MS) has benefited such study in the area of lipidomics; MS-based lipidomics (will be referred to simply as lipidomics from now onward) is capable of simultaneously targeting hundreds of lipid molecular species with high sensitivity, thus improving the speed and accuracy of the analysis over the traditional approaches. In addition, lipidomics also enables the detection of lipids that have never been analyzed by traditional methodologies, vastly improving the coverage of lipid species. Through lipidomics, we can gain a novel understanding of the relationship between drug and lipids with the advantages in speed, sensitivity, accuracy, and coverage of lipid species. The use of lipidomics has led to significant improvement in drug development process through the meticulously produced quality and quantity of information.

Many human diseases such as coronary heart disease, diabetes, Alzheimer's disease, multiple sclerosis, inflammatory diseases, and cancer have been linked to lipids in many scientific studies. In cardiovascular disease, abnormality in lipoprotein profile, omega-3/omega-6 fatty acid contents, and oxidized lipid accumulation in blood vessels have been reported [1–5]. In Alzheimer's disease, associations between disease states and sphingolipid contents or abnormality of cell membrane component both in brain and in cholesterol metabolism have been reported [6–11]. For cancer, prostaglandins and leukotrienes regulated tumor proliferation have

Lipidomics, First Edition. Edited by Kim Ekroos.

been suggested [12, 13] and as a result inhibitors for prostaglandin synthesis have been proposed as anticancer agents [14, 15]. Similarly, breast cancer and prostate cancer could also be explained in part by the imbalance of steroid hormones [16–19]. The use of lipidomics in these studies can, therefore, provide a deeper understanding of the mechanisms of disease progression. In summary, lipidomics is considered to be an indispensable technology for both biomarker discovery and drug target generation.

16.2 Biomarkers for Pharmaceutical Research

Drug discovery is now in a transition stage transcending a past strategy due to a decline of success rate in developing new drugs despite a huge expense on lengthy research and development. During this period, pharmaceutical research largely relied on a genomics-driven drug target generation, which is considered to be the best solution for the challenging issue. Indeed, genomics has succeeded in revealing many drug targets that have never been generated by traditional methodologies. There is, however, limited success in the development of drugs for targets generated by genomics, and it is generally agreed that the impact of genomics on drug development is limited [20]. Proteomics, once considered to be a promising technology in postgenomics era, has been primarily preoccupied with technology development and facing an inextricable situation as with genomics.

Several strategic changes have been made for the drug evaluation process in the past decade; these include high-throughput drug screening (HTS), drug metabolism and pharmacokinetics (DMPK) research, and drug formulation technologies. However, the evaluation of pharmacological efficacy, the most important part of drug development process, has not experienced any significant change. Innovative changes in the processes during preclinical and clinical stages are, therefore, urgently needed, and there are three approaches that can be considered. The first approach is streamlining proof-of-concept (PoC) studies in clinical studies by developing evaluation methods that can predict the efficacy in the early stage of clinical trials. The second approach is to improve the process for patient stratification. Since human being is such a diverse creature with wide genetic background, a drug may show efficacy only for a subpopulation of patients. Therefore, an optimal and more efficient clinical study could be performed if there is a process that can select patient population that responds to the drug tested. The third approach is to improve the process in understanding the mechanism(s) of drug action in preclinical studies. Proving drug efficacy for human by using disease model animals and cultured cells is quite challenging. Without doubt, a detailed understanding of mechanism(s) of drug action in these preclinical studies will help translating findings from animal to human and support the selection of the best drug candidates. Recent advance in technologies has provided new capabilities for the investigation of mechanism(s) of drug action and thus an improved drug evaluation system can be developed with more reliable methodologies for selection of good drug candidates.

The common key word for the three approaches mentioned above is "biomarkers." The Critical Path Initiative (CPI) led by US Food and Drug Administration (FDA) has declared the development of new biomarkers as a requirement during drug development. The establishment of biomarkers is, therefore, considered indispensable and key to successful drug development programs [21].

Biomarkers can be classified into several categories. Target engagement markers are used to confirm primary drug action and providing information regarding whether the drug is acting on its expected target. Proof-of-mechanism (PoM) makers are used to check drug-induced changes in the downstream pathways and cellular functions. Proof-of-principle (PoP) and PoC markers are used to predict and confirm whether the drug can truly show efficacies via key biological processes. These markers all play significant roles during drug development and help to demonstrate that a drug compound does what it is supposed to do *in vivo*.

Here follows a few biomarker examples to demonstrate their classifications. Many approved drugs or drugs still under development derive their efficacies by altering lipid profile. Here, the lipid species whose profiles have changed as a result of the mechanism of a drug could be considered as the PoM markers. Notable examples for PoM markers are prostaglandins for cyclooxigenase (COX) inhibitors and cholesterol for 3-hydroxy-3-methylglutaryl coenzyme A (HMG-CoA) reductases. In the case of peroxisome proliferator-activated receptor alpha (PPAR alpha) activators, wide modulation of lipid profile such as activation of lipid beta-oxidation and alteration of lipoprotein profile has been observed. Here, many of these downstream lipids can be used as potential PoM markers for the drugs. It is important to note that the availability of PoM markers sometimes is still not sufficient as indicators for efficacy prediction. Additional markers that can estimate the clinical benefit of drugs are also needed. This is especially true for diseases that require long-term PoC studies. Moreover, the availability of PoP markers that enable prediction of efficacy in early clinical study becomes critical. However, the development of PoP and PoC efficacy prediction markers is rather challenging. Although it remains to be validated, several reports indicated that cholesterol and triacylglycerol levels could be indicators for a risk of cardiovascular disease [22–27]. Another recent study has suggested that increased acylcarnitine levels in plasma could be used as possible indicators for type-2 diabetes and obesity [28]. These biomarkers exemplified above are still traditional indicators and, therefore, require further improvement. The discovery of these potential PoP and PoC markers was possible in part because of the strong relationship between lipids and diseases but, more importantly, because of the advancement in lipid analysis technologies. It is highly expected that lipidomics will advance this to a new level, as novel lipid details are revealed.

16.3 Strategy for Biomarker Discovery

A common difficulty in biomarker discovery research is the limited access to clinical tissue samples. For the development of biochemical biomarkers, samples using noninvasive methods, such as blood and urine, should be considered even though the drug actually targets solid organs such as liver, muscle, and brain. The reason for this is that blood and urine are much easier to obtain from living study subjects. For example, plasma lipids often reflect tissue state by both active/passive secretions. Despite other methodologies such as positron emission tomography (PET) and magnetic resonance imaging (MRI) that are available and can be applied for the evaluation of drugs in clinical trial, their limited accessibility and prohibitive cost have prevented these expensive imaging instruments from being widely used. As a result, biochemical-based biomarker is still the preferable method for clinical study where lipidomics is one major technology alongside other "omics" approaches.

There are mainly two approaches to develop biomarkers involving lipidomics technologies: (1) targeted and (2) unbiased (e.g., shotgun lipidomics). Targeted approach uses technologies focusing on specific molecules or molecular groups of similar chemical characteristics so that these molecules can be detected with high sensitivity and specificity. The targeted approach is also useful to study a specific lipid synthesis pathway if the drug is expected to modulate one or more lipid molecules within the pathway. The drug effects on the entire pathway can, therefore, be studied in detail to reveal its true pharmacological process. The eicosanoids [29–31] and steroids [32, 33] are representative molecular groups for targeted lipidomics because of their structure and chemical character similarities. Moreover, although they are normally present in trace amounts in the body, their detectability is not hampered due to the inherently high sensitivity of this approach.

Unbiased approach on the other hand is designed to analyze a wide range of molecules. Unlike targeted approach, where only a set of known molecules are programmed into the method, the molecular species that can be analyzed by the unbiased approach are unlimited. Despite this seemingly overwhelming advantage, the detection sensitivity of the unbiased approach in general (based on current technologies) is markedly less than in the targeted approach. As a result, the unbiased approach is actually "biased" toward high abundant lipid species in a body such as phospholipids, neutral lipids, and sphingolipids as these lipid species are often found in studies using unbiased lipidomics [34, 35]. Due to the tremendous amount of data generated from unbiased lipidomics studies, the data analysis and identification of molecular structures can be a time-consuming process. Despite such drawbacks, the unbiased lipidomics approach is important because the method allows exploration of novel findings. The ability to detect unexpected lipids, typically missed by the targeted approaches, is indispensable in biomarker discovery.

As discussed in other chapters, liquid chromatography and mass spectrometry play a critical role in the lipidomics field. A chromatographic separation is critical to obtain high sensitivity, high specificity, and precise quantitation. There are several types of mass spectrometers with characteristics unique to each instrument and they can be selected according to their strengths and features to develop platforms specifically for targeted and unbiased approaches. Triple quadrupole (QQQ) mass spectrometers are frequently selected for targeted lipidomics because this type of MS can perform selected reaction monitoring (SRM) and multiple reaction monitoring (MRM) scans. These scan functions are unique to QQQ instruments with the advantage of high detection sensitivity and fast simultaneous analysis of multiple molecules. The QQQ instruments are ideal for the simultaneous analysis of, for example, multiple lipid species within a pathway that are present at trace amounts.

For unbiased approaches, mass spectrometers that have the ability to obtain accurate molecular mass data and high-resolution spectra are preferred. These include time-of-flight (TOF) type mass spectrometers and Fourier transform (FT) type mass spectrometers. The ability to scan in high-resolution mode and obtain accurate molecular mass information is critical to distinguish molecules that have close structures and molecular weights. With these instrumentations, it is possible to distinguish the small differences in molecular masses of lipid molecules belonging to the same molecular group. For example, the general structure of the lipid class phosphatidylcholine (PC) consists of two fatty acyl chains by various combinations in the number of carbons and double bonds. The analysis of PC from plasma often results in about 70 different distinct signals representing various PC species with molecular mass close to one another depending on the attached fatty acids. There are at least two forms of structure in PC, acyl-form (PC) and ether-form (PC-O), and mass spectrometry is able to distinguish these structures. The ability to evaluate molecules that have minor differences in mass allows the evaluation of these PC lipids in a different perspective even though they are from the same molecular group: A typical 2D LC/MS data of unbiased lipidomics from plasma is shown in Figure 16.1a. Combining both the separation power of chromatography and the mass accuracy of mass spectrometer in positive and negative ion modes, lipid species from various classes and within the same class can be readily detected and identified, thus providing a comprehensive overview of all lipids present in the plasma sample (Figure 16.1b). Figure 16.2 shows the comparison profiles of triacylglycerol (TG) species, detecting more than 200 signals, in two distinct samples. The TG profiles of these two samples are apparently different particularly in long-chain fatty acid-containing TG species. Individual evaluation of each lipid species even within the same group can reveal new insights into the function of lipid regulatory systems, while differential analysis in the lipid levels between these two samples shows the various TG lipid population in the two sample states.

The targeted approach is usually the first in-line approach for biomarker research in pharmaceutical field because the drug development process is typically based on expected changes in a certain molecular profile, cellular function, and the resulting efficacy. Drug evaluation is generally performed by checking whether drug candidates follow defined or assumed critical events. Therefore, the first step of biomarker discovery is to survey peripheral pathways that are predicted by a presumed mode of actions. Starting from blood samples, the targeted approach platform optimized for each molecular group, called modules (Figure 16.3), will be best suited for this stage of the process. If an endogenous molecule within the pathways investigated is found to explain the drug effects based on the mechanisms of the drug

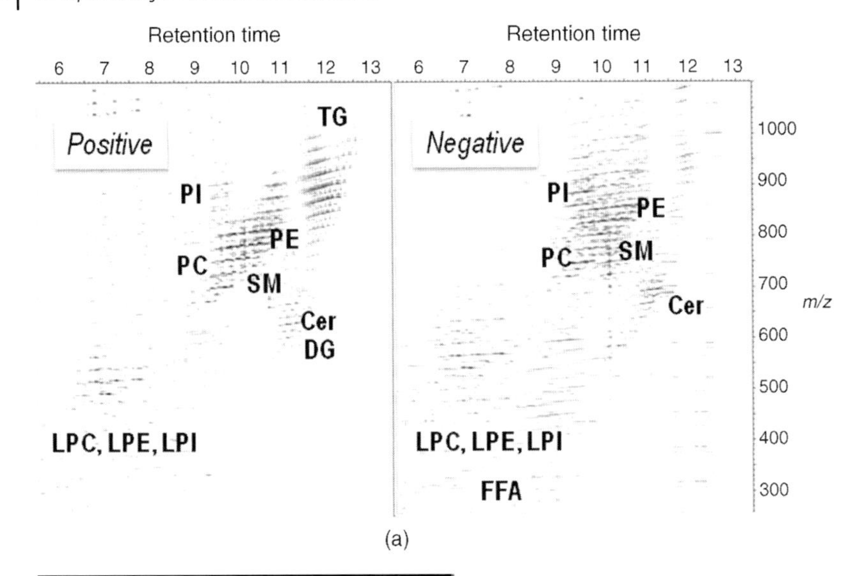

(a)

Class	Polarity	Number of molecules
LPC	+/-	100
LPE	+/-	50
LPI	+/-	10
PC	+/-	100
PE	+/-	20
PI	+/-	5
Cer	+/-	10
SM	+/-	10
TG	+	250
FFA	-	12

(b)

Figure 16.1 Plasma lipidomics 2D map.

and if there is true correlations between target tissues and blood, this molecule can be nominated as a prime biomarker candidate. To increase the chance of finding true biomarker candidates, multiple molecular pathways are to be considered. Therefore, several types of modules representing multiple molecular pathways

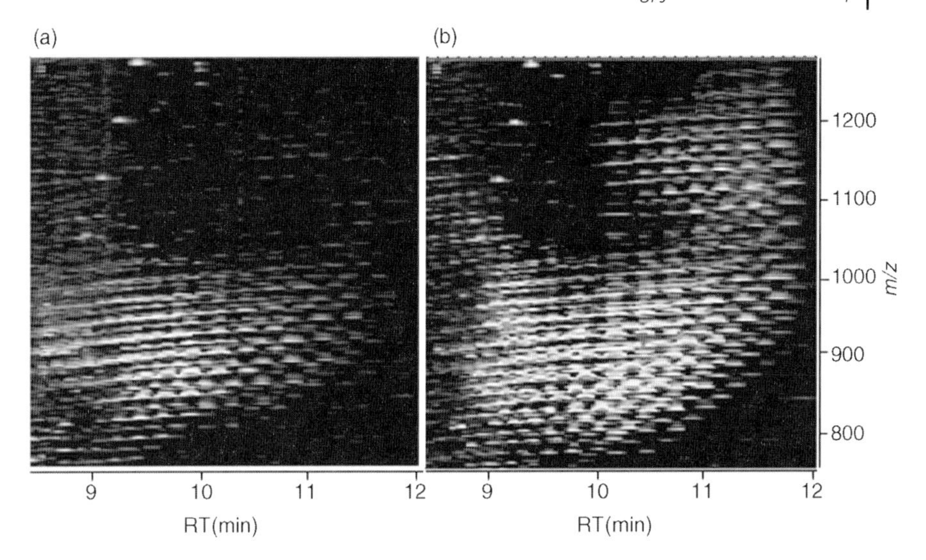

Figure 16.2 Triacyglycerol profile in different conditions.

should be included during the design of the targeted lipidomics method. For lipid metabolism-related drug and biomarker discovery projects, one might want to consider a combination of free fatty acid module, acylcarnitine module, and acyl-CoA module to use as the initial targeted lipidomics method. As mentioned earlier, the drawback for the targeted approach, which is mostly driven by hypotheses, can focus only on specific molecules of interest and is sometimes too limited to explore

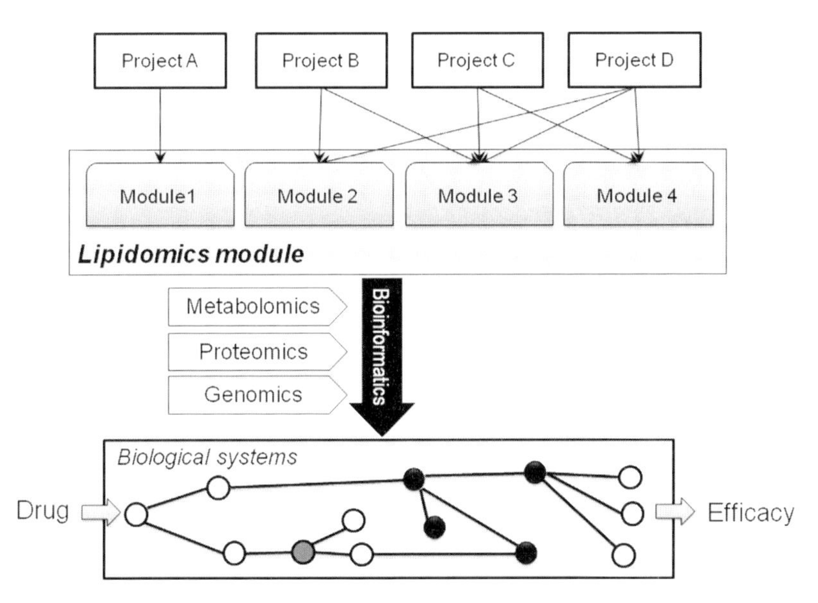

Figure 16.3 Module based lipidomics.

new or unexpected molecular species. It is often useful to incorporate the unbiased approach into the workflow complementing the targeted approach in order to obtain wider coverage of the molecular landscape and in this way improve the chance of finding new mechanisms of a drug. Candidate biomarkers discovered using the unbiased approach can later be followed up through further investigation using targeted modules established based on the new hypotheses.

16.4 Conclusions

Another application of lipidomics in pharmaceutical field is the discovery of drug targets. Targeted lipidomics can provide detailed information on what is taking place in a particular lipid pathway in both diseased and healthy people. A report by Kihara and coworkers have recently demonstrated the utilization of targeted eicosanoid analyses for discovery of new drug targets for multiple sclerosis [36]. The unbiased approach on the other hand can also offer many opportunities to discover novel insights of a disease, including understanding the toxic factors and pathways that connect to diseases. As mentioned, traditionally drug target generation mostly relies on genomics. The functionality of a body of every living being, however, is carried out by proteins and lipids. Proteomics and lipidomics should, therefore, be the focus areas for pharmaceutical research. The ultimate goal for lipidomics technology is to uncover a complete quantitative picture of the lipidome, and to use the knowledge for connecting all lipid-related biological processes. In general, “omics” technologies have to this point been able to detect only changes in certain parts of the underlying biological processes, therefore, producing scattered results often hard to interpret. Targeted and unbiased lipidomics could cover many parts of lipidome, but there is still a significant gap between these two approaches. Establishing technologies that are able to bridge the gap between the two is inevitable for the next-generation pharmaceutical research.

References

1 Nordestgaard, B.G., Chapman, M.J., Ray, K., Borén, J., Andreotti, F., Watts, G. F., Ginsberg, H., Amarenco, P., Catapano, A., Descamps, O.S., Fisher, E., Kovanen, P.T., Kuivenhoven, J.A., Lesnik, P., Masana, L., Reiner, Z., Taskinen, M.R., Tokgözoglu, L., and Tybjñrg-Hansen, A. (2010) Lipoprotein(a) as a cardiovascular risk factor: current status. *Eur. Heart J.*, **31** (23), 2844–2853.

2 Harris, W.S., Miller, M., Tighe, A.P., Davidson, MH., and Schaefer, E.J. (2008) Omega-3 fatty acids and coronary heart disease risk: clinical and mechanistic perspectives. *Atherosclerosis*, **197** (1), 12–24.

3 Chattipakorn, N., Settakorn, J., Petsophonsakul, P., Suwannahoi, P., Mahakranukrauh, P., Srichairatanakool, S., and Chattipakorn, S.C. (2009) Cardiac mortality is associated with low levels of omega-3 and omega-6 fatty acids in the heart of cadavers with a history of coronary heart disease. *Nutr. Res.*, **29** (10), 696–704.

4 Manicke, N.E., Nefliu, M., Wu, C., Woods, J.W., Reiser, V., Hendrickson, R.C., and Cooks, R.G. (2009) Imaging of lipids in

atheroma by desorption electrospray ionization mass spectrometry. *Anal. Chem.*, **81** (21), 8702–8707.

5 Shui, G., Cheong, W.F., Jappar, I.A., Hoi, A., Xue, Y., Fernandis, A.Z., Tan, B.K., and Wenk, M.R. (2011) Derivatization-independent cholesterol analysis in crude lipid extracts by liquid chromatography/mass spectrometry: applications to a rabbit model for atherosclerosis. *J. Chromatogr. A*, **1218** (28), 4357–4365.

6 Di Paolo, G. and Kim, T.W. (2011) Linking lipids to Alzheimer's disease: cholesterol and beyond. *Nat. Rev. Neurosci.*, **12** (5), 284–296.

7 Mathew, A., Yoshida, Y., Maekawa, T., and Sakthi, K.D. (2011) Alzheimer's disease: cholesterol a menace? *Brain Res. Bull.*, **86** (1–2), 1–12.

8 Mielke, M.M., Haughey, N.J., Ratnam Bandaru, V.V., Schech, S., Carrick, R., Carlson, M.C., Mori, S., Miller, M.I., Ceritoglu, C., Brown, T., Albert, M., and Lyketsos, C.G. (2010) Plasma ceramides are altered in mild cognitive impairment and predict cognitive decline and hippocampal volume loss. *Alzheimers Dement.*, **6** (5), 378–385.

9 Han, X., Rozen, S., Boyle, S.H., Hellegers, C., Cheng, H., Burke, J.R., Welsh-Bohmer, K.A., Doraiswamy, P.M., and Kaddurah-Daouk, R.. (2011) Metabolomics in early Alzheimer's disease: identification of altered plasma sphingolipidome using shotgun lipidomics. *PLoS One*, **6** (7), e21643.

10 Hejazi, L., Wong, J.W., Cheng, D., Proschogo, N., Ebrahimi, D., Garner, B., and Don, A.S. (2011) Mass and relative elution time profiling: two-dimensional analysis of sphingolipids in Alzheimer's disease brains. *Biochem. J.*, **438** (1), 165–175.

11 Piomelli, D., Astarita, G., and Rapaka, R. (2007) A neuroscientst's guide to lipidomics. *Nat. Rev. Neurosci.*, **8** (10), 743–754.

12 Wang, D. and Dubois, R.N. (2010) Eicosanoids and cancer. *Nat. Rev. Cancer*, **10** (3), 181–193.

13 Greene, E.R., Huang, S., Serhan, C.N., and Panigrahy, D. (2011) Regulation of inflammation in cancer by eicosanoids. *Prostaglandins Other Lipid Mediat.*, **96** (1–4), 27–36.

14 Reckamp, K.L., Krysan, K., Morrow, J.D., Milne, G.L., Newman, R.A., Tucker, C., Elashoff, R.M., Dubinett, S.M., and Figlin, R.A. (2006) A phase I trial to determine the optimal biological dose of celecoxib when combined with erlotinib in advanced non-small cell lung cancer. *Clin. Cancer Res.*, **12** (11), 3381–3318.

15 Reckamp, K., Gitlitz, B., Chen, L.C., Patel, R., Milne, G., Syto, M., Jezior, D., and Zaknoen, S. (2011) Biomarker-based phase I dose-escalation, pharmacokinetic, and pharmacodynamic study of oral apricoxib in combination with erlotinib in advanced nonsmall cell lung cancer. *Cancer*, **117** (4), 809–918.

16 Haslam, S.Z. and Woodward, T.L. (2003) Host microenvironment in breast cancer development: epithelial-cell–stromal-cell interactions and steroid hormone action in normal and cancerous mammary gland. *Breast Cancer Res.*, **5** (4), 208–215.

17 Okoh, V., Deoraj, A., and Roy, D. (2011) Estrogen-induced reactive oxygen species-mediated signalings contribute to breast cancer. *Biochim. Biophys. Acta.*, **1815** (1), 115–133.

18 Soronen, P., Laiti, M., Törn, S., Härkönen, P., Patrikainen, L., Li, Y., Pulkka, A., Kurkela, R., Herrala, A., Kaija, H., Isomaa, V., and Vihko, P. (2004) Sex steroid hormone metabolism and prostate cancer. *J. Steroid Biochem. Mol. Biol.*, **92** (4), 281–286.

19 Lange, C.A., Gioeli, D., Hammes, S.R., and Marker, P.C. (2007) Integration of rapid signaling events with steroid hormone receptor action in breast and prostate cancer. *Annu. Rev. Physiol.*, **69**, 171–199.

20 Hall, J., Dennler, P., Haller, S., Pratsinis, A., Säuberli, K., Towbin, H., Walther, K., and Woytschak, J. (2011) Genomics drugs in clinical trials. *Nat. Rev. Drug. Discov.*, **9** (12), 988–989.

21 Trist, D.G. (2011) Scientific process, pharmacology and drug discovery. *Curr. Opin. Pharmacol.*, **11** (5), 528–533.

22 Okamura, T., Kokubo, Y., Watanabe, M., Higashiyama, A., Ono, Y.,

Miyamoto, Y., Yoshimasa, Y., and Okayama, A. (2010) Triglycerides and non-high-density lipoprotein cholesterol and the incidence of cardiovascular disease in an urban Japanese cohort: the Suita study. *Atherosclerosis*, **209** (1), 290–294.

23 NCEP (2002) Third Report of the National Cholesterol Education Program (NCEP) Expert Panel on Detection, Evaluation, and Treatment of High Blood Cholesterol in Adults (Adult Treatment Panel III): final report. *Circulation*, **106** (25), 3143–421.

24 NCEP (2001) Executive summary of the Third Report of The National Cholesterol Education Program (NCEP) Expert Panel on Detection, Evaluation, and Treatment of High Blood Cholesterol in Adults (Adult Treatment Panel III). *JAMA*, **285** (19), 2486–2497.

25 Wilson, P.W. and Grundy, S.M. (2003) The metabolic syndrome: a practical guide to origins and treatment: Part II. *Circulation*, **108** (13), 1537–1540.

26 Iso, H., Naito, Y., Sato, S., Kitamura, A., Okamura, T., Sankai, T., Shimamoto, T., Iida, M., and Komachi, Y. (2001) Serum triglycerides and risk of coronary heart disease among Japanese men and women. *Am. J. Epidemiol.*, **153** (5), 490–499.

27 Miller, M. (2000) Differentiating the effects of raising low levels of high-density lipoprotein cholesterol versus lowering normal triglycerides: further insights from the Veterans Affairs High-Density Lipoprotein Intervention Trial. *Am. J. Cardiol.*, **86** (12A), 23L–27L.

28 McNamara, J.R., Shah, P.K., Nakajima, K., Cupples, L.A., Wilson, P.W., Ordovas, J. M., and Schaefer, E.J. (1998) Remnant lipoprotein cholesterol and triglyceride reference ranges from the Framingham Heart Study. *Clin. Chem.*, **44** (6 Pt 1), 1224–1232.

29 Mihalik, S.J., Goodpaster, B.H., Kelley, D.E., Chace, D.H., Vockley, J., Toledo, F.G., and DeLany, J.P. (2010) Increased levels of plasma acylcarnitines in obesity and type 2 diabetes and identification of a marker of glucolipotoxicity. *Obesity*, **18** (9), 1695–1700.

30 Gomolka, B., Siegert, E., Blossey, K., Schunck, W.H., Rothe, M., and Weylandt, K.H. (2011) Analysis of omega-3 and omega-6 fatty acid-derived lipid metabolite formation in human and mouse blood samples. *Prostaglandins Other Lipid Mediat.*, **94** (3–4), 81–87.

31 Masoodi, M., Mir, A.A., Petasis, N.A., Serhan, C.N., and Nicolaou, A. (2008) Simultaneous lipidomic analysis of three families of bioactive lipid mediators leukotrienes, resolvins, protectins and related hydroxy-fatty acids by liquid chromatography/electrospray ionisation tandem mass spectrometry. *Rapid Commun. Mass Spectrom.*, **22** (2), 75–83.

32 Dumlao, D.S., Buczynski, M.W., Norris, P. C., Harkewicz, R., and Dennis, E.A. (2011) High-throughput lipidomic analysis of fatty acid derived eicosanoids and N-acylethanolamines. *Biochim. Biophys. Acta.*, **1811** (11), 724–736.

33 Keski-Rahkonen, P., Huhtinen, K., Poutanen, M., and Auriola, S. (2011) Fast and sensitive liquid chromatography-mass spectrometry assay for seven androgenic and progestagenic steroids in human serum. *J. Steroid Biochem. Mol. Biol.*, **127** (3–5), 396–404.

34 Griffiths, W.J. and Wang, Y. (2009) The importance of steroidomics in the study of neurodegenerative disease and ageing. *Comb. Chem. High Throughput Screen.*, **12** (2), 212–228.

35 Han, X. and Gross, R.W. (2003) Global analyses of cellular lipidomes directly from crude extracts of biological samples by ESI mass spectrometry: a bridge to lipidomics. *J. Lipid Res.*, **44** (6), 1071–1079.

36 Kihara, Y., Matsushita, T., Kita, Y., Uematsu, S., Akira, S., Kira, J., Ishii, S., and Shimizu, T. (2009) Targeted lipidomics reveals mPGES-1-PGE2 as a therapeutic target for multiple sclerosis. *Proc. Natl. Acad. Sci. U.S.A.*, **106** (51), 21807–21812.

Index

Lipidomics, First Edition. Edited by Kim Ekroos.